BIOCHEMICAL MARKERS FOR CANCER

Edited by T. Ming Chu

Professor and Director
Department of Diagnostic
Immunology Research
and Biochemistry
Roswell Park Memorial Institute
Buffalo, New York

MARCEL DEKKER, INC.   New York and Basel

Library of Congress Cataloging in Publication Data
Main entry under title:

Biochemical markers for cancer.

(Clinical and biochemical analysis ; 11)
Includes index.
1. Cancer—Diagnosis. 2. Tumor proteins. 3. Tumor
lipids. I. Chu, T. Ming, [date]. II. Series. [DNLM:
1. Antigens, Neoplasm—Analysis. 2. Neoplasm proteins—
Analysis. W1CL654 v.11 / QZ 200 B614]
RC270.B56   616.99'40756   81-19601
ISBN 0-8247-1535-7   AACR2

COPYRIGHT © 1982 by MARCEL DEKKER, INC. ALL RIGHTS RESERVED

Neither this book nor any part may be reproduced or transmitted in any form
or by any means, electronic or mechanical, including photocopying, micro-
filming, and recording, or by any information storage and retrieval system,
without permission in writing from the publisher.

MARCEL DEKKER, INC.
270 Madison Avenue, New York, New York 10016

Current printing (last digit):
10   9   8   7   6   5   4   3   2   1

PRINTED IN THE UNITED STATES OF AMERICA

BIOCHEMICAL MARKERS FOR CANCER

CLINICAL AND BIOCHEMICAL ANALYSIS

A series of monographs and textbooks

EDITORS

Morton K. Schwartz
Chairman, Department of Biochemistry
Memorial Sloan-Kettering Cancer Center
New York, New York

Thomas Whitehead
Department of Clinical Chemistry
Queen Elizabeth Medical Centre
Birmingham, England

1. Colorimetric and Fluorimetric Analysis of Organic Compounds and Drugs, *M. Pesez and J. Bartos*
2. Normal Values in Clinical Chemistry: Statistical Analysis of Laboratory Data, *Horace F. Martin, Benjamin J. Gudzinowicz, and Herbert Fanger*
3. Continuous Flow Analysis: Theory and Practice, *William B. Furman*
4. Handbook of Enzymatic Methods of Analysis, *George G. Guilbault*
5. Handbook of Radioimmunoassay, *edited by Guy E. Abraham*
6. The Hemoglobinopathies, *Titus H. J. Huisman and J. H. P. Jonxis*
7. Automated Immunoanalysis (in two parts), *edited by Robert F. Ritchie*
8. Computers in the Clinical Laboratory: An Introduction, *E. Clifford Toren, Jr. and Arthur A. Eggert*
9. The Chromatography of Hemoglobin, *Walter A. Schroeder and Titus H. J. Huisman*
10. Nonisotopic Alternatives to Radioimmunoassay: Principles and Applications, *edited by Lawrence A. Kaplan and Amadeo J. Pesce*
11. Biochemical Markers for Cancer, *edited by T. Ming Chu*

ADDITIONAL VOLUMES IN PREPARATION

PREFACE

Biochemical differences between normal and tumor cells have been long recognized. Malignant transformations of normal cells usually result in synthesis and secretion of quantitatively or qualitatively abnormal biochemical substances in tumor-bearing hosts. Much effort has been directed toward the identification and characterization of these tumor-associated substances in recent years, which include antigens, antibodies, enzymes, hormones, and other proteins. Quantitation of these biochemical parameters has been employed as an adjuvant tool in diagnosis and treatment of cancer patients. This volume contains selected topics on biochemical markers which have been utilized in clinical oncology and describes a possible approach to the study of the basic mechanism of the marker at the molecular level, which is a logical extension of investigations in the area of biochemical markers for cancer.

Although the contributors to this volume are pioneer investigators in their respective fields, there is no single expert on all aspects of tumor markers. It is not our intention to cover the entire spectrum of cancer biochemical markers in a single volume. However, it is our hope that this book will be of interest both to practicing clinicians and laboratory scientists, as well as to many students of medicine and other disciplines of science. With their work and interest, continuing progress can be made in this rapidly expanding and important area of oncology.

T. Ming Chu

CONTRIBUTORS

ELLIOT ALPERT, M.D.[1]   Associate Professor of Medicine, GI Unit, Harvard Medical School, Boston, Massachusetts, Massachusetts General Hospital, Boston, Massachusetts

PATRICIA A. BOYD-LEINEN, Ph.D.[2]   Research Associate, Department of Molecular Medicine, Mayo Clinic, Rochester, Minnesota

CHUNSHING CHEN, Ph.D.   Research Associate, the Ben May Laboratory for Cancer Research and the Department of Biochemistry, University of Chicago, Chicago, Illinois

T. MING CHU, Ph.D.   Professor and Director, Department of Diagnostic Immunology Research and Biochemistry, Roswell Park Memorial Institute, Buffalo, New York

JAMES T. EVANS, M.D.[3]   Cancer Research Surgeon, Department of Surgical Oncology, Roswell Park Memorial Institute, Buffalo, New York

WILLIAM H. FISHMAN, Ph.D.   President and Chief Executive Officer, La Jolla Cancer Research Foundation, La Jolla, California

RONALD B. HERBERMAN, M.D.   Chief, Laboratory of Immunodiagnosis, National Cancer Institute, Bethesda, Maryland

HIDEMATSU HIRAI, M.D., Ph.D.   Professor, Department of Biochemistry, Hokkaido University School of Medicine, Sapporo, Hokkaido, Japan

---

[1] Present affiliation: Professor of Medicine and Chief of Gastroenterology, Department of Medicine, Baylor College of Medicine, Houston, Texas
[2] Present affiliation: Biochemical Editor, Chemical Abstracts Service, Columbus, Ohio
[3] Present affiliation: Associate Professor of Surgery, Department of Surgery, Erie County Medical Center, State University of New York at Buffalo, Buffalo, New York

E. DOUGLAS HOLYOKE, M.D.   Chief, Department of Surgical Oncology, Roswell Park Memorial Institute, Buffalo, New York

CARL S. KILLIAN, Ph.D.   Cancer Research Scientist, Department of Diagnostic Immunology Research and Biochemistry, Roswell Park Memorial Institute, Buffalo, New York

RANDALL H. HOTTEL, Ph.D.[1]   Staff Cell Biologist, La Jolla Cancer Research Foundation, La Jolla, California

CHING-LI LEE, Ph.D.   Senior Cancer Research Scientist, Department of Diagnostic Research and Biochemistry, Roswell Park Memorial Institute, Buffalo, New York

WALTER M. LEWKO, Ph.D.[2]   Assistant Professor of Biochemistry, Department of Biochemistry, Health Sciences Center, University of Louisville, Louisville, Kentucky

SHUTSUNG LIAO, Ph.D.   Professor, the Ben May Laboratory for Cancer Research and the Department of Biochemistry, University of Chicago, Chicago, Illinois

JAN LINDGREN, M.D.   Assistant Teacher, Department of Bacteriology and Immunology, University of Helsinki, Helsinki, Finland

RUEY MING LOOR, Ph.D.[3]   Research Associate, the Ben May Laboratory for Cancer Research and the Department of Biochemistry, University of Chicago, Chicago, Illinois

RAJENDRA G. MEHTA, Ph.D.[4]   Cancer Research Associate, Department of Biochemistry, Health Sciences Center, University of Louisville, Louisville, Kentucky

---

[1] Present affiliation: Microbiology Instructor, Department of Biology, Mesa College, San Diego, California
[2] Present affiliation: Laboratory of Pathophysiology, National Cancer Institute, Bethesda, Maryland
[3] Present affiliation: Senior Cancer Research Scientist, Department of Diagnostic Immunology Research and Biochemistry, Roswell Park Memorial Institute, Buffalo, New York
[4] Present affiliation: Senior Biochemist, Life Sciences Research, Illinois Institute of Technology Research Institute, Chicago, Illinois

Contributors

ARNOLD MITTELMAN, M.D. Associate Chief of Surgical Oncology, Department of Surgical Oncology and Program Director of Colon-Rectal Service, Roswell Park Memorial Institute, Buffalo, New York

GERALD P. MURPHY, M.D., D.Sc. Institute Director, Roswell Park Memorial Institute, Buffalo, New York

A. MUNRO NEVILLE, Ph.D., M.D., M.R.C. Path. Professor, Unit of Human Cancer Biology, Ludwig Institute for Cancer Research (London Branch), Royal Marsden Hospital, Sutton, Surrey, United Kingdom

DANIEL C. PARK, Ph.D. Research Fellow, Department of Biochemistry, Health Sciences Center, University of Louisville, Louisville, Kentucky

DAVID PRESSMAN, Ph.D.[1] Professor and Director, Department of Immunology Research, Roswell Park Memorial Institute, Buffalo, New York

DIANE HADDOCK RUSSELL, Ph.D. Professor, Department of Pharmacology, University of Arizona Health Sciences Center, Tucson, Arizona

EEVA-MARJA RUTANEN, M.D.[2] Resident, Department of Bacteriology and Immunology, University of Helsinki, Helsinki, Finland

MORTON K. SCHWARTZ, Ph.D. Professor and Chairman, Department of Biochemistry, Memorial Sloan-Kettering Cancer Center, New York, New York

BEN K. SEON, Ph.D. Principal Cancer Research Scientist, Department of Immunology Research, Roswell Park Memorial Institute, Buffalo, New York

MARKKU SEPPÄLÄ, M.D., Ph.D.[2] Professor and Chairman, Department of Obstetrics and Gynecology, University of Helsinki, Helsinki, Finland, and Department of Reproductive Physiology, St. Bartholomew's Hospital, London, United Kingdom

B. I. SAHAI SRIVASTAVA, Ph.D. Principal Cancer Research Scientist, Department of Experimental Therapeutics, Grace Cancer Drug Center, Roswell Park Memorial Institute, Buffalo, New York

---

[1] Deceased
[2] Present affiliation: Department of Obstetrics and Gynecology, University Central Hospital, Helsinki, Finland

TORSTEN WAHLSTRÖM, M.D. Chief Pathologist, Department of Obstetrics and Gynecology and Associate of Pathology, University of Helsinki, Helsinki, Finland

MING C. WANG, Ph.D. Associate Cancer Research Scientist, Department of Diagnostic Immunology Research and Biochemistry, Roswell Park Memorial Institute, Buffalo, New York

JAMES L. WITTLIFF, Ph.D. Professor and Chairman, Department of Biochemistry, Health Sciences Center, University of Louisville, Louisville, Kentucky

CONTENTS

PREFACE		iii
CONTRIBUTORS		v
1.	IMMUNOLOGICAL APPROACH TO THE BIOCHEMICAL MARKERS FOR CANCER Ronald B. Herberman	1
	I. Introduction	1
	II. Potential Clinical Applications	3
	III. Some Examples of Immunological Tests for Human Tumor-Associated Markers	12
	IV. Conclusions	20
	References	21
2.	ALPHA FETOPROTEIN Hidematsu Hirai	25
	I. Introduction	26
	II. Physiological Concentration of AFP in the Serum	27
	III. Physico- and Immunochemical Properties of AFP	28
	IV. Methods for Detection of AFP	32
	V. Physiology of AFP	33
	VI. Experimental Study of Hepatocarcinogenesis	34
	VII. AFP in Patients with Liver Diseases	36
	VIII. AFP in Germ-Cell Tumors	42
	IX. Association of Elevated Serum AFP With Gastrointestinal Tumors Other than Hepatoma	43
	X. AFP in Other Nonmalignant Diseases	44
	XI. AFP in Pregnancy and Fetal Distress	44
	XII. Immunological Effect of Anti-AFP Serum on AFP-Producing Cells	45
	XIII. Concluding Remarks	48
	References	49

3. CARCINOEMBRYONIC ANTIGEN AS A TUMOR MARKER 61
E. Douglas Holyoke, T. Ming Chu, James T. Evans, and
Arnold Mittelman

    I. Introduction 61
   II. The Clinical Usefulness of CEA 64
      A. Introduction 64
      B. Colon and Rectal Cancer 64
  III. Conclusion 76
      References 77

4. ENZYMES AS TUMOR MARKERS 81
Morton K. Schwartz

     I. Introduction 81
    II. Lysozyme 82
   III. γ-Glutamyl Transpeptidase 83
   IV. Ribonuclease 84
    V. Glycosyltransferases 85
   VI. Lactic Dehydrogenase 88
  VII. Aryl Sulfatase 89
 VIII. Histaminase 89
   IX. Hexosaminidase 90
    X. Terminal Deoxynucleotidyl Transferase 90
      References 90

5. DEVELOPMENTAL ALKALINE PHOSPHATASES AS
BIOCHEMICAL TUMOR MARKERS 93
Randall H. Kottel and William H. Fishman

    I. Introduction 93
   II. Developmental (Placental) Alkaline Phosphatases 98
  III. Developmental Alkaline Phosphatase Isoenzymes in
      Human Cancer 100
  IV. Summary 108
      References 109

6. PROSTATIC ACID PHOSPHATASE IN HUMAN PROSTATE
CANCER 117
T. Ming Chu, Ming C. Wang, Ching-Li Lee, Carl S. Killian,
and Gerald P. Murphy

    I. Introduction 117
   II. Biochemical Nature of Prostatic Acid Phosphatase 118

	III.	Immunological Specificity of Prostatic Acid Phosphatase	120
	IV.	Radioimmunoassay for Prostatic Acid Phosphatase	120
	V.	Counterimmunoelectrophoresis for Prostatic Acid Phosphatase	122
	VI.	Fluorescent Immunoassay for Prostatic Acid Phosphatase	126
	VII.	Solid-Phase Immunoadsorbent Assay	126
	VIII.	Clinical Evaluation of Immunoassays for Prostatic Acid Phosphatase	127
	IX.	Bone Marrow Acid Phosphatase	129
	X.	Summary	131
		References	132
7.	BIOCHEMICAL MARKERS FOR THE DIFFERENTIAL DIAGNOSIS OF LEUKEMIAS		137
	B. I. Sahai Srivastava	
	I.	Introduction	137
	II.	Purine and Pyrimidine Metabolizing Enzymes	139
	III.	DNA Polymerases	140
	IV.	Cell Membrane Enzymes	141
	V.	Acid Hydrolases	142
	VI.	Esterases	144
	VII.	Drug and Radiation Sensitivity of Leukemic Cells	145
	VIII.	Miscellaneous Markers	145
	IX.	Terminal Deoxynucleotidyl Transferase (TdT)	147
	X.	Protease and DNase Activity	162
	XI.	Conclusion	162
		References	163
8.	STEROID-RECEPTOR INTERACTIONS IN NORMAL AND NEOPLASTIC MAMMARY TISSUES		183
	James L. Wittliff, Patricia A. Boyd-Leinen, Rajendra G. Mehta, Walter M. Lewko, and Daniel C. Park	
	I.	Introduction	184
	II.	Characteristics of Steroid Receptors from Normal Mammary Gland	188
	III.	Characteristics of Steroid Receptors in Experimental Mammary Tumors	211
	IV.	Estrogen Receptors in Human Breast Carcinoma and Their Clinical Significance	217
	V.	Summary	223
		References	224

9. ISOFERRITINS: ALTERATION IN HUMAN MALIGNANCY AND
   CLINICAL SIGNIFICANCE   229
   Elliot Alpert

   I.   Introduction   229
   II. Structure   229
   III. Isoferritins   230
   IV. Tumor Isoferritins   231
   V.   Serum Ferritin   233
   VI. Serum Ferritin in Cancer   233
   VII. Conclusion   235
        References   235

10. POLYAMINES AS BIOCHEMICAL MARKERS OF TUMOR
    GROWTH PARAMETERS   241
    Diane Haddock Russell

    I.   Introduction   241
    II. Polyamine Accumulation in Tumor Cells   243
    III. Intra- and Extracellular Polyamines During Animal
         Tumor Regression   274
    IV. Extracellular Polyamine Levels in Diagnosed Cancer
         Patients   248
    V.   Polyamines as Markers of Response to Cancer Therapy   252
    VI. Urinary Polyamine Values and Disease Activity   259
    VII. Summary   260
         References   262

11. ECTOPIC HORMONE PRODUCTION BY TUMORS   267
    A. Munro Neville

    I.   Introduction   267
    II. Definition of Ectopic Hormone Production   268
    III. Evidence Required to Establish Ectopic Hormone
         Production by a Tumor   268
    IV. Range of Ectopic Hormonal Manifestations   280
    V.   Clinicopathological Applications   284
    VI. Biological Significance   287
    VII. Summary   290
         References   291

12. IMMUNOGLOBULINS AS CANCER MARKERS IN HUMANS   301
    Ben K. Seon and David Pressman

    I. Introduction   301
    II. Bence Jones Protein (Free Light Chain) and Light-Chain
        Fragment   302

	III.	Monoclonal and Biclonal Gammopathies	310
	IV.	Heavy-Chain Disease Proteins and Atypical Immunoglobulins	314
	V.	Conclusion	316
		References	316

13. MULTIPLE MARKERS IN THE MANAGEMENT OF CANCER PATIENTS — 321
    Markku Seppälä, Eeva-Marja Rutanen, Jan Lindgren, and Torsten Wahlström

	I.	Introduction	321
	II.	Trophoblastic Tumors	322
	III.	Nontrophoblastic Tumors of the Reproductive System	331
	IV.	Summary	345
		References	347

14. PROTEIN MARKERS FOR ANDROGENIC ACTIVITY IN RAT VENTRAL PROSTATE — 351
    Ruey Ming Loor, Chunshing Chen, and Shutsung Liao

	I.	Introduction	351
	II.	Androgen-Receptor Protein	351
	III.	Nuclear-Acceptor Protein	357
	IV.	Spermine-Binding Protein	360
	V.	Prostate α-Protein	363
	VI.	Concluding Remarks	365
		References	366

INDEX — 371

1

IMMUNOLOGICAL APPROACH TO THE BIOCHEMICAL MARKERS FOR CANCER

RONALD B. HERBERMAN / Laboratory of Immunodiagnosis, National Cancer Institute, Bethesda, Maryland

I. Introduction   1
II. Potential Clinical Applications   3
   A. Screening   6
   B. Aid in Diagnosis of Patients with Signs or Symptoms   7
   C. Aid to Histopathological Evaluation of Tumors   7
   D. Aid in Staging of Cancer Patients   8
   E. Localization of Tumor and Detection of Metastases   10
   F. Serial Monitoring To Determine Efficacy of Therapy and To Detect Recurrence or Metastases   11
III. Some Examples of Immunological Tests for Human Tumor-Associated Markers   12
   A. Alpha Fetoprotein (AFP)   12
   B. Carcinoembryonic Antigen (CEA)   14
   C. Ectopic Hormones   15
   D. Organ or Tissue TAAs   17
   E. Antigen-Antibody Complexes   19
IV. Conclusions   20
   References   21

I. INTRODUCTION

During the past 15 years there has been a rapid increase in interest and information on the existence and potential clinical value of human tumor-associated antigens. This has led to extensive efforts at immunotherapy and immunodiagnosis of cancer, and at immunological monitoring of cancer patients.

Despite considerable progress in this field, recently there has been a substantial amount of skepticism about these applications, and much of this can be traced to uncertainties about the very existence of human tumor-associated antigens. Part of this uncertainty can be attributed to problems in semantics. Early workers in tumor immunology looked for and described "tumor-specific antigens," i.e., antigens on tumor cells that were qualitatively different from antigens on any normal cells. The complete specificity required by this definition is quite difficult to prove and, with few exceptions, may not be correct. The main problem is to demonstrate sufficiently that a tumor antigen is entirely absent from normal tissues at any time in development. Many of the currently used techniques are relatively insensitive and could fail to detect the presence of small amounts of antigen. Furthermore, some antigens may occur only in some types of normal cells or in cells at one stage of differentiation.

In general it would seem most satisfactory to examine closely normal cells of the same organ and of the same histological type as the cancer cells. In addition it may frequently be necessary to study precursors of mature differentiated cells, either from embryos or from sites of active regeneration. Several of the tumor antigens which will be discussed below were originally thought to be tumor-specific or present only in tumor cells and embryonic cells. However, more concern for this issue and more sensitive techniques have shown that many antigens are present, at least in low amounts, in some normal cells after birth.

Because of these difficulties in demonstrating that an antigen is tumor-specific, the term "tumor-associated" has become popular. The term "tumor-associated antigen" (TAA) is essentially an operational definition for a substance that can be found on a tumor cell but is undetectable in the cells of a normal adult individual. By this definition an antigen detectable only in embryonic cells and tumor cells, a so-called oncofetal antigen, would be considered tumor-associated. Tumor-associated is often a provisional designation for an antigen that has not been fully characterized, or whose controls for specificity have not been, or for practical reasons cannot be, complete.

Beyond this retreat from tumor-specific to tumor-associated antigens, there is an increasing impression that if one looks hard enough, and with sufficiently sensitive techniques, most TAAs will be found to be present in some normal cells. This may be particularly true for TAAs that are common to a number of tumors of the same type. In animal tumor systems one category of common antigens that do appear to be restricted to tumor cells are those induced by oncogenic viruses. Even some of those may be present on morphologically normal cells infected by the viruses, long before malignant transformation occurs [1,2]. In the absence of a virus association with most human tumors, it is rather difficult to account for common, truly tumor-associated antigens. Of the common human TAAs whose specificities have been very extensively examined, the following have been shown to be the actual distribution of the antigens: (1) antigen with large quantitative differences in expression in tumor cells versus normal cells; (2) antigen well represented in normal tissues, but only in those of a particular type; (3) antigen expressed only or predominantly in normal cells at a particular stage in differentiation; (4) antigen that is detected in the circulation only of cancer patients, but it is actually a precursor molecule or a degradation product of a normal serum protein. Examples of antigens found in some of these categories will be discussed later.

It should be noted that tumor antigens which are ultimately shown to also be on normal cells may still be quite useful clinically. For example, for immunodiagnostic tests it must be possible to detect some consistent difference between cancer and noncancer. It is desirable but not necessary that the difference be qualitative. Quantitative differences between cancer patients and controls could also be sufficient. It would just be necessary to determine carefully the range of values in normal individuals and in patients without cancer.

To measure or localize TAAs or other immunological markers in cancer patients, most studies have relied on antibodies produced against these antigens in heterologous species. Unless highly purified antigens were used for immunization, the antisera usually reacted against many normal antigens and had to be absorbed extensively to render them specific for tumor-associated antigens. This should not be a problem for immune sera from cancer patients, and one might expect such sera to be more practical reagents for immunodiagnosis. In fact, however, the antitumor reactivity of patients' sera has usually been relatively weak, and the sera have not lent themselves well to development of radioimmunoassays or similar sensitive techniques for quantification of antigen concentrations.

In this chapter I will discuss the types of clinical applications for tests for TAAs and the issues and problems related to each of the possible uses.

II. POTENTIAL CLINICAL APPLICATIONS

Before discussing any of the particular markers, it is important to consider the possible objectives for use of these tests and the issues and problems related to each type of application. Table 1 summarizes some of the major potential applications for these assays. It is important to note at the outset that only a few tests for human TAAs have been definitely shown to have a place in clinical oncology. As will be discussed below, results with a number of assays are quite promising, and one major reason for the lack of detailed clinical information is that it has been difficult to satisfactorily transfer the technology from the research laboratory to the bedside. Many of the problems are not unique to tests for TAAs but are also true for other types of laboratory diagnostic tests, including some that have been available for many years and some that have been incorporated into widespread use without real validation or objective assessment of utility.

TABLE 1 Potential Clinical Applications of Immunological Tests for Tumor Markers

1. Detection: screening of populations and high-risk groups
2. Aid in diagnosis of patients with signs or symptoms suggestive of cancer
3. Aid to histopathological evaluation of Tumors
4. Aid in staging of cancer patients
5. Localization of tumor and detection of metastases
6. Serial monitoring to determine efficacy of therapy and to detect recurrence or metastases

---

In regard to diagnostic applications of these tests, the following can be listed as general criteria for useful tumor markers:

1. A first, obvious point is that the test needs to be able to detect some consistent qualitative or quantitative difference between cancer and noncancer.

2. A good diagnostic test should have a high degree of specificity; there should be very few false positives, i.e., individuals without cancer who have tests indicating cancer. In this sense the percent specificity of a test may be defined as:

(1-Incidence of false positive tests) × 100,

where

$$\text{Incidence of false positive tests} = \frac{\text{no. positive tests among individuals without cancer}}{\text{total noncancer individuals tested}}$$

3. Also, the test should be very sensitive and have few false negative results, i.e., it should be able to detect cancer in a large proportion of cancer patients. The percent sensitivity of a test may be defined as:

(1-Incidence of false negatives) × 100,

where

$$\text{Incidence of false negatives} = \frac{\text{no. negative tests among cancer patients}}{\text{total cancer patients tested}}$$

A particularly useful test for a tumor marker would be one that was positive even in cancer patients with localized tumors or small, metastatic deposits which were asymptomatic and undetectable by conventional diagnostic tests.

4. Some tests for tumor markers give only qualitative results, i.e., they are either positive or negative. However, a test is much more valuable if it provides quantitative information on the levels of the marker. Tests that have a large quantitative range between clinically detectable tumors and absence of tumors are particularly useful, since they offer the possibility of closely monitoring changes with tumor burden, and since they

are most likely to provide indications of small amounts of tumor. Since TAAs are products of tumor cells themselves, if they are released into the circulation their levels would be expected to depend on the mass of tumor. However, a variety of factors may influence the levels of tumor products:

(a) The number of tumor cells present.
(b) The proportion of tumor cells synthesizing the antigen and the synthetic rate per cell. Only certain cells within a tumor may make the TAA, and production may vary with the phase in the cell cycle and with the stage of differentiation of the cell.
(c) The location of the marker within the tumor cell and the mechanism for release from cells and entry into circulation. Some TAAs are cell membrane constituents or secretory products and may be shed or released from viable cells. Other TAAs may be intracellular constituents which would only be released when the tumor cells lose viability. Some TAAs released from solid tumors might enter the circulation in appreciable quantities only after invasion of blood vessels. With such antigens levels in the region of the tumor or in directly contiguous body fluids or excretions might be much higher than in the circulation, and testing of these might be more useful than tests on serum.
(d) The half-life of the antigen in the circulation can also vary considerably, depending on the size and nature of the substance. If the circulating antigen is immunogenic to the host and antigen-antibody complexes are formed, it is likely to be cleared much more rapidly than would a nonimmunogenic marker. When markers are not tumor products but rather are produced in response to tumor growth, the relationship between levels of response and the mass and extent of spread of the tumor might be quite different. It is not possible to set down any general principles for such reactive markers.

5. It would be very helpful if a test to be used for initial detection or diagnosis could provide information about the tumor type and location. One of the main concerns about detecting occult, clinically undetectable cancer is the difficulty in determining what type of cancer it is and where it is located. Clearly, at present more information than a diagnosis of "cancer, type and site unknown" would be needed for rational therapy. Therefore specificity of the TAA for a particular organ site or histologic type of cancer is an important factor initially, but it is not as essential for monitoring previously diagnosed patients.

A further issue concerns not so much tests for TAAs themselves, but the design of the studies to evaluate the usefulness of the marker. It is essential that the measurements and the data analyses be performed objectively, without knowledge of the clinical diagnosis or status of the patient. To accomplish this it is important that the laboratory receives coded specimens, without any identifiers as to source or type of donor. Beyond this it is necessary to design appropriate studies for the particular clinical application for which the tumor marker will be used. Adequate studies of this type have been performed with very few of the available tests for TAAs, and even then only for some of the possible clinical applications. The oncologist can and should play a central role in designing good studies, to provide solid information on the clinical value of a test.

Another point to emphasize in regard to the usefulness of various assays for one or more of the clinical applications is that a single test may not have sufficient specificity or sensitivity, but that the simultaneous use of several tests may provide highly discriminatory data. It is possible that assays for two or more immunological or biochemical markers would have additive or synergistic effects for improving the sensitivity or specificity of detection of tumor cells.

A. Screening

The use of tests to screen for cancer is probably the most difficult of the various potential applications to bring to fruition. If a test has been shown objectively to discriminate well between cancer and control groups, it then has to be evaluated for its use in screening by a study with an appropriate design. Despite the large number of tests for TAAs which have been developed, only a few have been directly evaluated for their use in screening. Many factors can affect the feasibility of a particular test for screening purposes. Most of these factors need to be extensively considered before a study on the possible usefulness of screening can be initiated.

1. It is particularly important that the assay be relatively simple and practical for application to testing of very large numbers of specimens or individuals. The procedures must be sufficiently well developed and standardized, so that reproducible results can be obtained over time and in many laboratories.

2. A suitable population must be available for study. For rapid identification of a useful test, it is very helpful to identify populations or families at high risk of developing cancer. It is very important to have sufficient access to the population, to permit retesting as appropriate, and to perform extensive clinical evaluations, particularly of the test-positive individuals. Furthermore, most of the individuals in the population must be available for clinical follow-up over a period of several years to determine which initially disease-free individuals, among both the test-positive and test-negative individuals, subsequently develop cancer.

3. Of fundamental concern is the specificity of the assay. It is difficult to make general statements about the acceptable levels of specificity

for screening tests. However, it would be important for a new screening test to be shown to have better specificity than currently available detection techniques, or to lead to improved specificity when used in conjunction with these procedures.

4. A further important issue is that the test should be very sensitive. To be useful as a screening procedure, the test should be able to detect asymptomatic individuals bearing small, localized tumors at a time when the disease is treatable and has not yet metastasized. The longer the lead time that a test provides (i.e., the interval between test positivity and clinical detection of disease), the more likely it is that the test will contribute to better response to therapy and to survival. It should be noted that, in order to determine accurately the lead time for a particular assay, tests need to be performed repeatedly to establish the point when the test first becomes positive. The amount of lead time a test might provide, and the likely clinical benefits to be accrued by early detection of disease, are likely to be determined considerably by the rate of tumor growth. Screening tests are more likely to be useful for slowly growing tumors with long latent periods than for explosively growing tumors that metastasize early.

B. Aid in Diagnosis of Patients With Signs or Symptoms

Some of the issues discussed above for screening also apply to the application of tests for TAAs as adjuncts to the diagnosis of cancer in patients with signs and/or symptoms suggestive of cancer. However, the concerns regarding the sensitivity and specificity of the assay are somewhat different. In regard to specificity in this setting, one is not concerned with identifying the small proportion of individuals with cancer out of a very large population of normal individuals. Rather it is necessary to discriminate among a group of individuals with disease, to determine which have malignant versus benign diseases. This presents particular problems since most forms of cancer arise in older populations with a variety of underlying benign chronic diseases that often affect the same organ system as the cancer. However, if a major cause of false positive results is a benign condition or conditions not involving the same organ system as the cancer, e.g., the liver, it may be possible to perform the relevant test to rule out those other conditions.

In regard to sensitivity, the test would not have to be able to detect very small, asymptomatic lesions. However, to be useful as a diagnostic adjunct, it should still be capable of detecting most resectable or otherwise treatable, localized tumors.

C. Aid to Histopathological Evaluation of Tumors

Recently the value of evaluating tumor specimens for the presence of various TAAs and markers has begun to be increasingly appreciated. This analysis may provide assistance in the histopathological classification of the tumor. The presence or the amount of a marker may provide useful prognostic information, since these factors may reflect the state of differentiation, immunogenicity, or metastatic potential of the tumor.

Another very important aspect of marker evaluation in tumors is to provide needed information for the subsequent monitoring of the patient. Identification of one or more markers within the tumor would provide a solid basis for using the assays for those markers to follow the course of disease. Although similar information might be obtained by testing for the markers in a pretherapy serum sample, direct examination of the tumor is likely to be more sensitive and specific.

In patients with small, localized tumors, the serum levels before therapy are often in the normal range despite active production of the markers by the tumor. Therefore the failure of the patient to have an initial elevated marker level should not rule out the possible later use of that marker. In fact, with some of the markers discussed below, such a disparity between tumor and circulatory levels has been noted. In contrast, failure to detect a marker within the tumor would make it much less likely to be subsequently detectable. However, although there is little evidence for this, it remains possible that some primary tumors would be negative for a TAA while the metastases might become positive. This could be envisioned if production of the antigen were more likely in a less differentiated or more aggressively growing tumor cell. This possibility needs to be directly explored.

Very recently, improved methods have been developed for examining markers within tumors. In addition to the usual studies of intact cells or tissue sections by immunofluorescence, it is now possible to look for the distribution of markers in fixed and stained tissue sections, using conventional light microscopy. This has been made possible by the development of immunoperoxidase staining techniques. Using this procedure, one can now accurately determine both the presence of the antigen and its location within various cell types in the tumor.

D. Aid in Staging of Cancer Patients

Tests for TAAs in newly diagnosed cancer patients, before any therapy or after primary surgical removal of tumor, might be very useful as an aid in assessing the stage of disease. Since the circulating levels of some TAAs or the degree of immunologic reactivity has been found to depend on the overall tumor burden and on the extent of spread of the tumor, tests on preoperative specimens can provide useful prognostic information. For example, elevated pretherapy levels of TAAs, particularly when quite high, may suggest the presence of metastases and poor prognosis. The test results might also reflect the state of differentiation of the tumor and its inherent aggressiveness.

Most studies on this aspect of marker utilization have been performed on groups of patients to determine the overall relationship between marker level and extent of tumor, as determined at surgery and by subsequent clinical course. To supplement information obtainable by other means, the marker data should be able to make prognostic discriminations among patients with the same clinical or histopathological stage of disease. If this is possible, then the use of markers in conjunction with clinical and other laboratory information could provide the basis for improved staging of patients and for the administration of therapy appropriate to the assessed extent of disease.

It should be noted that the type of data gathered in this area provides information on the probable prognosis of patients within a group, rather than on the clinical status of the individual patient. As with other staging criteria, the results of a test for TAA could not be taken as definitive evidence for occult metastases versus localized tumor.

One of the major challenges to the field of immunodiagnosis is to make the transition from population studies to a study in which statements can be made about individual patients, and on which therapy or other important clinical decisions may be based with some degree of certainty. One promising approach is the testing for levels of a circulating TAA after primary therapy, especially surgery, to determine whether all of the tumor was removed or eradicated. Persistence of elevated levels might provide strong evidence for residual tumor at the primary site or for regional or distant metastases. For example, after mastectomy for breast cancer, such a finding could indicate the need for axillary lymph node dissection and removal, or for radiotherapy or chemotherapy. On the other hand, the fall of elevated TAA levels into the normal range might provide some assurance of curative surgery, and might obviate the need for further therapeutic maneuvers. As will be discussed below, some of the tests are already being examined for this application.

To base important clinical decisions on marker data, a number of factors need to be considered: The assays for markers should be quantitative, and it would be desirable for them to be very sensitive so that they could reflect changes in levels over a wide range, hopefully 10-fold or more differences in concentration between the initial elevated values and the normal levels.

It must be emphasized that markers should not be expected to disappear immediately after tumor removal. The level at a particular time after therapy would reflect the original level and the length of time that the markers would remain in circulation. Thus, when elevated markers are detected after surgery, their significance can be evaluated only by obtaining additional specimens over a period of time. Furthermore, if the test is for reactive markers rather than for circulating levels of tumor-produced antigen, the disappearance after curative resection might not be expected. In fact, with some assays of reactive markers, the levels might increase after tumor removal.

E. Localization of Tumor and Detection of Metastases

Tests for TAAs could be very helpful in detecting the location of primary tumors and metastases, by at least two different approaches. The first, the determination of changes in levels of TAA in the blood in various regions of the body, is currently practical for some markers. The other approach, by in vivo localization of tumors by antibodies to TAA labeled with a radioisotope or heavy metal, is theoretically very appealing and potentially a powerful tool.

To detect the occult source of elevated levels of TAA, one can measure the differences in concentration in arterial and venous blood (A/V difference) in different regions where the metastasis is likely to be found. The utility of this approach depends on several factors. Of particular importance are the survival time in the circulation and the fraction of blood volume drained by the venous return being sampled, with more ready detection of A/V differences of short-lived antigens and of areas with low blood volume.

A related approach to localization would involve the measurement of TAA levels in extravascular fluids in the region of the tumor. For example, this might be particularly useful in detecting recurrent breast tumors by testing pleural fluid, or in detecting cerebral metastases by testing cerebrospinal fluid.

The possibility of localization of tumors by labeled antibodies to TAAs is a very attractive one. The rationale behind this approach is straightforward. If antibodies are specifically directed against antigens on a tumor, and if these antigens are absent or in low amounts on normal tissues or are inaccessible to systemically inoculated antibodies, one would expect to have localization of the antibodies at the site of tumor growth. Radiolabeling the antibodies, or labeling them with heavy metals, would allow the accumulation of antibodies to be detected by isotopic or radiologic scanning equipment.

In spite of these simple principles, little progress has been made in this area, and application at the clinical level is just now being attempted (see the discussion below on localization with antibodies to carcinoembryonic antigens). Some of the practical problems to be faced include:

1. Location of marker in tumor cell. One would anticipate that this procedure would be more easily applied to markers that are on the tumor cell surface than to those present only inside viable tumor cells.
2. Adequacy of blood flow to the tumor.
3. The levels of circulating marker. One would expect the inoculation of labeled antibodies into the circulation of a patient with high serum levels of marker would result in rapid antigen-antibody complex formation and nonspecific clearance of the label.

1 / Biochemical Markers for Cancer   11

    4. The proximity of the tumor to the major sites of nonspecific uptake of the labeled material. For example, it might be much more difficult to detect tumor within, or in the region of, the liver than it would to localize tumor elsewhere in the abdomen.
    5. The proportion of specific antibody to nonspecific immunoglobulins in the labeled reagent. The more purified the antibody to the TAA, and the less denatured or less likely the preparation is to bind nonspecificically, the more likely it will be to detect small foci of tumors.

F. Serial Monitoring To Determine Efficacy of Therapy and To Detect Recurrence or Metastases

Tests for TAAs have at least two important applications to clinical oncology: following the response of patients to chemotherapy or other forms of therapy, and monitoring patients without evidence of disease for early detection of recurrence or metastases. For these purposes the following are some of the requirements or desirable features for tests measuring levels of TAAs: (1) The assay should be sensitive and quantitative, and levels should be proportional to tumor mass. As pointed out earlier, tests for circulating antigens that vary over a wide range of concentrations are more likely to be useful. (2) For monitoring of patients after primary surgical resection of tumor, the test should become negative or return to normal levels after tumor removal. This could then provide a good low baseline for detection of rising levels. For this application the repeated levels of the patient himself provide the best baseline, and progressive increases in levels, even when still within the range of normal, could provide a meaningful indication of time recurrence. (3) The amount of fluctuation from one test to another must be considered. A variety of non-tumor related factors, including technical variation, therapy itself, and benign inflammatory or other transient diseases, could cause transient fluctuations or elevations in the assay. For almost all tests, therefore, it would be important to obtain repeated specimens over a period of time and to observe persistent or progressive elevation in antigen levels before the data should be taken as indicative of recurrence. (4) For accurate monitoring it is important that the antigen be produced by all or most of the tumor cells, so that recurrence of nonproducing tumors would be infrequent. If this is likely to be a problem, measurement of two or more antigens to detect each of the major types of tumor cells would be important. (5) The assay should be suitable for frequent, repeated testing. Assays requiring only small volumes of blood would obviously be advantageous. (6) For following the efficacy of therapy for metastatic or unresectable tumors, most of the above principles apply. However, the issues are somewhat different in that one would be starting with elevated levels and usually looking for a decrease. Here it would be important to establish a baseline of elevated levels by repeat sampling before therapy. Shortly after chemotherapy or radiotherapy, the levels of antigen might transiently rise rather than fall, due to destruction of tumor cells and consequent release of antigens into the circulation. Although this phase must be avoided during the usual monitoring, the detection of increase in levels immediately after therapy might actually be useful as an indication of some responsiveness of the tumor to the therapy. Failure to see a rise might rapidly predict the failure of the particular therapy being used. This approach appears useful for polyamines, but has not yet been sufficiently well investigated for TAAs. Significant and progressive decreases in marker levels following therapy could provide a reliable indication of response to therapy, and could even provide quantitative information on the degree of decrease in tumor burden and the extent of residual tumor.

In making a decision on whether to base therapy on elevated levels of markers in the absence of any other indications of tumor recurrence, one needs to consider the predictive accuracy of the marker and the possible benefits to be gained from the therapy versus the risks to the patient from the additional therapy.

III. SOME EXAMPLES OF IMMUNOLOGICAL TESTS FOR HUMAN TUMOR-ASSOCIATED MARKERS

Many investigators have inoculated rabbits or other animals with human tumor cells or extracts and obtained antisera with selective reactivity against antigens in tumors. This approach is leading to the development of an increasing number of potentially useful immunodiagnostic tests for cancer. Some of the antigens detected by heterologous antisera are common to many human tumors and can be detected on tumor cells and/or in the circulation of cancer patients. The main categories of such antigens are oncofetal antigens, ectopically produced placental or other hormones, and TAAs associated with tumors of a particular organ or tissue type. Oncofetal and placental antigens are normally present in the tissues of the fetus or placenta and then disappear or are much reduced in number after birth. Two oncofetal antigens, alpha-fetoprotein and carcinoembryonic antigen, have been extensively evaluated for most of the possible clinical applications discussed above. These are discussed in detail in other chapters, and therefore I will deal only briefly with them.

A. Alpha Fetoprotein (AFP)

Alpha fetoprotein (AFP) was the first of the oncofetal proteins to be widely studied and to be shown to be useful clinically. When AFP was first studied, by agar gel precipitation tests, it appeared to be quite specific, being undetectable in the sera of normal adults or of those with benign liver diseases. However, with the development of radioimmunoassays and other assays with high sensitivity, it was found that AFP was not a true TAA, since many normal individuals had low levels (1-40 ng/ml) of circulating antigen [3,4].

Hepatomas are usually diagnosed clinically at a very late stage, when effective therapy is no longer possible. Therefore the possible usefulness of AFP for early detection of hepatoma led to several large-scale screening studies in areas with relatively high incidences of this disease.

There have also been several studies on the usefulness of AFP for differential diagnosis of hepatoma in patients with liver disease [5-7]. Rapidly increasing AFP levels were closely associated with hepatocellular carcinoma, but in most cases in which carcinoma was detected, the individuals had advanced, untreatable disease. Therefore measurement of AFP has thus far been found to have only limited value in early diagnosis of hepatomas.

Circulating AFP levels, in combination with a radioimmunoassay for human chorionic gonadotropin (hCG), have also been measured in many patients with germ cell tumors of the testis. For such studies adequate detection of elevated AFP levels in most patients required the use of a sensitive radioimmunoassay. Since 89% of patients with germ cell tumors had elevations in one or both of these tumor markers [4], these procedures were evaluated for their possible usefulness in the differential diagnosis of testicular masses [8]. Most patients in this study had benign lesions, and some of these had elevated levels of AFP or hCG. Most but not all of those with malignant testicular tumors had elevated levels. Despite the good discrimination, it seems unlikely that these tests will be very useful for initial diagnosis, since some patients without elevated AFP or hCG levels might still have malignant lesions that would have to be diagnosed by orchiectomy.

The radioimmunoassays for AFP and hCG have had more clinical value as aids in the staging of patients with germ cell tumors of the testis. When these tests were performed after orchiectomy but before lymph node dissection, persistently elevated levels were shown to be closely associated with nodal involvement [9]. The immunological results yielded fewer errors (13% false negatives) than did clinical and lymphangiographic assessment (39% false negatives or positives).

Measurement of AFP and hCG in testicular tumors by an immunoperoxidase technique has also been shown to be helpful in histopathological evaluation [10]. AFP was found in embryonal cells and endodermal sinus tumor cells, whereas hCG was restricted to syncytiotrophoblastic giant cells. When some tumors which were histologically considered to be pure seminomas were examined, they were found to have small nests of germ cells detectable by the sensitive immunological procedure.

Measurement of circulating AFP levels has been particularly useful in monitoring therapy in patients with hepatoma or germ cell tumors. With hepatomas, which are refractory to most chemotherapeutic agents, careful monitoring of AFP levels has helped to identify somewhat effective drugs [11]. With testicular germ cell tumors, for which there are more effective chemotherapeutic agents, monitoring AFP levels has been even more useful [12, 13]. A decline in elevated levels to normal was associated with clinically effective therapy, but because of mixed tumor cell elements in many tumors, it was necessary to measure both AFP and hCG levels.

B. Carcinoembryonic Antigen (CEA)

In 1969 a very sensitive radioimmunoassay was developed for carcinoembryonic antigen (CEA). It detected circulating levels of CEA in the plasma of most patients with colorectal carcinoma and some other types of gastrointestinal cancer, but in virtually no specimens from other types of malignancies, benign diseases, or normal adults [14]. However, extensive research by many investigators has made it apparent that CEA is not specific for gastrointestinal cancers. Elevated circulating levels of CEA have been described for many cancer patients with a variety of histological types (including carcinomas of the breast and lung) and for some patients with nonneoplastic diseases, particularly those involving inflammation (e.g., cirrhosis, ulcerative colitis, hepatitis, and cirrhosis [15]). These data may be explained by recent findings that CEA is also not a truly tumor-associated antigen, since it is detectable in normal tissues and secretions [16]. A further shortcoming of this test is that it is insufficiently sensitive to detect localized gastrointestinal or other cancers in many patients [15].

Despite the limitations in the specificity and sensitivity of the radioimmunoassay for CEA, this assay has found its way into widespread, almost general use. A radioimmunoassay for CEA is the only assay for cancer that has been licensed in the United States by the Food and Drug Administration, and it is estimated that over half a million tests for CEA are being performed in the United States each year.

Initially there was much hope that testing for CEA would be a good screening procedure for early detection of gastrointestinal cancers. However, probably because of the recognized problems with specificity and sensitivity, there have as yet been no reports of prospective screening studies. Two retrospective studies on previously collected serum specimens failed to show a sufficient correlation between elevated CEA levels and subsequent development of cancer [17, 18]. There was a substantial number of false positive and false negative results, and therefore these studies have provided little encouragement for the usefulness of CEA in the early detection of cancer.

Radioimmunoassays for CEA have also yielded somewhat disappointing results in regard to differential diagnosis of malignant and benign gastrointestinal and other diseases. For some time this application was considered to be particularly valuable, but most of the data came from studies on selected populations. In a large, well-designed prospective study of the possible usefulness of CEA in the initial diagnosis of gastrointestinal cancer in a symptomatic group of patients, there was too much overlap between the results with cancer patients and those with benign diseases [19].

In contrast to the major problems in applying CEA assays to detection or diagnosis, a number of recent studies have demonstrated a possible value of CEA in indicating prognosis and monitoring therapy. The most extensive data have been accumulated in studies of patients with colorectal cancer, and these are discussed in detail in Chapter 3.

In lung and breast cancer, measurement of CEA has appeared promising in assessing prognosis and detecting recurrences; the relevant data are also discussed in Chapter 3.

As a further potentially important clinical application, Goldenberg et al. [20] have recently shown that radiolabeled antibodies to CEA can selectively localize in tumor deposits. In a few instances metastatic lesions detected by this method were not found by clinical or conventional laboratory diagnostic procedures. Somewhat surprisingly, in view of the theoretical considerations discussed above, lesions were frequently detected by the radioantibodies despite rather high circulating levels of CEA. These results are quite encouraging, and it will be of considerable interest to determine the actual clinical value associated with the more accurate localization of tumors by this procedure.

C. Ectopic Hormones

It has been known for many years that some human tumors are associated with ectopic production of a variety of hormones. The development of radioimmunoassays for very sensitive detection of hormones has led to findings that a substantial proportion of some types of tumors produce ectopic hormones. The pattern of distribution of some of the ectopically produced hormones warrants their consideration as TAAs. Some hormones, like hCG and placental lactogen, are normally produced by the placenta but are virtually undetectable in the circulation of adults who are not pregnant and are tumor-free. Calcitonin, although produced in small amounts by normal C cells of the thyroid, is present in negligible amounts in the circulation of most tumor-free individuals, even after administration of stimulating agents [21]. The precursor form of adrenocorticotropic hormone (ACTH), termed pro- or big ACTH, normally does not enter the circulation and therefore represents an interesting category of tumor-associated proteins, being produced by a large proportion of lung cancers [22].

The details of the clinical applications of ectopic hormones are discussed in Chapter 11. For the present discussion I wish only to emphasize the demonstrated usefulness of detection of calcitonin for diagnosing medullary carcinoma of the thyroid, and of hCG for diagnosing and monitoring choriocarcinoma. Medullary carcinoma of the thyroid represents one disease for which an immunological test is an accepted, and probably the most sensitive, method for diagnosis. The radioimmunoassay for thyrocalcitonin is being used in many centers for diagnosis. Since the disease can be inherited within some families as an autosomal dominant trait, a very high-risk population has been available for screening. Tashjian and co-workers [23] screened a large number of members of five kindreds and performed 38 thyroidectomies on asymptomatic individuals, on the basis of positive calcitonin assays. Almost 90% of these individuals were found to have medullary carcinoma of the thyroid. Particularly encouraging was the finding that 40% of them had localized disease, which has a much better prognosis than the more frequent occurrence of extension beyond the gland.

Although these data are very impressive, their significance cannot be fully determined. It is not yet clear what the expected incidence of fatal thyroid cancer is among such families, and whether all cases of asymptomatic localized disease progress to invasion. Further, some of the individuals with positive tests were found to have only C-cell hyperplasia and no sign of malignancy. Although C-cell hyperplasia has been postulated to represent a preinvasion lesion [24], there is no direct information about its malignant potential.

Elevated calcitonin levels are not completely pathognomonic for medullary carcinoma of the thyroid, since some patients with subacute thyroiditis have had high basal levels [25], as have some patients with oat cell carcinoma of the lung [26] and other types of cancer [27]. Tests for basal levels of calcitonin do not appear to be sufficiently sensitive to detect thyroid cancer in many patients. However, infusion of calcium and particularly of pentagastrin [28] has improved the diagnostic accuracy of the test considerably.

It remains to be determined whether these procedures can detect all patients with medullarly carcinomas of the thyroid, and what the impact on survival will be. The very success of the calcitonin screening test has made it difficult to plan studies to answer these questions.

Elevated levels of human chorionic gonadotropin (hCG) can be detected by radioimmunoassay in the circulation of almost all patients with choriocarcinoma, most patients with germ cell tumors of the testis and ovary, and a small proportion of patients with other types of cancer [29]. Measurement of hCG levels can play an important role in differentiating choriocarcinoma from normal pregnancy [30].

Determination of serum and urinary hCG levels is also very useful in the staging of patients with choriocarcinoma. Very high values place the patient in a high-risk category, with an appreciably lower response rate to therapy with single agent chemotherapy [31,32]. Serial monitoring of hCG has become an established part of the management of patients with choriocarcinoma, since changes in levels provide an excellent reflection of tumor burden. This assay has allowed clinicians to determine accurately the most effective chemotherapy and how long to treat patients, and has led to considerable success in controlling the disease.

D. Organ or Tissue TAAs

Many investigators have immunized heterologous species with tumors of various organs. After absorption, some of the antisera have appeared to have selective reactivity with tumors of the same histological type or from the same organ. This has raised some hope that a battery of tests, each able to detect tumors of a particular organ, could be developed. Such tests would have obvious clinical value in screening and diagnosis. There has been some progress in this area, particularly with some tumor types, but various problems have been encountered. In addition to the frequent findings that an apparent organ or tissue TAA is actually a normal tissue constituent, most of the antigens have not been found to be restricted exclusively to one type of tumor.

There have been extensive studies of antigens associated with human leukemias [33]. Greaves and his associates prepared antibodies with the most impressive specificity by immunizing rabbits with acute lymphocyte leukemia (ALL) cells, precoated with antitonsil antibodies in an attempt to block recognition of normal lymphocyte antigens. Their reagent was quite specific for ALL cells lacking T-cell markers and for acute undifferentiated leukemias. The antigen was undetectable on fetal cells, activated lymphocytes, enzyme-treated lymphocytes, normal bone marrow cells, and virus-infected cells. Such good specificity was even seen by immunofluorescence, allowing the examination of individual cells.

These extensive studies suggested that the antiserum was clearly detecting a TAA and so held much promise for diagnosis. However, further recent studies have revealed this antigen to be present in normal fetal liver cells and, of particular concern, in regenerating bone marrow cells [33]. It therefore appears that this leukemia antigen is actually an antigen associated with a particular phase of normal differentiation of bone marrow stem cells into lymphocytes, and that the ALL cells have a tendency to be arrested at this point in differentiation. Greaves [33] has suggested that many, if not all, common human TAAs are actually differentiation antigens.

Gastrointestinal tumors have also been the subject of much study, following the pioneering studies on AFP and CEA. Because of the problems of specificity and sensitivity with the usual radioimmunoassays for CEA, many investigators have searched for other markers or have attempted to isolate more purified forms of CEA or fragments of CEA, which might provide the basis for assays with greater specificity for colorectal cancer or other gastrointestinal malignancies.

Edgington et al. [34] described an isomeric form of CEA, termed CEA-S. In their initial report on the use of radioimmunoassay to CEA-S, most patients with gastrointestinal cancer had elevated values, but very few patients with other types of cancer or with benign diseases had elevations. Therefore this assay appeared to have high tumor-site specificity along with good sensitivity.

In a further evaluation with coded sera from the National Cancer Institute, the assay for CEA-S detected a somewhat higher portion of patients with gastrointestinal cancer than did the CEA assay, and there was a remarkably low incidence of false positives in patients with other types of cancer and in non-tumor-bearing individuals [35]. Disappointingly, it has not yet been possible to standardize such an improved assay for large-scale clinical use. Leung et al. [36] performed studies with a fragment of CEA-M, prepared by cleavage with a reducing agent and termed TA. Preliminary results with an assay to TA suggested improved specificity relative to CEA.

Pusztaszeri et al. [37] recently produced antisera in rabbits against a high molecular weight antigen, the zinc glycinate marker (ZGM). ZGM was found in half of the colon carcinomas examined, but not in normal colon, and it appeared to be distinct from a variety of previously described antigens. Recently a radioimmunoassay to ZGM has been developed, but the clinical usefulness of this assay in detecting circulating ZGM remains to be determined.

Pant et al. [38] have described another new antigen associated with colon cancer, termed CSAp. It appeared to be mainly restricted to colon tumors but has also been found in inflammatory colon tissues, and therefore it is probably a normal tissue associated antigen rather than a TAA.

Fetal sulfoglycoprotein (FSA) was detected by Häkkinen using rabbit antisera to a glycoprotein fraction of gastric juice from patients with gastric carcinoma [39]. This has provided the basis for a screening test for FSA in gastric juice, since this antigen appeared to be absent from gastric secretions of normal individuals. Many people in Finland, which has a relatively high incidence of gastric cancer, have been screened, and the results to date illustrate some of the potential problems with screening. The overall false positive rate was only 3.3%, but among those with positive tests, only 1% were found to have gastric cancer detectable by gastroscopy.

Several groups have made considerable efforts to develop a test for a TAA of pancreatic cancer, since this is such a highly malignant and frequent type of cancer and is very difficult to diagnose. Gelder et al. [40] have described a pancreatic oncofetal antigen (POA) and measured it in the sera of patients by a quantitative rocket immunoelectrophoresis assay. Although this assay has shown some ability to discriminate about half of the patients with pancreatic cancer from controls, there have been some elevated values in patients with other types of cancer or with benign pancreatic or biliary disease. In preliminary studies serial determination of POA levels showed some correlation with clinical course.

Prostatic acid phosphatase has long been known to be a marker for advanced prostate cancer, and recently several investigators have prepared heterologous antibodies against it and have shown that levels of the circulating isoenzyme have the properties of a tumor-associated marker. Particularly encouraging has been the observation that a sensitive radioimmunoassay could detect elevated levels of the antigen in the sera of an appreciable proportion of patients with localized prostatic cancer. The actual clinical usefulness of this assay is now being intensively evaluated (see Chapter 6).

Several investigators have described TAAs in carcinoma of the cervix and have developed tests for circulating levels of antigens [41]. Kato and Torigoe [42], using a radioimmunoassay to a cervical TAA, was able to discriminate well between sera from cervical cancer patients and sera from controls. They have also obtained some evidence for the clinical value of the assay for monitoring the course of disease.

In the past few years several groups have docused their attention on antigens associated with ovarian cancer [43]. Bhattacharya and Barlow [44] described a mucoprotein antigen, OCAA, associated with cystadenocarcinomas, and have been able to develop a radioimmunoassay to detect this antigen in the circulation [43]. Despite current problems with low sensitivity in the assay, it has shown some promise for measuring the response to therapy of patients with advanced serous or mucinous cystadenocarcinomas of the ovary.

Several investigators have also tried to produce antisera to melanoma-associated antigens [45]. After extensive absorption some of these reagents appeared to detect common melanoma antigens [45,46]. However, the specificity of the reactivities has not been completely documented, and clinically useful tests have yet to be developed.

A number of attempts have been made to produce heterologous antisera to lung TAAs, but thus far there have been very few clinical applications. Recently Braatz et al. [47] isolated a lung-tumor-associated antigen and used the purified antigen to set up a radioimmunoassay [48]. Although there have been problems with cross-reactivity with a normal serum protein, the current assay appears to be able to discriminate fairly well between sera from tumor-bearing lung cancer patients and sera from heavy smokers, normal controls, and patients with benign lung diseases.

E. Antigen-Antibody Complexes

If antibodies are being produced by cancer patients against circulating TAAs or other antigens, then antigen-antibody complexes would be expected to be formed. Recently, several sensitive tests have been developed to detect such circulating immune complexes, and these procedures have been used to examine sera from cancer patients and controls. Several investigators detected elevated levels of complexes in the sera of an appreciable number of patients with cancers of various types [49-54]. Some studies have indicated a correlation between levels of circulating complexes and clinical course [49-51, 55]. In the studies by Carpentier and his associates [49, 50], circulating complexes were associated with extensive disease, and in patients with acute leukemia they were correlated with poorer response to chemotherapy and short survival.

Analysis of the antigens in the immune complexes of cancer patients might be expected to be a useful new way to identify immunogenic TAAs. But thus far there have only been limited efforts in this direction [53].

IV. CONCLUSIONS

A large body of evidence now clearly indicates that many human tumors have characteristic antigens. A central issue remaining after all these studies is the nature of the detected antigens. It has been particularly difficult in these clinical studies to ascertain definitively the specificity of the tumor antigens. This is particularly true for the most frequently described tumor antigens, those common to tumors of more than one patient. A large number of controls and tests must be done to determine whether some antigens on tumors are organ-associated antigens, differentiation antigens, virus-associated antigens, or fetal antigens.

Despite these difficulties in defining and characterizing TAAs, many human TAAs have been described through a variety of immunological approaches. A number of these have provided some promising indications for clinical usefulness, particularly in the area of managing cancer patients: For example, tests for TAAs provide some help in staging, assessing prognosis, and monitoring for early detection of recurrence.

In addition to the value of a particular test, many investigators are gaining experience with using multiple tests for TAAs, which may increase sensitivity and/or specificity. The application of these tests for screening and initial diagnosis of cancer has been very limited thus far. Very few tests for TAAs have provided sufficient encouragement regarding their ability to discriminate small numbers of tumor-bearing individuals from the large background of individuals with benign disease or no disease at all. Only the radioimmunoassay for calcitonin, in conjunction with provocative tests, has been reasonably well documented as useful in detecting cancer and then only for the relatively rare familial medullary carcinoma of the thyroid.

The above discussions make it clear that there are many gaps in our knowledge of human TAAs. There is a pressing need for more sensitive, standardized assays and for purification and characterization of the antigens, to see which are really TAAs. It will also be necessary to proceed methodically from promising indications for the clinical usefulness of some of these tests to well-designed trials to document their value for various important applications.

REFERENCES

1. L. Old, E. A. Boyse, and E. Stockert, Cancer Res. 25:813 (1965).
2. B. Wahren, Exp. Cell Res. 42:230 (1966).
3. E. Ruoslahti, and M. Seppälä, Int. J. Cancer 8:374 (1971).
4. T. A. Waldmann and K. R. McIntire, Cancer 34:1510 (1974).
5. T. Koji, T. Munehisa, K. Yamaguchi, Y. Kusumoto, and S. Makamura, Ann. N.Y. Acad. Sci. 259:239 (1975).
6. F. G. Lehman, Ann. N.Y. Acad. Sci. 259:199 (1975).
7. K. Okuda, Y. Kubo, and H. Obata, Ann. N.Y. Acad. Sci. 259:248 (1975).
8. M. R. Moore, C. L. Vogel, K. N. Walton, P. Counts, K. R. McIntire, and T. A. Waldmann, Obstetrics 147:167 (1978).
9. P. T. Scardino, H. D. Cox, T. A. Waldmann, K. R. McIntire, B. Mittemeyer, and N. Javadpour, J. Urol. 118:994 (1977).
10. R. J. Kurman, P. T. Scardino, K. R. McIntire, T. A. Waldemann, and N. Javadpour, Cancer 40:2136 (1977).
11. D. C. Ihde, R. C. Kane, M. H. Cohen, K. R. McIntire, and J. D. Minna, Cancer Treat. Rep. 61:1385 (1977).
12. N. Javadpour, K. R. McIntire, and T. A. Waldmann, Natl. Cancer Inst. Monogr. 49:209 (1978).
13. P. H. Lange, K. R. McIntire, T. A. Waldmann, T. R. Hakala, and E. E. Fraley, J. Urol. 118:593 (1977).
14. D. M. P. Thomson, J. Krupey, S. O. Freedman, and P. Gold, Proc. Natl. Acad. Sci. U.S.A. 64:161 (1969).
15. N. Zamcheck, T. L. Moore, P. Dhar, and H. A. Kupchik, N. Engl. J. Med. 286:83 (1972).
16. V. L. W. Go, H. V. Ammon, K. H. Holtermuller, E. Krage, and S. F. Phillips, Cancer 36:2346 (1975).
17. D. P. Stevens, I. R. Mackay, and K. J. Cullen, Br. J. Cancer 32:147 (1975).
18. R. R. Williams, K. R. McIntire, T. A. Waldmann, M. Feinleib, V. L. W. Go, W. B. Kannel, T. R. Dawber, W. P. Castelli, and P. M. McNamara, J. Natl. Cancer Inst. 58:1547 (1977).
19. T. Hersh, C. Moore, M. S. Perkel, and E. C. Hall, in Compendium of Assays for Immunodiagnosis of Cancer (R. B. Herberman, ed.), Elsevier, North Holland/New York, 1979, pp. 209-210.
20. D. M. Goldenberg, F. Deland, E. Kim, S. Bennett, F. J. Primus, J. R. Van Nagell, Jr., N. Estes, P. Desimone, and P. Rayburn, N. Engl. J. Med. 298:1384 (1978).
21. H. Health, III and C. D. Arnaud, in Immunodiagnosis of Cancer (R. B. Herberman and K R. McIntire, eds.), Marcel Dekker, New York, 1978, pp. 409-420.

22. G. Gewirtz and R. S. Yalow, J. Clin. Invest. 53:1022 (1974).
23. A. H. Tashjian, Jr., H. J. Wolfe, and E. F. Voelkel, Am. J. Med. 56:840 (1974).
24. H. J. Wolfe, K. E. W. Melvin, S. J. Cervi-Skinner, A. A. Al-Saadi, J. F. Juliar, C. E. Jackson, and A. H. Tashjian, N. Engl. J. Med. 289:437 (1973).
25. S. J. Cervi-Skinner and B. Castleman, N. Engl. J. Med. 289:472 (1973).
26. O. L. Silva, K. L. Becker, A. Primack, J. Doppman, and R. H. Snider, N. Engl. J. Med. 290:1122 (1974).
27. R. C. Coombes, C. Hillyard, P. B. Greenberg, and I. MacIntyre, Lancet 1:1080 (1974).
28. S. A. Wells, D. A. Ontjes, G. W. Cooper, J. F. Hennessy, G. J. Ellis, H. T. McPherson, and D. C. Sabiston, Jr., Ann. Surg. 182:362 (1975).
29. G. D. Braunstein, J. L. Vaitukaitis, P. P. Carbone, and G. T. Ross, Ann. Intern. Med. 78:39 (1973).
30. G. D. Braunstein, in Immunodiagnosis of Cancer (R. B. Herberman and K. R. McIntire, eds.), Marcel Dekker, New York, 1978, pp. 383-409.
31. D. P. Goldstein, in Controversy in Obstetrics and Gynecology II (D. R. Reid and C. D. Christian, eds.), W. B. Saunders, Philadelphia, 1974, pp. 219-230.
32. G. T. Ross, D. P. Goldstein, R. Hertz, M. B. Lipsett, and W. D. Odell, Am. J. Obstet. Gynecol. 93:223 (1965).
33. M. F. Greaves, in Immunodiagnosis of Cancer (R. B. Herberman and K. R. McIntire, eds.), Marcel Dekker, New York, 1978, pp. 542-587.
34. T. S. Edgington, R. W. Astarita, and E. F. Plow, N. Engl. J. Med. 293:103 (1975).
35. T. S. Edgington, E. F. Plow, W. Go, R. Herberman, P. Burtin, I. Jordan, C. Chavkin, D. H. Deheer, and R. M. Nakamura, Bull. Cancer 63:613 (1976).
36. J. P. Leung, T. Y. Eshda, and V. T. Marchesi, J. Immunol. 119:664 (1977).
37. G. Pusztaszeri, C. A. Saravis, and N. Zamcheck, J. Natl. Cancer Inst. 56:275 (1976).
38. K. D. Pant, H. L. Dahlman, and D. M. Goldenberg, Immunol. Communic. 6:411 (1977).
39. I. Häkkinen, in Immunodiagnosis of Cancer (R. B. Herberman and K. R. McIntire, eds.), Marcel Dekker, New York, 1979, pp. 342-357.
40. F. Gelder, C. Reese, A. R. Moosa, and P. Hunter, in Immunodiagnosis of Cancer (R. B. Herberman and K. R. McIntire, eds.), Marcel Dekker, New York, 1978, pp. 357-368.

41. A. J. Nahmias and R. B. Ashman, in Immunodiagnosis of Cancer (R. B. Herberman and K. R. McIntire, eds.), Marcel Dekker, New York, 1978, pp. 857-874.
42. H. Kato and T. Torigoe, Cancer 40:1621 (1977).
43. M. Bhattacharya, in Immunodiagnosis of Cancer (R. B. Herberman and K. R. McIntire, eds.), Marcel Dekker, New York, 1979, pp. 632-643.
44. M. Bhattacharya and J. J. Barlow, Cancer 31:588 (1973).
45. S. Ferrone and M. A. Pellegrino, in Immunodiagnosis of Cancer (R. B. Herberman and K. R. McIntire, eds.), Marcel Dekker, New York, 1979, pp. 588-632.
46. J. C. Bystryn and J. R. Smalley, Int. J. Cancer 20:165 (1977).
47. J. A. Braatz, K. R. McIntire, G. L. Princler, K. H. Kortright, and R. B. Herberman, J. Natl. Cancer Inst. 61:1035 (1978).
48. K. R. McIntire, J. A. Braatz, S. A. Gaffar, G. L. Princler, and K. H. Kortright, in Carcino-embryonic Proteins: Chemistry, Biology, Clinical Applications (F.-G. Lehmann, ed.), Vol. II, Elsevier/North Holland Biomedical Press, Amsterdam, 1979, pp. 533-540.
49. N. A. Carpentier, G. T. Lange, D. M. Fiere, G. J. Fournie, P. H. Lambert, and P. A. Miescher, J. Clin. Invest. 60:874 (1977).
50. N. Carpentier, P. H. Lambert, and P. A. Miescher, in Current Trends in Tumor Immunology (R. Reisfeld, S. Ferrone, and R. B. Herberman, eds.), STPM Garland Press, New York, 1979, pp. 165-174.
51. D. G. Jose and R. Seshadri, Int. J. Cancer 13:824 (1974).
52. H. Teshima, H. Wanebo, C. Pinsky, and N. K. Day, J. Clin, Invest. 59:1134 (1977).
53. A. N. Theofilopoulos and F. J. Dixon, in Immunodiagnosis of Cancer (R. B. Herberman and K. R. McIntire, eds.), Marcel Dekker, New York, 1979, pp. 896-937.
54. A. Theofilopoulos, C. B. Wilson, and F. J. Dixon, J. Clin Invest. 57:169 (1976).
55. R. D. Rossen, M. A. Reisberg, E. M. Hersh, and J. U. Gutterman, J. Natl. Cancer Inst. 58:1205 (1977).

2

ALPHA FETOPROTEIN

HIDEMATSU HIRAI / Department of Biochemistry, Hokkaido University School of Medicine, Sapporo, Hokkaido, Japan

I. Introduction   26
II. Physiological Concentration of AFP in the Serum   27
   A. Serum AFP Levels in Human Serum   27
   B. Serum AFP in Other Species   27
III. Physico- and Immunochemical Properties of AFP   28
   A. Purification   28
   B. Physicochemical Properties   28
   C. Amino Acid Composition and Sequence   28
   D. Carbohydrate Composition and Electrophoretic Microheterogeneity   29
   E. Immunological Cross-reactivity of AFPs between Mammalian Species   31
IV. Methods for Detection of AFP   32
   A. Immunodiffusion in Gel   32
   B. Hemagglutination, Latex Agglutination, and Complement Fixation   32
   C. Radioimmunoassay   32
   D. Enzyme Immunoassay   33
V. Physiology of AFP   33
   A. Biosynthesis of AFP in Vivo   33
   B. AFP Synthesis in a Cell-free System   33
   C. Estrogen Binding with AFP   34
   D. Immunosuppressive Effect of AFP   34
VI. Experimental Study of Hepatocarcinogenesis   34
   A. Liver Cell Proliferation and AFP Synthesis   34
   B. AFP Synthesis in the Precancerous Stage in the Liver   35
VII. AFP in Patients with Liver Diseases   36
   A. AFP in Hepatoma   36
   B. Serum AFP Levels in Patients with Hepatitis and Liver Cirrhosis   39

C. Early Diagnosis, Prospective Study and a Retrospective Analysis
           of Hepatoma   40
 VIII. AFP in Germ-Cell Tumors   42
        A. AFP in Patients with Germ-Cell Tumors   42
        B. Human Chorionic Gonadotropin (HCG) and AFP   42
   IX. Association of Elevated Serum AFP With Gastrointestinal Tumors
        Other than Hepatoma   43
    X. AFP in Other Nonmalignant Diseases   44
        A. Tyrosinosis   44
        B. Ataxia Telangiectasia   44
   XI. AFP in Pregnancy and Fetal Distress   44
        A. AFP in Normal Pregnancy   44
        B. AFP in Fetal Distress   45
  XII. Immunological Effect of Anti-AFP Serum on AFP-Producing
        Cells   45
        A. Effect of the Antiserum on Tumor Cells in Cultures   45
        B. Effect of the Antiserum on Tumors Transplanted in Animals   46
        C. Effect of the Antiserum on DAB Hepatocarcinogenesis in
           Rats   46
        D. Formation of the Homologous Antibody to AFP and Its Effect on
           Hepatoma Growth   46
        E. Effect of Antibody on the Fetus in Pregnant Animals   47
        F. AFP-Antibody as Specific Vehicle to Target Tumor Cells   47
 XIII. Concluding Remarks   48
        References   49

I. INTRODUCTION

Alpha fetoprotein (AFP) is a typical oncodevelopmental protein. It is synthesized mainly in fetal life, and practically no production occurs in normal adult. However, the synthesis commences again when some adult cells become transformed to cancer cells. The AFP gene or genes that control AFP synthesis repressed in adult cells are activated again by unknown mechanisms in association with cancer. This phenomenon was designated by W. H. Fishman in 1976, as an "oncodevelopmental gene expression."

AFP production is specific for fetal liver and yolk sac or for tumors that arise from the liver or from yolk sac elements. Detection of AFP in patients is an important tool in diagnosing AFP-producing tumors and is indispensable to the antenatal diagnosis of birth defects. This chapter surveys the chemical and immunological nature of AFP, its behavior in the fetus and in cancer, and the detection of this protein in clinical medicine.

The pioneer works of Stevens (1959) and Pierce and Wallace (1971) have demonstrated the totipotential of mouse teratocarcinoma cells, and have shown that under appropriate conditions these cells differentiate into normal cells. The tumor cells and the embryo are similar, except that cancer cells have unconditioned growth. Synthesis of AFP in both types of cells is just one example of a shared property.

Anti-AFP antibody is cytotoxic to AFP-producing cells in vitro. Immunotherapy or immunoprevention of cancer by using antiserum to AFP, as well as detection of tumor localization with tagged AFP antibody, are important approaches that are being studied extensively.

II. PHYSIOLOGICAL CONCENTRATION OF AFP IN THE SERUM

A. Serum AFP Levels in Human Serum

In the early stage of gestation AFP is a main component of the fetal serum. The highest concentration occurs in 12-14 weeks of gestation, when it reaches 1-3 mg/ml. Subsequently, it decreases as gestation proceeds. At term the AFP concentration in cord blood shows a wide distribution, from 4 to 188 µg/ml (Seppälä and Ruoslahti, 1973).

Within 1 year after birth the serum AFP levels decrease to the normal adult level, 0-25 ng/ml, with an average of 10 ng/ml. The level is so low that it can be detected only by sensitive methods such as radioimmunoassay.

B. Serum AFP in Other Species

Massayeff et al. (1975) determined the serum AFP levels of SD rats 3 weeks to 2 years old. AFP fell sharply after birth from 500 ng/ml at the third week to 116 ng/ml at the 32nd day. The average level was 18 ng/ml for adult rats. Watabe et al. (1972) quantitated the serum proteins of fetuses and neonates of Donryu rats with electrophoresis and the immunoprecipitation method of Mancini. The serum at day 14 of fetal life was composed of about 50% AFP and 20% albumin, and only traces of γ-globulin. AFP decreased with age, whereas albumin increased. AFP concentration in serum determined by the Mancini method was 7.5 mg/ml for the 14-day-old fetus, 4 mg/ml at birth, and 1 mg/ml 28 days after birth. AFP in pregnant rats was detectable by the 13th day of gestation and increased to 60 µg/ml at term. It decreased rapidly to less than 5 µg/ml on the 3rd day after parturition.

Olsson et al. (1977) measured serum AFP of normal adult mice of 27 different strains. AFP in both sexes was 34-173 ng/ml. But one strain, BALB/c/J, showed an extremely high level of about 1000 ng/ml (250-2300 ng/ml). The results of some hybridization and back-cross experiments suggested that the AFP in the mouse was controlled by a single Mendelian gene, the Rαf gene (for regulation of AFP).

III. PHYSICO- AND IMMUNOCHEMICAL PROPERTIES OF AFP

A. Purification

Physicochemical properties of AFP are similar to those of serum albumin. Fetal sera or ascitic fluids of some tumor patients are usually good sources for AFP isolation.

The immunochemical method was applied to the purification of human AFP in 1970 by Nishi and Hirai. Specific antiserum which reacted with AFP was mixed with fetal serum or ascitic fluids, and the resulting antigen-antibody complex was separated by centrifugation. The complex thus separated was dissociated into AFP and the antibody by dissolving it in an acidic buffer (pH 2). AFP and antibody were then separated by a molecule-sieving column.

The immunoadsorbent technique was also efficient in purifying AFP (Nishi et al., 1972a; Rouslahti et al., 1974). Purified human AFP has been crystallized either in a concentrated ammonium sulfate solution or in pentane diol solution (Nishi et al., 1972a; Hirai et al., 1976).

B. Physicochemical Properties

Table 1 lists the physicochemical properties of human and of rat AFP (Hirai et al., 1973, 1976). Rat AFP and human AFP are similar in physicochemical properties. Human serum albumin (Schultze and Heremans, 1966) is similar to AFP. Conventional physicochemical methods for protein purification are thus less applicable to the isolation of AFP, and immunochemical procedures are essential for purification of AFP.

C. Amino Acid Composition and Sequence

Table 2 shows the amino acid composition of human and rat AFP (Hirai et al., 1976). Human fetal AFP and hepatoma AFP have an identical composition (Nishi, 1970). The composition of rat AFP is similar to that of human AFP (Watabe, 1974). Amino acid compositions of AFP from mouse (Watabe, 1974; Zimmerman et al., 1977), rabbit (Pihko et al., 1973), swine (Carlsson et al., 1976), cattle (Marti et al., 1976), sheep (Lai et al., 1977; Marti et al., 1976), and chicken (Lindgren, 1976) are also similar to each other.

The N-terminal amino acid sequence of human AFP has been determined by several authors (Ruoslahti et al., 1976a, 1976b; Aoyagi et al., 1977). Yachnin et al. (1976a) reported a high degree of homology between the amino acid sequence of AFP and that of serum albumin. However, antisera to albumin and AFP do not cross-react immunochemically. The complementary dynamics of AFP and albumin synthesis in fetal life appear to suggest that albumin and AFP are two related molecules.

TABLE 1 Physicochemical Properties of AFP

	Human AFP fetal hepatoma		Rat AFP hepatoma	Human albumin[a]
Sedimentation constant $s_{20,w}$ [b]	4.50	4.50	4.79	4.6
Diffusion constant $D_{20,w}$ [c]	6.18	–	5.68	6.1
Partial specific volume V_{20} [d]	0.726	0.726	0.727	0.733
Molecular weight	64,600[e]		75,100	69,000
Isoelectric point pI [f]	4.7	4.7	4.7	4.9
$E_1^{1\%}$ cm (278 mμ)[g]	5.30	5.26	4.27	5.8
	5.30	5.26		

[a] From Schultze and Heremans (1966).
[b] 10^{-13} sec (S).
[c] 10^{-7} cm^2 sec^{-1}.
[d] ml/g.
[e] Calculated from $s_{20,w}$ and $D_{20,w}$; 70,000 was obtained by SDS disc electrophoresis.
[f] pH.
[g] Optical density of a 1% solution in 1 cm cuvette at 270 mμ.
Source: From Hirai et al. (1973).

D. Carbohydrate Composition and Electrophoretic Microheterogeneity

AFP contains 3-4% carbohydrates, and its composition is shown in Table 3. The carbohydrate composition of rat AFP has been reported by Kerchkaert et al. (1977). AFP can be chemically differentiated from albumin since albumin contains practically no carbohydrate.

Electrophoretic microheterogeneity of AFPs has been reported. It is assumed that the difference in electrophoretic mobility of variants is due to the difference in sialic acid contents, although sialic acid seems to play no role in immunological reactivity.

The heterogeneity of AFP with respect to its binding with Con A and other lectins has been reported by many authors (Nunez et al., 1976; Kerchkaert et al., 1977; Gold et al., 1978). Some molecules of AFP bind with lectins, while some others do not. Galactosyl and mannosyl moieties probably are responsible for Con A binding.

TABLE 2 Amino Acid Composition of AFPs[a]

	Human AFP fetus	Human AFP hepatoma	Rat AFP hepatoma	Human albumin[b]
Asx	87	88	91	86[c]
Thr	65	63	52	46
Ser	66	64	76	39
Glx	197	186	153	134[d]
Pro	38	39	40	41
Gly	47	48	44	20
Ala	90	87	82	102
Cys	20	24	27	27
Val	48	51	41	67
Met	8	10	24	8
Ile	44	46	46	13
Leu	94	96	99	99
Tyr	28	29	30	26
Phe	48	52	48	49
Lys	64	63	74	92
His	22	21	32	25
Arg	31	30	35	38
Trp	3	3	6	1.6

Units in mol/1000 mol amino acids.
Asx: aspartic acid + asparagine.
Glux: glutamic acid + glutamine.
Amide of albumin (asparagine + glutamine) is 63 mol.
[a] AFPs isolated from serum of human fetuses, ascites of hepatoma patients, and ascites of hepatoma-bearing rats.
[b] Data from Schultze and Heremans (1966) recalculated by the author as moles per 1000 mol amino acid residues.
[c] Asparic acid only.
[d] Glutamic acid only.
Source: From Hirai (1976).

The binding between AFP and fatty acid was demonstrated by Parmelee et al. (1978), and this should be taken into account in the electrophoretic microheterogeneity of AFP. Parmelee has speculated that the fatty acids in the AFP molecule has an effect on the immunosuppressive property of AFP.

E. Immunological Cross-reactivity of AFPs Between Mammalian Species

AFPs in fetal sera from 12 mammalian species were studied by Gitlin and Boesman (1967b). The fetal sera of squirrel, dog, cat, seal, sheep, and armadillo were shown to cross-react with a rabbit antiserum against human AFP. Nishi and Hirai (1972a) examined the cross-reactivity of AFP using antisera produced in rabbits, horses, and chickens against human, rat, and mouse AFP. Interestingly, the rabbit antiserum to human AFP reacted with the fetal serum of rabbit, which indicated that rabbits produced antibodies not only to human AFP but also the AFP of the rabbit itself. This phenomenon was interpreted as the breakdown of tolerance to AFP (Nishi and Hirai, 1972b). In general, AFPs of mammals cross-react with each other in spite of the barrier of species specificity. On the other hand albumin is quite species-specific. Therefore, from the evolutional point of view, AFP is a more primitive molecular species.

TABLE 3 Carbohydrate Composition of AFP

	Human		Rat hepatoma	Human albumin*
	Hepatoma	Fetus		
Nitrogen	14.9	14.7	13.1	16.0
Sulfur	1.8	1.7	2.0	1.5
Carbohydrate	3.4	–	5.3	0.08
Mannose	0.68 (3)			
Galactose	0.49 (2)			
Glucosamine	0.89 (4)			
Fucose	±			
Sialic acid	1.3 (3)			

Numbers in parentheses indicate moles per 70,000 g AFP.
Nitrogen and sulfur were determined at the Element Analysis Center, Hokkaido University. Carbohydrates were analyzed by gas chromatography.
Source: From Hirai et al. (1973).
*This column was reprinted by permission of the publisher from Schultze, A. E., and Heremans, J. P., Molecular Biology of Human Proteins, p. 182. Copyright 1966 by Elsevier North Holland, Inc.

IV. METHODS FOR DETECTION OF AFP

A. Immunodiffusion in Gel

Ouchterlony double immunodiffusion is a qualitative method with a sensitivity of 10 μg/ml. The single radial immunodiffusion (Mancini's technique) is a quantitative, simple, and reliable method, although its sensitivity also is 10 μg/ml. Electroimmunodiffusion (counterimmunoelectrophoresis, electrosyneresis, etc.) is a sensitive semiquantitative method; 1000 ng AFP/ml can be detected.

The so-called rocket method (Laurell, 1966) has also been applied to AFP determination. With crossed immunoelectrophoresis, a method refined by Axelsen et al. (1973) in which radiolabeling of AFP was successfully applied to increasing sensitivity, 100 ng/ml or less may be determined. The method is applicable not only to quantitation but also to characterization of antigenic structure.

B. Hemagglutination, Latex Agglutination, and Complement Fixation

Hemagglutination has been applied to AFP determination by using antibody-coated sheep erythrocytes (Nishi and Hirai, 1973a; Lehmann and Lehmann, 1974). It is as sensitive (10-20 ng/ml) as radioimmunoassay. However, nonspecific agglutination may occasionally be caused by some unknown serum factors. It is recommended that serum samples be diluted 10 to 20 times to avoid this nonspecific agglutination. The prozone phenomenon may also take place.

A latex agglutination technique was developed by Cahill and Cohen (1974) and by Alpert et al. (1973). Latex particles coated with antibody fraction were agglutinated by AFP present in samples. AFP at 500 ng/ml was detectable. Nonspecific agglutination by unknown serum factors was again observed.

Complement fixation technique has been applied to AFP determination by Kahan and Levine (1971) with a sensitivity of 50 ng/ml.

C. Radioimmunoassay

At present, radioimmunoassay (RIA) is the most sensitive and commonly used technique (Nishi and Hirai, 1971, 1973b, 1974; Ruoslahti et al., 1972). RIA is capable of detecting 1 ng/ml, but it is suggested that AFP determination of less than 20 ng/ml is not always reproducible by RIA.

Solid-phase RIA (the sandwich technique) has also been applied to AFP determination. Nishi et al. (1974) and Nishi and Hirai (1976) employed filter paper disks coated with a monospecific anti-AFP antibody. This method is simple, rapid, and sensitive enough to detect 0.1 ng/ml of AFP with a range of 1-1000 ng/ml, 10 times greater than that of the competitive RIA technique.

D. Enzyme Immunoassay

In enzyme immunoassay (EIA) enzymes are used in place of radioactive isotopes of RIA; therefore specific equipment for counting radioactivity is not necessary (Belanger et al., 1973a; Hibi, 1978). Enzymes such as alkaline phosphatase, peroxidase, and glucose oxidase have been applied.

V. PHYSIOLOGY OF AFP

A. Biosynthesis of AFP in Vivo

A very small number of dividing cells are present in the adult liver, and these proliferating liver cells may be responsible for the production of trace amount of AFP in the adult. Significant synthesis of AFP takes place only in the fetus. The liver, yolk sac, and gastrointestinal tracts have been found to be main sites for its synthesis in fetuses of mammals (Gitlin and Boesman, 1967a, 1967b).

AFP synthesis is observed most frequently in tumors that originate from liver (hepatoma) and germ cells (yolk sac tumors), and occasionally in primary tumors of the stomach. These organs are the sites of AFP synthesis in the fetus, and when cancer develops in these organs in the adult, the synthesis of AFP is resumed. This phenomenon forms the basis for development of cancer as a result of "dedifferentiation" or "retrodifferentiation."

B. AFP Synthesis in a Cell-Free System

AFP synthesis in mice at the level of messenger RNA has been investigated by Tamaoki et al. (1976) and Koga et al. (1974a, 1974b). Messenger-RNA-rich fractions (polysomal RNA) and ribosomal RNAs were isolated from both fetal and adult livers of mice and added to a protein-synthesizing system. AFP synthesis occurred overwhelmingly with fetal messenger RNA fraction, while synthesis of albumin occurred dominantly with adult messenger RNA fractions. No difference was observed between the ribosomal RNA fractions from fetal and adult livers. The data suggest that mRNA for AFP is enriched in fetal liver, that mRNA for albumin is enriched in adult liver, and that regulation of synthesis of these proteins takes place at the level of mRNA (transcription), but not at the level of ribosomal RNA (translation). It was also observed that, as the gestation proceeds, mRNA for albumin increased parallel to the synthesis of this protein.

C. Estrogen Binding with AFP

Nunez and his colleagues (1971) have observed that rat fetal plasma showed somewhat different properties than adult serum in binding with estrogens, and they identified the binder as AFP in fetal rat serum. Rat AFP binds especially tightly with estrone, estradiol, and diethylstilbestrol, but only loosely with estriol, and practically not at all with androgens. AFP binding with estrogens is so tight and specific that it has been applied to histochemical demonstration and to affinity chromatographic purification of rat AFP using estrogen as a ligand (Arnon et al., 1973). The dissociation constant of rat AFP-estradiol complex was 10^{-8} to 10^{-9} M^{-1} (Benassayag et al., 1975; Aussel et al., 1974). Mouse AFP also was found to bind estrogens, while human AFP showed practically no binding, if any (Nunez et al., 1976).

The physiological significance of estrogen binding of AFP remains unknown, although attempts have been made to correlate estrogen binding and the immunosuppressive activity of AFP.

D. Immunosuppressive Effect of AFP

In 1975 Murgita and Tomasi reported an immunosuppressive effect of mouse AFP. Their report attracted the attention of many investigators, since the physiological function of AFP has not yet been demonstrated at all. The immunosuppressive effect of AFP postulated by Tomasi's group was confirmed by some investigators (Keller et al., 1976; Zimmerman, 1977; Yachin, 1976b; Lester et al., 1976), but not by others (e.g., Belanger et al., 1976).

To determine whether AFP is truly immunosuppressive, it is necessary to exclude carefully any participation of other factors. This is absolutely necessary because in some instances the normal serum often contains immunosuppressive factors (Cooperband et al., 1968). Nevertheless, this phenomenon is interesting and potentially important.

This author does not doubt the possible immunosuppressive activity of AFP in some experimental systems, but the evidence to date is not complete enough to support the conclusion that the observed immunosuppression is due solely to the biological function of AFP.

VI. EXPERIMENTAL STUDY OF HEPATOCARCINOGENESIS

A. Liver Cell Proliferation and AFP Synthesis

The AFP gene seems to be expressed generally in association with cell proliferation, such as in the liver and yolk sac in the fetus, in regenerating liver cells, or in some tumor cells. The elevation of serum AFP in patients with hepatitis or liver cirrhosis has also been frequently observed, where regenerating and proliferating liver cells seem to be responsible for the AFP synthesis. The rate of AFP synthesis correlates to the extent of regeneration (Pihko and Ruoslahti, 1974). Although AFP synthesis occurs generally in association with cell proliferation, Watanabe et al. (1976) have postulated that in rat serum AFP synthesis is not necessarily dependent on the liver cell destruction with the subsequent proliferation of liver cells. The rate of synthesis of DNA and AFP were not always proportional, and in some conditions AFP synthesis occurred without DNA synthesis.

Alpert anf Feller (1978) measured serum AFP in a large series of patients who had undergone partial hepatectomy. None of the serum samples showed significant elevation of the serum AFP concentration. These observations in animal experiments and patients seem to indicate that the AFP gene is usually expressed when the competent cells proliferate, but in some cases the gene may be expressed without cell division.

B. AFP Synthesis in the Precancerous Stage in the Liver

An important phenomenon was observed by Watabe (1971) and Hirai et al. (1973) in studying hepatocarcinogenesis in rats with p-dimethylaminoazobenzene (DAB). Hepatoma generally develops by feeding rats with a diet containing 0.06% DAB for at least 12 weeks. However, AFP began to appear in the second and third week after the onset of DAB feedings. The appearance of AFP was transient, and AFP decreased and was undetectable in the 10th to 11th week, but then followed an enormous elevation of AFP in association with hepatoma development. The transient appearance of AFP in the early stage was given the name "primary reaction." This phenomenon was later confirmed by several investigators (Kitagawa et al., 1972; Kroes et al., 1972; deNechaud and Uriel, 1973; Doležalová et al., 1974; Kelleher et al., 1976). Histological changes in the primary reaction had been fully investigated by Onoé and his colleagues independently from Watabe's observation (Iwasaki et al., 1972; Onoé et al., 1975).

The oval cells that appear in the period of primary reaction seem to be involved in synthesizing AFP. AFP has been demonstrated in oval cells by the immunofluorescence technique (Fujita et al., 1975; Sakamoto et al., 1975; Kuhlman, 1978). Cells corresponding to oval cells in rat liver were observed in the livers of patients with hepatitis (Sakamoto et al., 1976).

Oval cells are considerably smaller than normal hepatocytes, have large nuclei, and are considered to be regenerating liver cells. When DAB feeding was stopped before the peak of the primary reaction (the seventh week), the oval cells disappeared and the histology of the liver returned entirely to normal. However, when DAB feeding was continued until the end of the primary reaction (the 10th to 11th week), almost all of the rats developed hepatoma. Therefore some irreversible changes in the cells took place in the period between the 7th and 10th week which resulted in the transformation of hepatocytes into hepatoma cells. Whether oval cells themselves become transformed directly into cancer cells or whether some other intermediate cells are transformed remains unknown. The changes of liver cells that take place in the primary reaction might correspond to the changes that occur in human hepatitis and liver cirrhosis. It should be noted that human hepatitis is caused frequently by viral infection, while experimental data in animals have been obtained mostly through chemical agents. Therefore a direct comparison of these two systems may not be proper.

Recently a mouse model system for spontaneous hepatoma development was studied by Becker et al. (1977) and Jalenko et al. (1978). They detected a slight elevation of AFP in these cancer-prone mice for a short period just before the appearance of tumors. Some premalignant histological changes were observed in the livers of mice with an elevated AFP. The AFP elevation and histological changes in the precancerous stage were similar to those observed in chemical carcinogenesis.

The dynamics of AFP synthesis in animals treated with hepatocarcinogenic chemical or viral agents, and in animals prone to spontaneous hepatoma development, provide valuable information toward the understanding of human hepatoma and may be beneficial in the early diagnosis of hepatoma patients or in screening high-risk groups.

VII. AFP IN PATIENTS WITH LIVER DISEASES

A. AFP in Hepatoma

1. Serum AFP Levels in Patients with Hepatoma

Tartarinov (1964a, 1964b) discovered elevated AFP levels in patients with hepatoma independently from the Abelev's (1963) observation of elevated AFP in mice bearing transplanted hepatomas. These discoveries of Russian researchers have been the basis for the current concept of oncodevelopmental gene expression.

Table 4 shows the serum AFP positivity in Japanese patients (Hirai et al., 1973). Approximately 80% of patients with histologically diagnosed hepatoma gave positive reactions (>10,000 ng/ml). Data from 38 reports up to 1970 summarized by Abelev (1971) are shown in Table 5. Approximately 40-80% of the patients' sera were positive. It appears that the positivity is higher in non-Caucasian countries, which might indicate some ethnic differences in the AFP positivity.

The data in Table 6 are arranged by this author from a summarized table in a review of Lamerz (1975), where AFP was measured by RIA. Ethnic differences were again observed between Japan and the Western countries, although the extent of the difference was smaller than that presented in Table 5.

Conclusions cannot be made on whether these discrepancies really represent a racial difference, since investigators used different methods

TABLE 4 Prevalence of AFP in Japan (>10,000 ng/ml)

Diseases	Number of cases	Positivity (%)	Remarks
Primary liver cancer	254/352	72	–
Hepatoma	178/227	78	Histologically confirmed cases only
Cholangioma	1/14	–	Histologically confirmed cases only
Metastatic liver cancer	5/218	3.3	–
Liver cirrhosis	6/350	–	–
Miscellaneous diseases	7/2360 [a]	–	Seven cases are teratoma of testis
Healthy adults	0/6009 [b]	–	–

[a] Serum samples were from a clinical laboratory excluding liver cancer patient samples.
[b] Donor blood was from a blood bank.
Source: Data up to September 1970. The data were collected from 14 institutions in Japan. AFP was detected by immunodiffusion in agar (sensitivity: >10,000 ng/ml).

to measure AFP. Also, differences may have originated from the different etiologies of the hepatomas.

2. AFP-Producing and AFP-Nonproducing Hepatoma

Hepatoma may be classified into two groups, AFP-producing and AFP-nonproducing hepatoma. Tumor cell cloning experiments have demonstrated that rat hepatomas consist of these two kinds of tumor cells. AFP levels in hepatoma patients are most likely decided by the ratio of the number of AFP-producing cells to the number of AFP-nonproducing cells. No difference in histological findings and clinical symptoms between AFP-producing and nonproducing hepatoma has been found so far.

3. Mosaic Structure of Hepatoma in Respect to AFP Synthesis

Immunohistochemical examination of AFP in hepatoma has demonstrated that only a fraction of the tumor cells were stained positively.

TABLE 5 Serum AFP Positivity in Patients with Hepatoma (>10,000 ng/ml)

	No. positive/total	Percent
England	18/41	43
France	18/30	60
Greece	18/35	51
USA	36/87	41
USSR	86/112	79
South Africa	178/223	80
Mozambique	29/56	52
Senegal	219/304	72
Uganda	31/58	53
Indonesia	87/100	87
Singapore	24/32	75
Hong Kong	27/42	64
Japan	153/228	67
Total	924/1348	69

AFP was detected by immunodiffusion in agar gel. Values greater than 10 μg/ml were taken as positive.
Source: Adapted from Abelev (1971).

TABLE 6 Serum AFP Positivity in Patients with Hepatoma Determined by Radioimmunoassay (>20 ng/ml)

Countries	No. of total cases tested	Percent positivity	Reference
Japan	78	84	Nishi et al. (1973b)
	107	93	Ishii (1973)
	100	90	Hirai et al. (1973)
	228	91	Hirai (1976)
England	22	77	Chayuialle et al. (1973)
Finland	14	86	Ruoslahti et al. (1972b)
USA	130	73	Bloomer et al. (1975)

Levels higher than 20 ng/ml were taken as positive.

Tsukada et al. (1973a, 1974a) cultured a rat ascites hepatoma AH66, a high AFP producer, from which 19 clones were separated by the soft agar method. These clones showed various degrees of AFP production including clones that were practically non-AFP-producers. Yoshida sarcoma has been known to be a typical nonproducing tumor cell line. However, Isaka et al. (1973, 1975) cloned Yoshida sarcoma and were successful in isolating some clones that produced AFP. These cloning experiments have clearly demonstrated the mosaic structure of hepatoma in relation to AFP-productivity, although some other factors should be taken into account. Tsukada and Hirai (1975) made a synchronized culture of two clones of the AH66 cell line; one was a high AFP producer and another was a high albumin producer. It is possible that AFP is detectable in cells only in a certain period of the cell cycle.

B. Serum AFP Levels in Patients with Hepatitis and Liver Cirrhosis

Serum AFP rises in patients with hepatitis and liver cirrhosis relatively frequently, but generally only at the RIA level. Tables 7 and 8 are rearranged from the data in reviews by Lamerz (1975) and Hirai (1976). Most positive cases had values under 400 ng/ml. According to the experience of this author, when the AFP levels exceed 400 ng/ml one should consider that the patient has a high risk of developing hepatoma. The ethnic difference in AFP levels seen in hepatoma patients appears to be observed also in

TABLE 7 Serum AFP Positivity in Patients with Hepatitis

Hepatitis	Countries	Total no.	Positive (%)	Reference
Acute	Japan	134	35	Ishii (1973)
	Canada	128	20	Silver et al. (1974)
	Finland	19	16	Ruoslahti et al. (1973)
Acute + subacute	Japan	759	41	Hirai (1976)
Subacute	USA	50	12	Bloomer et al. (1975)
Classical	USA	100	27	Bloomer et al. (1975)
Drug-induced	USA	16	31	Bloomer et al. (1975)
Alcoholic	USA	40	15	Bloomer et al. (1975)

Determined by RIA.
The values greater than 20-30 ng/ml were taken as positive.

TABLE 8 Serum AFP Positivity in Patients with Liver Cirrhosis

Countries	Total no.	Positive (%)	Reference
Japan	218	44	Hirai (1976)
Japan	139	27	Ishii (1973)
Japan	38	47	Nishi et al. (1973b)
USA	84	17	Bloomer et al. (1975)
Finland	13	15	Ruoslahti et al. (1972b)
Germany	318	32	Lehmann (1976)
Germany	43	12	Lamerz et al. (1975)

Determined by RIA.
The values greater than 20-30 ng/ml were taken as positive.

patients with hepatitis and cirrhosis. In the Japanese population the positivity is 27-47%, while in the Caucasian countries it is 12-13%.

Bloomer et al. (1975) analyzed over 200 cases of hepatitis and found that the positivity was approximately 30% in so-called classical and drug-induced hepatitis, and only 15% in alcoholic hepatitis. If Caucasian cases involve more patients with alcoholic hepatitis, the lower positivity of AFP levels in Western countries can be explained. The same rationale may apply to a lower positivity rate in liver cirrhosis in Western countries.

One hypothesis of how hepatoma develops is that it arises from regenerating liver cells which are caused by some liver injury such as hepatitis. The sequential development of hepatitis, cirrhosis, and hepatoma may be presumed, although no solid evidence supports such a conclusion. An association of HB antigen with hepatoma has been emphasized by several investigators (Purves et al., 1973b; Bloomer et al., 1975; Vogel et al., 1972), but discounted by others (Ruoslahti et al., 1973; Silver et al., 1974).

C. Early Diagnosis, Prospective Study, and a Retrospective Analysis of Hepatoma

A prospective study of early diagnosis of hepatoma by AFP measurement involved 9864 male adults in Daker (Leblanc et al., 1973). AFP was measured by the immunodiffusion technique. Nine cases of hepatoma were identified, of which one was surgically treated. Purves et al. (1973a) screened patients for hepatoma in a large southern African population with RIA; they found retrospectively that the hepatoma-sensitive population had high AFP levels.

With the cooperation of this author, Koji et al. (1975) have screened patients for hepatoma in a mass program in a town where the HB antigen or antibody incidence was greater than 10%. Approximately 1000 apparently healthy individuals were examined, and two who were AFP-positive (over 200 ng/ml) were detected. One patient had a successful partial hepatectomy to resect a tumor, 2-3 cm in diameter. The patient is alive and well 5 years after the operation.

A large mass screening assay survey was done in the People's Republic of China on 343,999 individuals using the Ouchterlony technique or counter-immunoelectrophoresis (Co-ordinating Group for the Research of Liver Cancer, 1974). The study identified 149 AFP-positive cases, and 129 (88.4%) were firmly diagnosed as having hepatoma. In a group of 54 AFP-positive patients, 20 had obvious signs of hepatoma, whereas the remaining 34 showed these signs 1-10 months (averaging 3.1 months) after the AFP screening. In a few cases the AFP determination led to curative resection of the tumors.

Lehmann (1976) conducted a prospective study in West Germany involving 318 patients with various kinds of liver cirrhosis and discovered 15 "clinically unknown" hepatomas. He reported that when patients with liver cirrhosis showed a serum AFP over 400 ng/ml, a high risk of hepatoma should be considered. Miyaji (1973) analyzed autopsy data of 10,303 cases of liver cirrhosis over 10 years (1958-1967) in Japan and found that the rate of association with hepatoma was 27%.

A retrospective analysis performed in patients in the Boston City Hospital (1917-1968) by Purtilo et al. (1973) showed that 35 of 98 subjects dying of hepatoma had fatty nutritional cirrhosis, whereas 22 of 98 had postnectotic cirrhosis. Chronic alcoholism was the pathogenic agent implicated in the production of fatty nutritional cirrhosis, hemochromatosis, and postnectotic cirrhosis in 56 of 98 patients dying of hepatoma. Eight years elapsed between the onset of alcoholic cirrhosis and the detection of hepatoma in 23 of these patients. The patients who had postnecrotic cirrhosis and/or hemochromatosis also had a high risk of developing a hepatoma (24%).

Okuda et al. (1975) suggested that patients with chronic hepatitis who undergo rapid transition to cirrhosis may be a high-risk group for the development of hepatoma. Following the serum AFP levels of such high-risk patients for a number of years may make it possible to diagnose the hepatoma early enough for the tumor to be resected.

VIII. AFP IN GERM-CELL TUMORS

A. AFP in Patients with Germ-Cell Tumors

Abelev (1967) first reported that 8 of 18 patients with teratoblastoma in the testes had elevated AFP levels according to the Ouchterlony method, whereas only 1 of 8 patients with seminoma and choriocarcinoma had an elevated AFP. Subsequently, many reports have substantiated the association of AFP with testicular and ovarian tumors.

1. Testicular Tumors

AFP levels of 153 patients with testicular tumor were reported by Grigor et al. in 1977. None of the 35 patients with pure seminoma showed raised AFP levels (over 25 ng/ml), whereas patients with active teratoma did have raised AFP. Most patients with teratoma actually had mixture of cell types including teratoma, yolk sac tumor, seminoma, and trophoblastic tumor elements. The AFP serum levels varied over a wide range from 25 ng/ml to 130,000 ng/ml. In the 74 cases of active teratoma, 56 (76%) were AFP positive, and all of the 10 cases with pure yolk sac tumor showed elevated AFP concentrations. Lange et al. (1976, 1977) observed 67 patients with testicular germ-cell tumor, of whom 27 had nonseminomatous tumors and all showed raised AFP. Sakashita et al. (1976, 1977) collected 36 cases of testicular tumor patients. None of 12 pure seminoma patients showed elevated AFP, whereas 20 of 24 patients with tumors containing yolk sac elements were AFP-positive.

2. Ovarian Tumors

Much as it does in testicular tumors, AFP appears in ovarian tumors that contain yolk sac tumor elements. Talerman et al. (1978) observed 14 cases of ovarian tumor, and 9 of the tumors contained yolk sac tumor elements; all had elevated AFP. The AFP level correlated well with surgery for the tumors and with the clinical course of the patients: Once the AFP-producing tumor was removed, the serum AFP concentration decreased to the normal level within 5-7 weeks. AFP rose again with recurrence of the tumor.

B. Human Chorionic Gonadotropin (HCG) and AFP

Concomitant synthesis of AFP and HCG in germ-cell tumors has been observed. Newlands et al. (1976) measured AFP and HCG in 44 patients with testicular tumors and showed a good concordance between these two markers. He has emphasized the importance of simultaneous determination of both markers. Kurman et al. (1976) histochemically examined HCG and AFP in 15 cases of ovarian tumor, and HCG was demonstrated in 10 cases.

Of these 10, seven showed the concomitant presence of AFP. However, there are some reports stating that AFP and HCG levels are not always parallel.

IX. ASSOCIATION OF ELEVATED SERUM AFP WITH GASTROINTESTINAL TUMORS OTHER THAN HEPATOMA

AFP synthesis is also observed in fetal stomach and intestine. Therefore, the re-expression of the AFP gene in cancers of these organs may also occur. Kuriyama et al. (1972) investigated 189 stomach cancer patients in Japan and found only one with an elevated serum AFP according to immunodiffusion (>10,000 ng/ml). Fifty patients were clinically diagnosed to be free of liver metastasis, and only 5 of 50 showed elevated serum AFP levels over 20 ng/ml.

McIntire et al. (1975) used RIA to examine 426 patients with gastrointestinal malignancy, including hepatoma. Elevation of AFP levels in these patients is shown in Table 9. They found 14 patients with stomach cancer in whom AFP levels were over 40 ng/ml. Of these, nine had liver metastases, and AFP levels over 100 ng/ml. One patient had an AFP greater than 10,300 ng/ml. Five patients with colon cancer also had an AFP over 40 ng/ml; three of them had liver metastases and an AFP level over 100 ng/ml, with one level at 60,000 ng/ml. Eleven patients with pancreas cancer had an elevated AFP; six of them had an AFP greater than 100 ng/ml, and five of these six had liver metastases.

TABLE 9 Elevated Serum AFP (>40 ng/ml) in Patients with Gastrointestinal Neoplasm

Primary tumor sites	No. of patients	No. of elevated	Percent elevated
Esophagus	4	0	0
Stomach	95	14	15
Small bowel	10	0	0
Colon-rectum	191	5	3
Pancreas	45	11	24
Biliary tract	8	2	25
Liver	73	51	70

Source: From McIntire et al. (1975).

X. AFP IN OTHER NONMALIGNANT DISEASES

A. Tyrosinosis

High AFP levels in patients with tyrosinosis have been reported (Belanger et al., 1973a, 1973b; Buffe et al., 1974; Guillouzo et al., 1976). Belanger observed 25 cases of children with tyrosinosis. AFP levels were 5000-500,000 ng/ml in most cases, and the highest level was 1 mg/ml. A positive correlation was observed between the AFP level, the extent of hepatic regeneration, and the circulating methionine level.

B. Ataxia Telangiectasia

Ataxia telangiectasia, an autosomal-recessive disorder characterized by cerebellar ataxia and oculocutaneous telangiectasia, is often accompanied by abnormal liver function and immunodeficiency. In most patients a deficiency in both humoral and cellular immunity has been observed. Absence of normal development of the thymus, and decreased levels of IgG and IgE are especially characteristic.

Waldman et al. (1972) studied 20 patients, 4-19 years old, with ataxia telangiectasia, and found an elevated serum AFP from 44 to 2800 ng/ml (normal: < 30 ng/ml) in all cases. The elevation was not observed in siblings and parents of the patients. They also examined patients with agammaglobulinemia and IgA and/or IgE deficiency, but all of these patients had normal levels of AFP. Sugimoto et al. (1977) examined five patients with ataxia telangiectasia who were 5-10 years old. The serum AFP levels were elevated in all cases to 97-283 ng/ml. Decreased IgA and IgE levels were also noted.

XI. AFP IN PREGNANCY AND FETAL DISTRESS

A. AFP in Normal Pregnancy

1. AFP in Maternal Serum

AFP appears in the serum of pregnant women and derives from the fetus via placenta and amniotic fluid. AFP appears in maternal sera as early as the seventh week of gestation (Seppälä and Ruoslahti, 1972) and reaches the highest level in the third trimester with a mean level of 150-250 ng/ml. After delivery maternal serum AFP decreases with a half-life of 4-5 days. AFP in cord blood at term (35-42 weeks gestation) ranges from 4-188 µg/ml (Seppälä, 1975).

2. AFP in Amniotic Fluids

AFP in amniotic fluids is 1/10-1/300 of that in fetal sera, and higher than that in maternal sera. AFP in amniotic fluid decreases as gestation proceeds. AFP in amniotic fluids is considered to be derived mostly from fetal urine.

B. AFP in Fetal Distress

AFP measurement in maternal sera and in amniotic fluids has become an important tool for antenatal detection of birth abnormalities. Brock and Sutcliffe (1972) discovered an extreme elevation of AFP in amniotic fluids when there was a neural tube defect in the fetus. Independently from Brock, Hino et al. (1972), working in this author's laboratory, found a highly elevated AFP level in the maternal serum of a woman who subsequently gave birth to an anecphalic baby. Previews on this subject are available elsewhere (Brock, 1977; Seppälä, 1977).

Open neural tube defects in the fetus produce elevations of AFP concentration in amniotic fluids from the second trimester with very few exceptions. In almost all cases of anecephaly, the AFP concentration in amniotic fluid was elevated. The elevation of AFP in amniotic fluids most likely is the result of a leakage of fetal cerebrospinal fluids through the open·neural tube defect. Reflecting high concentrations of AFP in amniotic fluids, the AFP levels in maternal serum are often elevated over the normal range. A United Kingdom collaborative study (Report, 1977) reported that maternal serum AFP levels were elevated in 88% of cases with anencephaly and 79% of cases with open spina bifida. Therefore screening for high-risk pregnancy is now possible by measurement of AFP levels in maternal sera.

An elevated AFP in amniotic fluids has also been reported in the following conditions: encephalocele, meningocele and/or myelocele, hydrocephalus, esophageal atresia, congenital nephrosis, intrauterine fetal death, and severe Rh immunization (Seppälä, 1975). The serum AFP levels in mothers with toxemia are low (Hino et al., 1972; Seppälä, 1975). When there is more than one fetus, maternal serum AFP levels are always higher than in a single pregnancy (Garoff and Seppälä, 1973).

XII. IMMUNOLOGICAL EFFECT OF ANTI-AFP SERUM ON AFP-PRODUCING CELLS

A. Effect of the Antiserum on Tumor Cells in Cultures

A cytotoxic and growth inhibitory effect of horse antisera against rat AFP on rat ascites hepatoma cells has been demonstrated in a culture system (Tsukada et al., 1973b, 1974b) using AH66 rat ascites hepatoma, a high AFP-producing cell line. Twenty percent of the tumor cells were killed within 2 hours after incubation with antiserum. The cytotoxicity was not complement-dependent. Two clones were separated from the AH66 cell line (Tsukada et al., 1973a, 1974a), a high AFP-producer and a low AFP-producer. As expected, anti-AFP antiserum had a strong effect on the high AFP-producer and a relatively weak effect on the low AFP-producer.

Hepatomas usually consist of a mixture of AFP-producing and non-producing tumor cells. It is expected that once killing of AFP-producing cells by antiserum takes place in a tumor, the AFP-nonproducing cells may be destroyed as well.

B. Effect of the Antiserum on Tumors Transplanted in Animals

Compared with in vitro experiments, only a slight effect has been observed on tumors transplanted in animals (Tsukada et al., 1973b). However, negative results have also been reported by Parks et al. (1974) and Sell et al. (1976).

C. Effect of the Antiserum on DAB Hepatocarcinogenesis in Rats

Kanada et al. (1978) observed a striking inhibitory effect of the AFP antiserum on the development of hepatoma in rats fed with 3'-Me-DAB. Histological examination of the rats treated with anti-AFP antiserum revealed a strong vacuolar degeneration of hepatoma cells. Administration of antiserum at the time of completion of the primary reaction (see p. 35) appeared to be of importance. Some putative precancerous cells seem to be ready for their transformation to hepatoma cells at the end of primary reaction in the 10th week, and these cells are most likely to be affected by anti-AFP antiserum. One-shot injection of the antiserum at the end of primary reaction was remarkable. This suggests the possible clinical use of heterologous antiserum.

D. Formation of the Homologous Antibody to AFP and Its Effect on Hepatoma Growth

An interesting phenomenon was discovered by Nishi et al. (1972b, 1973c). When they immunized animals (rabbits, rats, horses) with a heterologous (human) AFP the animals produced antibodies to both the heterologous and the homologous AFPs. The formation of homoantibody can be evoked by immunization not only with heterologous AFP, but also with chemically modified homologous AFP (Ruoslahti et al., 1975). These data may be interpreted either as a result of formation of cross-reactive antibody or as the "breaking of tolerance to the self-AFP."

Can the homoantibody affect tumor cells that produce homologous AFP? Ruoslahti et al. (1976c) immunized mice with rat AFP and observed the formation of both hetero- and homoantibodies in mice. These mice were then challenged with BW7756 mouse hepatoma, but no difference in tumor incidence was observed when compared to control mice. However, Goussev et al. (1974) and Yazova et al. (1976) demonstrated some effect of homoantibody formed in rats immunized with mouse AFP. After being challenged with Zajdele ascites hepatoma, all four control animals died with a tumor on day 8, but 5 of 10 homoantibody-producing rats were still alive on day 14.

Wahren et al. (1976) attempted the induction of homoantibody into patients. Two patients with testicular embryonal carcinoma exhibiting elevated serum AFP were given 1 mg of rabbit AFP once a month. The serum AFP decreased following the immunization. It was suggested that formation of homoantibody in patients caused the decrease of circulating AFP. Some clinical improvements were observed in these patients (Wahren, personal communication).

E. Effect of Antibody on the Fetus in Pregnant Animals

The embryocytotoxicity of anti-AFP antibody has been demonstrated by several authors (Smith, 1972; Nishi et al., 1973c; Slade, 1973; Mizejewski and Grimley, 1976). AFP antibody released into fetus via placenta may affect yolk sac or liver cells where AFP is produced, resulting in death of the fetus or abortion, although negative data have been reported (Leung et al., 1974). The placental barrier certainly should be taken into account.

F. AFP-Antibody as Specific Vehicle to Target Tumor Cells

Radiolabeled antitumor antibody can be used in the immunodetection of tumors. Goldenberg and his group have successfully localized the CEA-producing tumors by computerized photoscan, using radioiodine-labeled anti-CEA antibody both in animal models (Goldenberg et al., 1974) and in patients (Goldenberg et al., 1978). Over 90% of the patients with cancer of the colon and the lung were positively detected by this technique.

Anti-AFP antibody labeled with radioiodine has been used to detect hepatoma in rats in our laboratory (Koji et al., 1980). Rats transplanted with hepatoma or with primary hepatoma developed by feeding DAB were given radioiodinated anti-AFP antibody or its (Fab')2 fragment. A significant accumulation of labeled antibody was detected in tumor tissues, and the photoscan gave good images in some cases.

The use of AFP antibody as a carrier of chemotherapeutic agents to target tumor cells is a promising approach to the immunotherapy of cancer (Hurwitz et al., 1975; Rosenberg and Terry, 1977; Ghose et al., 1978).

XIII. CONCLUDING REMARKS

Alpha fetoprotein is produced in the fetus as well as in some cancer patients. It is also found in normal adults and in some patients with nonmalignant disease. The normal human serum level of AFP is less than 20 ng/ml. Production of this protein by cells of hepatoma, by teratocarcinoma, and by the fetal liver or yolk sac is extremely high. When serum AFP in patients older than 1 year is 400 ng/ml or more, a hepatoma or teratocarcinoma should be strongly suspected. Elevation of serum AFP is occasionally observed in patients with other malignancies, such as stomach cancer, and also in patients with nonmalignancies such as hepatitis and liver cirrhosis; in these cases the level of AFP is usually less than 400 ng/ml.

Not all patients who have hepatoma or teratocarcinoma have an elevated serum AFP. A normal level of AFP does not exclude the presence of these cancers. Patients with hepatoma often first present clinically with advanced disease demonstrating an elevated serum AFP. Therefore an early diagnosis of these cancers by serum AFP is not always expected, but moderate success in using AFP as a screening test has been reported.

Antiserum to AFP has been found to be cytotoxic to the AFP-producing cells. Studies to evaluate the therapeutic or preventive use of anti-AFP antibody should be performed, since antiserum to secretory-membrane-bound proteins may be a promising approach to preventing and treating cancer.

AFP has played an important role in our understanding of the mechanism of carcinogenesis, especially from the point of view of oncodevelopmental gene expression. Some functions of AFP have not been understood completely. Homology in chemical structure has been found for albumin and AFP.

Although confirmation is required, AFP has been reported to possess immunosuppressive properties which may play a role in causing the low immune reactivity of the fetus and the decreased immune capacity in cancer patients.

Rat and mouse AFP binds estrogen specifically and tightly, whereas human AFP binds only slightly. This phenomenon may have a physiological significance in rodents, although it cannot now be generalized for all mammals.

ACKNOWLEDGMENTS

The author expresses his thanks to Dr. Terry Wepsic of my laboratory (presently at the Veteran's Administration Hospital in San Diego, California) for his valuable suggestions in the preparation of this article. The author also thanks Dr. Yumiko Fukushi and Miss Masae Takada for their excellent technical assistance in collecting references.

REFERENCES

Abelev, G. I., Perova, S., Khramkoba, P. N., Postnikova, Z. A., and Irlin, I. S. (1963). Production of embryonal α-globulin by transplantable hepatomas. Transplantation 1:174-180.

Abelev, G. I., Assecritova, I., Kraevsky, N., Perova, S., and Perevodchikova, N. (1967). Embryonic serum α-globulin in cancer patients diagnostic value. Int. J. Cancer 2:551-558.

Abelev, G. I. (1971). Alpha-fetoprotein in ontogenesis and its association with malignant tumors. Adv. Cancer Res. 14:295-358.

Alpert, E., Coston, R. L., Cahill, J. F., and Cohen, H. (1973). Evaluation of a rapid latex-agglutination test for detection of α-fetoprotein. Tumor Res. 8:47-50.

Alpert, E., and Feller, E. (1978). Human AFP is not induced by normal liver regeneration. Scand. J. Immunol. 8:Suppl. 8.

Aoyagi, T., Ikenaka, T., and Ichida, F. (1977). Comparative chemical structures of human α-fetoproteins from fetal serum and from ascites fluid of a patient with hepatoma. Cancer Res. 37:3663-3667.

Arnon, R., Teicher, E., Bustin, M., and Sela, M. (1973). Preparation of antiserum to α-fetoprotéin making use of estradiol affinity column. FEBS Letters 32:(2)335-338.

Aussel, C., Ureil, J., and Mercien-Bodard, C. (1974). Isolement de l'α-fetoprotéine du rat (αFP) et étude de sa propriété de fixation d'oestrogènes. Proc. Int. Conf. "Alpha-Fetoprotein" (R. Masseyeff, ed.), INSERM 17-24, Nice.

Axelsen, N. H., Kroll, J., and Weeks, B. (eds.) (1973). A Manual of Quantitative Immunoelectrophoresis. Univerditetsforlaget, Oslo-Bergen-Tromso.

Becker, F. F., Stillman, D., and Sell, S. (1977). Serum α-fetoprotein in a mouse strain (C3H-Aryfb) with spontaneous hepatocellular carcinomas. Cancer Res. 37:870-872.

Belanger, L. J., Larochelle, J., Prive, L., and Belanger, M. (1973a). Tyrosinemie héréditarie. IV. Pathogénèse et traitment. Pédiatire 28:35-55.

Belanger, L., Slyvester, C., and Dufour, D. (1973b). Enzyme-linked immunoassay for AFP: Competitive and sandwich procedure. Clin. Chim. Acta 48:15-18.

Belanger, L., Daguillard, F., Waithe, W. I., and Dufour, D. (1976). Studies on the effect of alpha$_1$-fetoprotein on immune status of children with tyrosinosis and on mitogen-induced rat and human lymphocyte transformation in vitro. In Onco-Developmental Gene Expression (W. H. Fishman and S. Sell, eds.), Academic Press, New York, pp. 329-335.

Benassayag, C., Vallette, G., Cittanova, N., Nunez, E., and Jayle, M. F. (1975). Isolation of two forms of rat alpha-fetoprotein and comparison of their binding parameters with estradiol-17β. Biochim. Biophys. Acta 412:295-305.

Bloomer, J. R., Waldmann, T. A., McIntire, R., and Klatskin, G. (1975). Relationship of serum α-fetoprotein to the severity and duration of illness in patients with viral hepatitis. Gastroenterology 68:342-350.

Brock, D. J. H., and Sutcliffe, R. G. (1972). Alpha-fetoprotein in the antenatal diagnosis of anencephaly and spina bifidia. Lancet 29:197-199.

Brock, D. J. H. (1977). Prenatal diagnosis of neural tube defects. Eur. J. Clin. Invest. 7:465-472.

Buffe, D., and Rimbaut, C. (1974). α_1-Foetoprotéine dans les maladies métaboliques de l'enfant. Proc. Int. Conf. "Alpha-Fetoprotein" (R. Masseyeff, ed.), INSERM 191-195, Nice.

Cahill, J., and Cohen, H. (1974). A rapid screening test for detection of AFP as an indicator of fetal distress. Am. J. Obstet. Gynecol. 119:1095-1100.

Carlsson, R. N. K., Ingvarsson, B. I., and Karlsson, B. W. (1976). Isolation and characterization of alpha-fetoprotein from fetal pigs. Int. J. Biochem. 7:13-20.

Chayvialle, J. A. P., and Ganguli, P. C. (1973). Radioimmunoassay of alpha-fetoprotein in human plasma. Lancet 1:1355-1356.

Co-ordinating Group for the Research of Liver Cancer (1974). Application of serum alpha-fetoprotein assay in mass survey of primary carcinoma of the liver. J. Chinese Med. 2:241-245.

Cooperband, S. R., Bondevik, H., Schmid, K., and Mannick, J. A. (1968). Transformation of human lymphocytes: Inhibition by homologous alpha-globulin. Science 159:1243-1244.

DeNechaud, B., and Uriel, J. (1973). Antiènes cellulaires transitories du foie de rat. III. Mode de réapparition de l'α-fetoprotéine au cours de l'hépatocarcinogénèse chimique. Int. J. Cancer 11:104-115.

Doležalová, V., Šimíčková, M., Kočent, A., and Feit, J. (1974). Dynamics of α-fetoprotein production compared to α_2-macroglobulin during induction of primary hepatoma with 4-dimethylaminoazobenzene in rats. Neoplasms 21:369-380.

Fujita, S., Ishizuka, H., Kamimura, N., Kaneda, H., and Ariga, K. (1975). The α-fetoprotein-producing cells in the early stage of the experimental liver cancer. Ann. N.Y. Acad. Sci. 259:217-200.

Garoff, L., and Seppälä, M. (1973). Alpha fetoprotein and human placental lactogen levels in maternal serum in multiple pregnancies. J. Obstet. Gynecol. 80:695-700.

Ghose, T., Path, F. R. C., and Blain, A. H. (1978). Antibody-linked cytotoxic agents in the treatment of cancer. Current status and future prospects. J. Natl. Cancer Inst. 61:657-675.

Gitlin, D., and Boesman, M. (1967a). Sites of serum α-fetoprotein synthesis in the human and in the rat. J. Clin. Invest. 46:1010-1016.

Gitlin, D., and Boesman, M. (1967b). Fetus-specific serum proteins in several mammals and their relations to human α-fetoprotein. Comp. Biochem. Physiol. 21:327-336.

Gold, P., Labitan, A., Wong, H. C. G., Freedman, S. O., Krupey, J., and Shuster, J. (1978). Physicochemical approach to the purification of human α_1-fetoprotein from the ascites fluid of a hepatoma-bearing patient. Cancer Res. 38:6-12.

Goldenberg, D. M., Preston, D. F., and Hansen, H. J. (1974). Photoscan localization of GW-39 tumors in hamsters using radiolabeled anticarcinoembryonic antigen immunoglobulin G. Cancer Res. 34:1-9.

Goldenberg, D. M., DeLand R., Kim, E., Bennett, S., Primus, F. J., van Nagell, J. R., Ester, N., DeSimone, P., and Rayburn, P. (1978). Use of radiolabeled antibodies to carcinoembryonic antigen for the detection and localizing of diverse cancers by external photoscanning. N. Engl. J. Med. 298:1384-1388.

Goussev, A. I., and Yasova, A. K. (1974). Termination of natural tolerance to alpha-fetoprotein in rats: Its effect on hepatoma growth and course of pregnancy. Proc. Int. Conf. "Alpha-Fetoprotein" (R. Masseyeff, ed.), INSERM, Nice, 225-270.

Grigor, K. M., Detre, S. I., Kohn, J., and Neville, A. M. (1977). Serum alpha-fetoprotein levels in 153 male patients with germ cell tumors. Br. J. Cancer 35:52-58.

Guillouzo, A., Feldman, G., and Belanger, L. (1976). Localization of alpha-fetoprotein-containing liver cells in tyrosinosis and in newborn rat. In Onco-Developmental Gene Expression (W. H. Fishman and S. Sell, eds.), Academic Press, New York, pp. 647-654.

Hibi, N. (1978). Enzyme-immunoassay of human α-fetoprotein. GANN 69:67-75.

Hino, M., Nishi, S., and Yamashita, K. (1972). α-Fetoprotein in pregnant women. Igaku-no-Ayumi 82:512-513 (in Japanese).

Hirai, H. (1976). Carcinofetal proteins—4th Arne Tiselius Memorial Lecture. Protides Biol. Fluids (H. Peeters, ed., Pergamon Ltd.) 23:3-23.

Hirai, N., Nishi, S., Watabe, H., and Tsukada, Y. (1973). Some chemical, experimental and clinical investigations on α-fetoprotein. In Alpha-fetoprotein and Hepatoma (H. Hirai and T. Miyaji, eds.), GANN Monograph No. 14:19-34.

Hurwitz, E., Levy, R., Maron, R., Wilchek, M., Arnon, R., and Sela, M. (1975). The covalent binding of daunomycin and adriamycin to antibodies, with retention of both drug and antibody activities. Cancer Res. 35:1175-1181.

Isaka, H., Umehara, S., Hirai, H., Tsukada, Y., and Watabe, H. (1973). Isolation of α-fetoprotein producing cells from Yoshida ascites sarcoma and its clone. GANN 64:133-138.

Isaka, H., Umehara, S., and Tsukada, Y. (1975). α-Fetoprotein synthesis in cultured cells: Studies on Yoshida sarcoma and ascites hepatoma in vitro. Ann. N.Y. Acad. Sci. 259:74-79.

Ishii, M. (1973). Radioimmunoassay of α-fetoprotein. GANN Monograph on Cancer Res. University of Tokyo Press, Tokyo 14:89-98.

Iwasaki, T., Dempo, K., Kaneko, A., and Onoé, T. (1972). Fluctuations of various cell populations and their characteristics during azo-dye carcinogenesis. GANN 63:21-30.

Jalenko, H., Virtanen, I., Engvall, E., and Ruoslahti, E. (1978). Early increase of serum alpha-fetoprotein in spontaneous hepatocarcinogenesis in mice. Int. J. Cancer 21:453-459.

Kahan, B., and Levine, L. (1971). The occurrence of a serum alpha-1-protein in developing mice and murine hepatomas and teratomas. Cancer Res. 31:930-936.

Kaneda, H., Fujita, S., Ishizuka, H., Satoh, K., Honda, I., Nishi, S., Tsukada, Y., and Hirai, H. (1978). Effect of anti-rat AFP horse serum on the development of hepatoma in the rat by 3'-methyl-4-dimethyl aminoazobenzene. In Carcinoembryonic Proteins (F. G. Lehmann, ed.), vol. 1, Elseiver/North Holland, Amsterdam, pp. 539-544.

Kelleher, P. C., Nadworny, H. A., and Smith, C. J. (1976). Cyclic evaluation of pretumor rat serum alpha-fetoprotein during 3'-M-4-DAB hepatocarcinogenesis. In Onco-development Gene Expression (W. H. Fishman and S. Sell, eds.), Academic Press, New York, pp. 691-694.

Keller, R. H., Calvanice, N. J., and Tomasi, T. B., Jr. (1976). Immunosuppressive properties of AFP: Role of estrogens. In Onco-developmental Gene Expression (W. H. Fishman and S. Sell, eds.), Academic Press, New York, pp. 287-295.

Kerckaert, J. P., Bayard, B., Debray, H., Sautiere, P., and Biserte, G. (1977). Rat alpha-fetoprotein heterogeneity. Comparative chemical study of the two electrophoretic variants and their ricunus lectin-binding properties. Biochim. Biophys. Acta 493:293-303.

Kitagawa, T., Yokochi, T., and Sugano, H. (1972). α-Fetoprotein and hepatocarcinogenesis in rats fed 3'-methyl-4-dimethylaminoazo-benzene or N-2-fluoroenylacetamide. Int. J. Cancer 10:368-381.

Koga, K., O'Keefe, D. W., Ito, T., and Tamaoki, T. (1974a). Transcriptional control of α-fetoprotein synthesis in developing mouse liver. Nature 252:495-497.

Koga, K., and Tamoaki, T. (1974b). Developmental changes in the synthesis of α-fetoprotein and albumin in the mouse liver. Cell-free synthesis by membrane bound polyribosomes. Biochemistry 13:3024-3028.

Koji, T., Munehisa, T., Yamaguchi, K., Kusumoto, Y., and Nakamura, S. (1975). Epidemiological studies of α-fetoprotein and hepatitis B antigen in Tomie town, Nagasaki, Japan. Ann. N.Y. Acad. Sci. 259:239-247.

Koji, T., Ishii, N., Munehisa, T., Kusomoto, Y., Nakamura, S., Tamenishi, A., Hara, A., Kobayashi, K., Tsukada, Y., Nishi, S., and Hirai, H. (1980). Localization of radioiodinated antibody to alpha-fetoprotein in hepatoma transplanted in rats and a case report of AFP antibody in treatment of a hepatoma patient. Cancer Res. 40:3013-3015.

Kroes, R. W., Williams, G. M., and Weisburger, J. H. (1972). Early appearance of serum α-fetoprotein during hepatocarcinogenesis as a function of age of rats and extent of treatment with 3'-methyl-4-dimethylaminoazobenzene. Cancer Res. 32:1526-1532.

Kuhlmann, W. D. (1978). Localization of alpha$_1$-fetoprotein and DNA-synthesis in liver cell populations during experimental hepatocarcinogenesis in rats. Int. J. Cancer 21:368-380.

Kuriyama, N., Noguchi, S., Okamura, J., Murakami, F., and Jinnai, D. (1972). Alpha-fetoprotein in patients with gastric carcinoma with or without liver metastasis. Med. J. Osaka Univ. 23:133-140.

Kurman, R. J., and Norris, H. J. (1976). Endodermal sinus tumor of the ovary. A clinical and pathologic analysis of 71 cases. Cancer 38:4202-4219.

Lai, P. C. W., Hay, D. M., Peters, E. H., and Lorscheider, F. L. (1977). Immunochemical purification and characterization of ovine α-fetoprotein. Biochim. Biophys. Acta 493:201-209.

Lamerz, R., and Fateh-Moghadam, A. (1975). Carcinofetal antigen. I. Alpha-fetoprotein. Klin. Wochenschr. 53:147-169.

Lange, P. H., McIntire, K. R., Waldman, T. A., Hakala, T. R., and Fraley, E. E. (1976). Serum alpha-fetoprotein and human chorionic gonadotrophin in the diagnosis and management of nonseminomatous germ cell testicular cancer. N. Engl. J. Med. 295:1237-1240.

Lange, P. H., McIntire, K. R., Waldman, T. A., Hakala, T. R., and Fraley, E. E. (1977). Alpha-fetoprotein and human chorionic gonadotrophin in management of testicular tumors. J. Urol. 118:593-596.

Laurell, C. B. (1966). Quantitative estimation of proteins by electrophoresis in agarose gel containing antibodies. Anal. Biochem. 15:45-52.

Leblanc, L., Tuyns, A. J., and Masseyeff, R. (1973). Screening for primary liver cancer. Digestion 8:8-14.

Lehmann, F. G., and Lehmann, D. (1974). Hemagglutination method for the detection of α$_1$-fetoprotein. Colloq. Inst. Natl. Sante Rech. Med. 28:571-579.

Lehmann, F. G. (1976). Prognostic significance of alpha$_1$-fetoprotein in liver cirrhosis: Five-year prospective study. In Onco-developmental Gene Expression (W. H. Fishman and S. Sell, eds.), Academic Press, New York, pp. 407-415.

Lester, E. P., Miller, J. B., and Yachnin, S. (1976). Human alpha-fetoprotein as a modulator of human lymphocyte transformation:

Correlation of biological potency with electrophoretic variants. Proc. Natl. Acad. Sci. USA 73:4645-4648.

Leung, C. C. K., Watabe, H., and Brent, R. L. (1974). The effect of heterologous antisera to alpha-fetoprotein. Am. J. Anatomy 2:307-311.

Lindgren, J. (1976). Chicken alpha-fetoprotein: Molecular properties and expression. Protides Biol. Fluids (H. Peeters, ed., Pergamon Ltd.) 24:277-280.

Marti, J., Aliau, S., and Moretti, J. (1976). Studies on AFP from ovine and bovine fetuses. Protides Biol. Fluids (H. Peeters, ed., Pergamon Ltd.) 24:259-262.

Masseyeff, R., Gilli, J., Krebs, B., Calluaud, A., and Bonet, C. (1975). Evolution of α-fetoprotein serum levels throughout life in human and rats and during pregnancy in the rat. Ann. N.Y. Acad. Sci. 259:17-28.

McIntire, K. R., Waldmann, T. A., Moertel, C. G., and Go, V. L. W. (1975). Serum α-fetoprotein in patients with neoplasms of the gastrointestinal tract. Cancer Res. 35:991-996.

Miyaji, Y. (1973). The association of hepatocellular carcinoma with cirrhosis among autopsy cases in Japan during 10 years from 1958 to 1967. In Alpha-fetoprotein and Hepatoma (H. Hirai and T. Miyaji, eds.), GANN Monograph No. 14:179, University of Tokyo Press, Tokyo.

Mizejewski, G. J., and Grimley, P. M. (1976). Abortogeneic activity of antiserum to alpha-fetoprotein. Nature 259:222-224.

Murgita, R. A., and Tomasi, T. B., Jr. (1975). Suppression of the immune response by α-fetoprotein on the primary and secondary antibody response. J. Exp. Med. 141:269-286.

Newlands, E. S., Dent, J., Kardana, A., Searle, F., and Bagshawe, K. O. (1976). Serum α_1-fetoprotein and HCG in patients with testicular tumors. Lancet 2:744-745.

Nishi, S. (1970). Isolation and characterization of a fetal α-globulin from sera of fetuses and a hepatoma patient. Cancer Res. 30:2507-2513.

Nishi, S., and Hirai, H. (1971). Diagnosis of hepatoma with a radioimmunoassay of α-fetoprotein. Nuclear Medicine 8:233 (in Japanese).

Nishi, S., and Hirai, H. (1972a). Purification of human, dog and rabbit α-fetoprotein. Biochim. Biophys. Acta 278:293-298.

Nishi, S., Watabe, H., and Hirai, H. (1972b). Production of antibody to homologous α-fetoprotein in rabbits, rats and horses by immunization with human α-fetoprotein. J. Immunol. 109:957-960.

Nishi, S., and Hirai, H. (1973a). Hemagglutination test for α-fetoprotein. Tumor Res. 8:51-53.

Nishi, S., and Hirai, H. (1973b). Radioimmunoassay of α-fetoprotein in hepatoma, other liver diseases and pregnancy. In Alpha-fetoprotein and Hepatoma (H. Hirai and T. Miyaji, eds.), GANN Monograph No. 14:79-87, University of Tokyo Press, Tokyo.

Nishi, S., Watabe, H., and Hirai, H. (1973c). Species cross-reaction of α-fetoprotein and break-down of the tolerance to α-fetoprotein by immunization with heterologous α-fetoprotein. Tumor Res. 8:17-22.

Nishi, S., Hirai, H., Sekiguchi, K., and Kurata, K. (1974). Radioimmunoassay of tumor related antigens. Proc. 1st World Cong. Nucl. Med. (Tokyo) 169-172.

Nishi, S., and Hirai, H. (1976). A new radioimmunoassay of α-fetoprotein and carcinoembryonic antigen. Protides Biol. Fluids 23:303-307.

Nunez, E., Engelman, F., Benassayag, G., and Jayle, M. (1971). Identification et purification preliminaire de la foetoprotéine liant les oestrogènes dans le serum des rats nouveaux nés. C.R. Acad. Sci. Paris 273:242-245.

Nunez, E. A., Benassayag, C., Savu, L., Vallette, G., and Jayle, M. F. (1976). Purification and comparative estrogen binding properties of different forms of rat, mouse and human alpha$_1$-fetoprotein. In <u>Oncodevelopmental Gene Expression</u> (W. H. Fishman and S. Sell, eds.), Academic Press, New York, pp. 365-372.

Okuda, K., Kotoda, K., Obata, H., Hayashi, H., Hisamitsu, T., Tamiya, M., Kubo, Y., Yajushiji, F., Nagata, E., Junnouchi, S., and Shimokawara, Y. (1975). Clinical observations during a relatively early stage of hepatocellular carcinoma with special reference to serum α_1-fetoprotein levels. Gastroenterology 69:226-234.

Olsson, M., Lindalh, G., and Ruoslahti, E. (1977). Genetic controls of alpha-fetoprotein synthesis in the mouse. J. Exp. Med. 145:819-827.

Onoé, T., Kaneko, A., Dempo, K., Ogawa, K., and Minase, T. (1975). α-Fetoprotein and early histological changes of hepatic tissue in DAB-hepatocarcinogenesis. Ann. N.Y. Acad. Sci. 259:168-180.

Parks, L. C., Baer, A. N., Pollack, M., and Williams, G. M. (1974). Alpha fetoprotein: An index of progression or regression of hepatoma and a target for immunotherapy. Ann. Surg. 180:599-605.

Parmelee, D. C., Evenson, M. A., and Deutsch, H. F. (1978). The presence of fatty acids in human alpha-fetoprotein. J. Biol. Chem. 253:2114-2119.

Pierce, G. B., and Wallace, C. (1971). Differentiation of malignant to benign cells. Cancer Res. 31:127.

Pihko, H., Lindgren, J., and Ruoslahti, E. (1973). Rabbit α-fetoprotein. Immunochemical purification and partial characterization. Immunochemistry 10:381-385.

Pihko, H., and Ruoslahti, E. (1974). α-Fetoprotein production in normal and regenerating mouse liver. Colloq. Inst. Natl. Sante Rech. Med. 28:333-336.

Purtilo, D. T., Mazor, M. O., Usar, M. C., and Gottleib, L. S. (1973). Cirrhosis and hepatoma occurring at Boston City Hospital (1917-1968). Cancer Res. 32:458-462.

Purves, L. R., Manso, C., and Torres, F. D. (1973a). Serum alpha-fetoprotein levels in people susceptible to primary liver cancer in Southern Africa. In Alpha-fetoprotein and Hepatoma (H. Hirai and T. Miyaji, eds.), GANN Monograph No. 14:51-66, University of Tokyo Press, Tokyo.

Purves, L. R., Branch, W. R., and Geddes, E. W. (1973b). Serum alpha-fetoprotein. VII. The range of apparent serum values in normal people, pregnant women and primary liver cancer high risk. Cancer 31:578-587.

Report of the U.K. Collaborative study on alpha-fetoprotein in relation to neural tube defects (1977). Lancet 1:1323-1332.

Rosenberg, S. A., and Terry, W. D. (1977). Passive immunotherapy of cancer in animals and man. Adv. Cancer Res. 25:323-388.

Ruoslahti, E., Seppälä, M., Vuopio, P., Seksela, E., and Peltokallio, P. (1972). Radioimmunoassay of α-fetoprotein in primary and secondary cancer of the liver. J. Natl. Cancer Inst. 49:623-628.

Ruoslahti, E., Seppälä, M., Rosanen, J. A., Vuopio, P., and Helske, T. (1973). Alpha-fetoprotein and hepatitis B antigen in acute hepatitis and primary cancer of liver. Scand. J. Gastroenterol. 8:197.

Ruoslahti, E., Pihko, H., and Seppälä, M. (1974). Alpha-fetoprotein: Immunochemical properties. Expression in normal state and in malignant and non-malignant liver disease. Transplant. Rev. 20:38-60.

Ruoslahti, E., and Pihko, H. (1975). Effect of chemical modification of the immunogenicity of homologous α-fetoprotein. Ann. N.Y. Acad. Sci. 259:85-94.

Ruoslahti, E., and Terry, W. D. (1976a). α-Fetoprotein and serum albumin show sequence homology. Nature 260:804-805.

Ruoslahti, E., and Engvall, E. (1976b). Immunological cross-reaction between alpha-fetoprotein and albumin. Proc. Natl. Acad. Sci. USA 73:4641-4644.

Ruoslanti, E., Engvall, E., Jalenko, H., and Pihko, H. (1976c). Immunization of mice against autologous alpha-fetoprotein (AFP)—Reduction of serum AFP during tumorigenesis but lack of effect on incidence of transplanted hepatomas. In Onco-developmental Gene Expression (W. H. Fishman and S. Sell, eds.), Academic Press, New York, pp. 349-353.

Sakamoto, S., Yachi, A., Anzai, T., and Wada, T. (1975). α-Fetoprotein-producing cells in hepatitis and in liver cirrhosis. Ann. N.Y. Acad. Sci. 259:253-258.

Sakashita, S., Hirai, H., Nishi, S., Nakamura, K., and Tsuji, I. (1976). α-Fetoprotein synthesis in tissue culture of human testicular tumors and the examination of experimental yolk sac tumors in the rat. Cancer Res. 36:4232-4237.

Sakashita, S., Nishi, S., Hirai, H., and Tsuji, I. (1977). Synthesis of α-fetoprotein and some other serum proteins in testicular tumors. Invest. Urol. 15:2-4.

Schultze, H. E., and Heremans, J. P. (1966). In Molecular Biology of Human Proteins. Elsevier/North Holland, Amsterdam, p. 182.

Sell, S., Sheppard, H. W., Jr., Nickel, R., Stillman, D., and Michaelsen, M. (1976). Effect of anti-α_1-fetoprotein on α_1-fetoprotein-producing rat tumors in vivo and in vitro. Cancer Res. 36:476-480.

Seppälä, M., and Ruoslahti, E. (1972). Radioimmunoassay of maternal serum alpha fetoprotein during pregnancy and delivery. Am. J. Obstet. Gynecol. 112:208-212.

Seppälä, M., and Ruoslahti, E. (1973). AFP: Physiology and pathology during pregnancy and application to antenatal diagnosis. J. Perinatal Medicine 1:104-113.

Seppälä, M. (1975). Fetal pathophysiology of human α-fetoprotein. Ann. N.Y. Acad. Sci. 259:59-73.

Seppälä, M. (1977). Immunologic detection of alpha-fetoprotein as a marker of fetal pathology. Clin. Obstet. Gynecol. 20:737-757.

Silver, H. K., Deneault, J., Gold, P., Thompson, W. G., Shuster, J., and Freedman, S. O. (1974). The detection of α_1-fetoprotein in patients with viral hepatitis. Cancer Res. 34:244-247.

Slade, B. (1973). Antibodies to α-fetoprotein cause foetal mortality in rabbits. Nature 246:493.

Smith, J. A. (1972). Alpha-fetoprotein: A possible factor necessary for normal development of the embryo. Lancet 1:851.

Stevens, L. C. (1959). Embryology of testicular teratomas in strain 129 mice. J. Natl. Cancer Inst. 23:1249.

Sugimoto, T., Kodowako, T., Tozawa, M., Sawada, A., Kusunoki, T., and Yamaguchi, M. (1977). Blood levels of fetal proteins in patients with ataxia telangiectasia. Igaku-no-Ayumi 101:769-779 (in Japanese).

Talerman, A., Haije, W. G., and Baggerman, L. (1978). Serum alpha-fetoprotein (AFP) in diagnosis and management of endodermal sinus (yolk sac) tumor and mixed germ cell tumor of the ovary. Cancer 41:272-278.

Tamaoki, T., Miura, K., Lin, T., and Banks, P. (1976). Development changes in alpha-fetoprotein and albumin messenger RNAs. In Onco-developmental Gene Expression (W. H. Fishman and S. Sell, eds.), Academic Press, New York, pp. 115-122.

Tartarinov, Yu. S. (1964a). New data on the embryospecific antigenic components of human blood serum. Fed. Proc. 24:916-918.

Tatarinov, Yu. S. (1964b). Detection of embryonic specific α-globulin in the blood sera of patients with primary liver tumor. Vopr. Med. Klim. 10:90-91 (in Russian).

Tsukada, Y., and Hirai, H. (1973a). In vitro cloning of a rat ascites hepatoma cell line, with reference to alpha-fetoprotein synthesis. Tumor Res. 8:88-93.

Tsukada, Y., and Hirai, H. (1973b). Effect of anti-AFP-serum on rat hepatoma cells transplanted and cultured. Abstract of 32nd General Assembly of Jap. Cancer Association, 63.

Tsukada, Y., Mikuni, M., and Hirai, H. (1974a). In vitro cloning of a rat ascites hepatoma cell line, AH66, with special reference to alpha-fetoprotein synthesis. Int. J. Cancer 13:196-202.

Tsukada, Y., Mikuni, M., Watabe, H., Nishi, S., and Hirai, H. (1974b). Effects of anti-alpha-fetoprotein serum on some cultured tumor cells. Int. J. Cancer 13:187-195.

Tsukada, Y., and Hirai, H. (1975). α-Fetoprotein and albumin synthesis during the cell cycle. Ann. N.Y. Acad. Sci. 259:37-46.

Vogel, C. L., Anthony, P. P., Sadikalo, F., Barker, L. F., and Peterson, M. R. (1972). Hepatitis-associated antigen and antibody in hepatocellular carcinoma: Results of a continuing study. J. Natl. Cancer Inst. 48:1583-1588.

Wahren, B., Esposti, P., Galder, H., and Alpert, E. (1976). Marker antigens and immunization with heterologous AFP in germinal tumors of the testis. In Onco-developmental Gene Expression (W. H. Fishman and S. Sell, eds.), Academic Press, New York, pp. 147-154.

Waldman, T. A., and McIntire, K. R. (1972). Serum-alpha-fetoprotein levels in patients with ataxia telangiectasia. Lancet 2:1112-1115.

Watabe, H. (1971). Early appearance of embryonic α-globulin in rat serum during carcinogenesis with 4-dimethylaminoazobenzene. Cancer Res. 31:1192-1194.

Watabe, H., Hirai, H., and Satoh, H. (1972). Alpha-fetoprotein in rats transplanted with ascites hepatoma. GANN 63:189-199.

Watabe, H. (1974). Purification and chemical characterization of α-fetoprotein from rat and mouse. Int. J. Cancer 13:377-388.

Watanabe, A., Miyazaki, M., and Taketa, K. (1976). Prompt elevation of rat serum α-fetoprotein by acute liver injury following a single injection of ethionine. Int. J. Cancer 17:518-524.

Yachnin, S., Hsu, R., Heinrikson, R., and Miller, J. B. (1976a). Comparison of the physical properties and immunosuppressive effects of hepatoma and fetal human alpha-fetoprotein (AFP). Clin. Res. 24:483A.

Yachnin, S. (1976b). Determination of the inhibitory effect of human alpha-fetoprotein on in vitro transformation of human lymphocytes. Proc. Natl. Acad. Sci. USA 73(8):2857-2861.

Yazova, A. K., Suslov, A. P., and Brondz, B. D. (1976). Macrophage migration inhibition caused by T-cell in rats with terminated natural tolerance to alpha-fetoprotein (AFP). In <u>Onco-developmental Gene Expression</u> (W. H. Fishman and S. Sell, eds.), Academic Press, New York, pp. 547-553.

Zimmerman, E. F., Voorting-Hawking, M., and Michael, J. G. (1977). Immunosuppression by mouse sialylated α-fetoprotein. Nature 265(5592):354-356.

3

CARCINOEMBRYONIC ANTIGEN AS A TUMOR MARKER

E. DOUGLAS HOLYOKE, JAMES T. EVANS, and ARNOLD MITTLEMAN / Department of Surgical Oncology, Roswell Park Memorial Institute, Buffalo, New York

T. MING CHU / Department of Diagnostic Immunology Research and Biochemistry, Roswell Park Memorial Institute, Buffalo, New York

I. Introduction   61
II. The Clinical Usefulness of CEA   64
   A. Introduction   64
   B. Colon and Rectal Cancer   64
III. Conclusion   76
   References   77

I. INTRODUCTION

Seventy-five years ago during the initial phases of study of tumor immunity using animal models, because of a failure to appreciate the effects of transplantation per se on host resistance and tumor rejection, many investigators felt, as Mr. Shaw tell us in "A Doctor's Dilemma," that tumor immunity could be a real and a strongly operative force. Successful treatment by vaccination was anticipated as being very likely. Disillusion set in as the complexities and problems involved became apparent. However, interest was renewed in the 1950s when Foley, Kein, and others clearly demonstrated the presence of tumor antigenicity in syngeneic systems [1, 2]. Although most thought was focused on the possible therapeutic significance of this fact of true tumor antigenicity, it was obvious that this also implied the possibility of identification and isolation of tumor antigens which might provide a means of improved tumor diagnosis.

Using antitumor antiserum absorption and complement fixation techniques, Korosteleva demonstrated as early as 1957 that common tumor-specific antigens could be demonstrated in various types of malignant tumors obtained from different individuals [3]. Abelev and his co-workers identified an antigen component of newborn mouse serum and reported their findings in 1963 [4]. Tartarinov reported in 1964 on the presence of such an alpha-fetoprotein (AFP) in the sera of patients with primary hepatocellular cancer [5].

It was against this background that Gold and Freedman began looking for specific tumor antigens in man [6]. Colon was chosen as the target organ primarily because the usual resected specimen following surgery contains "normal" colon mucosa far enough away from the primary malignancy to assure that control tissue is available from the same individual for each tumor tested. Gold immunized rabbits with extracts of pooled colonic cancer and absorbed the antisera produced in this way with a pooled extract of normal human colon tissue and with human blood components. He also injected a group of newborn rabbits at birth with normal human materials in an effort to render them tolerant and later immunized them with pooled tumor material.

The antisera obtained by both methods were tested against tumor and normal tissue antigens by agar gel diffusion, immunoelectrophoresis, hemagglutination, passive cutaneous anaphylaxis, and immunofluorescence assay. Distinct antibody activity directed against two apparently qualitatively tumor-specific antigens or at least antigenic determinents were identified. Identical tumor-specific antigens were also demonstrated in a number of individual colonic cancers obtained from different human tumors. It is of interest that these workers went to some effort to assure that what they were identifying was not due to either bacterial products or to clotting factors, such as fibrin.

Gold and his fellow investigators then went on to test for the distribution of these antigens in other normal and malignant tissues. In adults identifiable antigenic material was present in all malignant tumors of the epithelium of the gastrointestinal tract and pancreas. These antigens were not found in other adult tissues by the immunodiffusion techniques mentioned above. Similar antigens were present in fetal gut, liver, and pancreas between 2 and 6 months of gestation. This led to the name "carcinoembryonic" antigen (CEA) for these materials. This name has proved to be somewhat of a misnomer in that subsequently developed and new more sensitive techniques for measuring small amounts of antigen were able to identify these materials in normal serum, in normal adult colon tissue, and in other nongastrointestinal malignancies such as those of the breast and lung, and neuroblastoma.

In 1960 Yalow and Berson described a technique of radioimmunoassay for insulin which can be applied to detect a variety of protein and polypeptide entities in small quantity [7]. Thompson and the group at Montreal [8] applied this technique using a modification of the co-precipitation-inhibition assay technique described by Farr [9]. Table 1, adopted from Schwartz [10], indicates how greatly radioimmunoassay has increased our ability to detect small amounts of suitable material. This progress in the measurement of trace materials is a major factor in our renewed search for valid and useful tumor markers.

TABLE 1 Method Sensitivity Alpha-fetoprotein

Method	Limit of detection (10^9 g/ml or ng/ml*)
1. Double gel immunodiffusion	1,000-10,000
2. Immunoelectrophoresis	5,000
3. Counterimmunoelectrophoresis	25
4. Immunoradiography	10
5. Radioimmunoassay	0.25

*Values may vary with the quality of antisera used.
Source: From Ref. 10.

A quick review of the table will show that sensitivity of measurement has been increased by radioimmunoassay up to 40,000 times over double gel immunodiffusion. The adaption of radioimmune assay to measure carcinoembryonic antigen levels in serum or plasma begun by this group is obviously a major event.

The method has the following steps. Preparation of purified antigen is first. This may be done from either primary or metastasic tumor tissue, but in fact it has usually been from the latter. Since colon cancer is the primary source of antigen, it is usually much easier to obtain substantial amounts of relatively solid tumor tissue from a moderate-sized metastasis than it is from the primary, which may be infiltrative and considerably ulcerated. Next the CEA has to be isolated and later labeled, usually by means of the chloramine T method modified after Hunter and Greenwood [11]. Antisera are usually prepared by immunization of goats with purified CEA. This is perhaps the procedure that implies the greatest opportunity for variance, both because of the idiosyncrasy of some goats in the antibody response and because of the problems of absorption to assure specific anti-CEA activity. The radioimmunoassay reported by Thompson was based on the coprecipitative inhibition technique described by Farr [9]. Figure 1 reports the initial classic results of Thompson et al. in their assay of patients with colon cancer.

The work reported in Figure 1 initiated a great deal of clinical and scientific study. Results began to accumulate in some quantity by 1971. Now, 10 years later, as of the time of this analysis and as we approach a decade of work, it is a good time to review our current concepts of, and the evidence of, any signficant clinical uses of CEA.

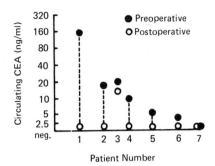

FIGURE 1 CEA and colon cancer. Seven of 11 patients were measured as shown, both preoperatively and postoperatively. Only one false negative result was recorded. (From Ref. 8.)

II. THE CLINICAL USEFULNESS OF CEA

A. Introduction

A biological marker capable of detecting the presence and growth of a malignant tumor of any type might play a role in screening, diagnosing, staging of disease, and monitoring disease for cure, regression, or progression in response to therapy. It may be used also to monitor occult disease progression, increasing our sensitivity to clinical recurrence with a view to earlier therapy of present disease.

CEA is elevated in patients with a number of different tumors [12, 13]. However, since the antigens reported by Gold were initially identified in malignant colon cancer, and it was initially suggested that identification of this antigen in serum might be useful in the screening, diagnosis, and monitoring of this disease, most clinical studies have involved populations of patients at risk for or suffering from colorectal cancer. More recently, several studies of patients with gynecologic malignancy have been reported [14, 15].

B. Colon and Rectal Cancer

1. CEA in Screening

All studies to date have indicated clearly that our assays for carcinoembryonic antigen lack both the sensitivity and specificity to be useful as a diagnostic screening technique for colorectal cancer in the general population [16-18]. This could have been anticipated because several relatively early studies indicated that CEA is quantitatively increased only in malignant colon tissue over normal colon mucosa; thus increased CEA in sera of patients with colon cancer represents a quantitative departure from normal, but is not a qualitative change [19-21]. Further studies have demonstrated that CEA in serum or plasma is elevated in serum in a variety of benign conditions, and also that CEA is not elevated in serum in a substantial percentage of patients with malignant disease [22-24].

Even using the radioimmunoassay for CEA in serum or plasma does not provide enough sensitivity or specificity to be useful for screening. It is possible in this type of assay to adjust sensitivity and specificity by adjusting our cutoff level. That is that level of antigen in serum which we define as distinguishing between benign and malignant disease. In our laboratory at Roswell Park, if we set a cutoff or diagnostic level higher than 2.5 ng/ml, toward 5 or 10 ng/ml, specificity is increased but sensitivity drops off precipitously [25]. At any level below 2.5 ng/ml, false positive values become a very severe problem, even though there are still a substantial number of patients with clinically treatable malignancy who do not show any elevation of CEA.

As pointed out by Herberman, it is difficult to make generalizations concerning acceptable levels of specificity (false positive results) or sensitivity (false negative results) for screening assays [26]. As he observed, the prevalence of the type of cancer in the populations to be screened must be considered. The lower the prevalence of a cancer in a population, the more critical specificity becomes. Given Herberman's example using a 2% false positive rate or a specificity of 98%, if a population being screened has a cancer incidence of 20 per 100,000 individuals tested, even if the test could detect all such people with the malignancy, 1980 of 2000 or 99 of 100 test-positive individuals would not have cancer. Our experience at Roswell Park as well as that of others indicates that CEA is indeed elevated above our usual cutoff in about 2% of a young, healthy control population. It is also elevated in a variety of benign conditions, some quite subtle, for instance in smokers, in a much larger percentage of individuals. On the other hand, in symptomless colon cancer it has a sensitivity of only 50% [27,28].

In corroboration, Chu and Murphy conducted a study of an older-aged group of business executives from cooperating organizations; 1800 individuals were tested [29]. Two previously unsuspected malignancies were detected, one in the pancreas and one in the colon. The cost of detection was about $20,000 per patient, and disease was too advanced for curative therapy at discovery. The test is simply not suitable for screening.

As we have pointed out before, when a test proves to be too insensitive for screening, it is often suggested that it may be useful in screening or monitoring high-risk groups. For solid tumors of the colon and rectum, high-risk groups that we can definitely identify represent a small percentage of the whole. These groups would include individuals with a history of colitis, multiple polyps, or previous neoplasm, or with a family history of colon cancer [30,31]. The problem in screening patients with chronic inflammatory disease of the colon for possible cancer in that CEA is elevated in so many individuals who turn out to have only inflammatory disease that it is not discriminating [32]. A study of the literature and our own experience with the presence of polyps or a family history of multiple polyposis indicates that the assay is also not useful in providing an earlier diagnosis of malignancy in these individuals if they do develop frank neoplasm [33,34]. Experience in monitoring individuals with a significant family history is not available. We will consider the possible benefits of monitoring patients with a previous history of malignancy later.

2. CEA as a Diagnostic Adjuvant

The primary problem with using CEA as a diagnostic adjuvant is again the rather high incidence of elevation of CEA in benign disease of poor specificity. Table 2 summarizes the experience of Costanza in this regard [35].

Others have reported similar findings [36,37]. However, if we use clinical and other laboratory evaluations and discount individuals known to have benign disease, the false positive rate can be significantly reduced without an adverse effect on the rate of negatives we encounter in those with tumor. However, sensitivity is still low. Several reports and our own experience indicate that if we use the Astler-Coller modifications of the Dukes classification, only 20 to 25% of colon or rectal cancer victims who are at Stage A or Stage B_1 are detected [28,38]. This is not high enough to permit the assay to be very useful in diagnosing early malignancy. Indeed, in our experience CEA is elevated in the large B_2 lesions that penetrate serosa, or in the large C_2 lesions that penetrate serosa and have nodes.

TABLE 2 Benign Disease and CEA

Disease	CEA elevated (%)
Collagen disease	53
Acute benign GI disease	44
Nonspecific colitis	36
Diabetes mellitus	27
Cardiovascular disease	18
Functional bowel disease	15

Source: From Ref. 35.

Using a TNM classification, it is generally in the T_3 or T_4 patients (i.e., those in whom the tumor perforates serosa or invades adjacent organs) that CEA is elevated clearly. This guideline overshadows the effect of the presence of a minimal number of lymph nodes on CEA [39,40]. When all of these constraints are taken into account, it is clear that CEA assay is at best a limited tool in making an early diagnosis.

It has been reported that CEA assay increased the accuracy of diagnosis on a series of symptomatic patients and was additive to barium enema as a diagnostic technique [41]. While we do not feel that barium enema in this study was as accurate as we have usually found it to be, and in spite of the limitations noted above, we still believe that the CEA assay should be included in the investigation of patients with symptoms of large bowel disease, especially those with negative or doubtful barium enema, and that (albeit in a small group of patients) this assay may increase our diagnostic success. However, we do not have enough information to state categorically that this is so.

3. Staging by CEA

When staging cancer accurately, the clinician is attempting to make a statement about the prognosis and anticipated course of the disease. Originally, staging began to have meaning as it developed as a tool to determine when surgery for local control was optimal and indicated. For colon cancer the primary determining measurements of the probable success or failure of surgery are depth of penetration and the presence of lymph nodes. Dukes originally presented a staging of A, B, and C, which allowed for graded penetration of the wall of the colon and the presence of nodes. This has been refined further. Many physicians have found that the Astler-Coller staging system with B_1 and B_2 lesions, the latter indicating penetration of the bowel wall, and the presence of possible lymph nodes are clinically meaningful stages.

We have found, as have many other investigators, that invasion of surrounding organs and the location and number of involved lymph nodes is important. In our studies we now employ a T, N, and M classification to help us characterize our patients as precisely as possible. This classification, as we have used it since 1971, is seen in Table 3. It follows the standard approach for a T, N, M classification and is used by us as a histological and pathological, but not clinical statement of extent of disease.

In addition to the distinctions described in Table 3, we feel it is important to emphasize the possible value of recording a ratio of positive to negative nodes. This is particularly true for group studies or for review of patients since it affords some measure of how diligently lymph nodes have been searched for.

The question of the use of CEA as a predictor is not yet settled completely after all of our studies, although we have reported [42], as have others including Zamcheck [43], and Mach [44], that it may indeed be

TABLE 3 Roswell Park T, N, M Staging of Colorectal Cancer

Tumor mass and invasion		
T_1-T_4	T_1	Mucosal or submucosal tumor
	T_2	Muscularis invasion by tumor
	T_3	Perforation by tumor
	T_4	Invasion of adjacent organs: Bladder, vagina, small bowel, etc.
N_0-N_3	N_0	No positive nodes[a]
	N_1	1-2 nodes near the lesion
M_0-M_1	M_0	No metastases
	M_1	Metastases present

[a] All nodes cleared.

predictive of length of free interval and survival. The question remains whether this parameter adds anything to the simple Dukes classification in staging patients with colon cancer. Chu has reviewed all our patients studied both prospectively and retrospectively from 1971, and his data indicate that CEA is additive to Dukes classification in prognosis [45]. However, after 18 months, although less than a sixth of our recurrences occur this late, a normal initial CEA does not protect against recurrence. We are continuing our prospective Roswell Park Memorial Study and CEA is predictive.

Evans and Mittelman have recently reviewed their data (Fig. 2) [46]. For patients with Dukes B lesions, CEA positivity increased the failure rate (F) from <10%, or 2 of 31, to 25% or 4 of 16. Similarly for Dukes C lesions, CEA positivity increased the incidence of recurrence from 33%, or 6 of 18, to 50% or 8 of 16. The results are for 81 patients who have been followed for 5 years, so the figure does not represent a projection but actual recurrence figures. Table 4 reports these data also.

Recently, McIntyre reviewed the studies of CEA taken one month after surgery and before randomization to treatment groups in the Gastrointestinal Tumor Study Group Colon Adjuvant Study [47]. CEA specimens were measured in participating hospitals, in institutions, or by Zamcheck's

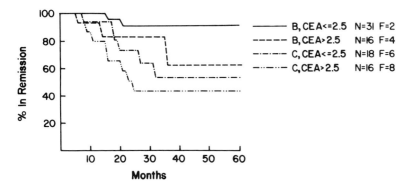

FIGURE 2 Recurrence of disease by Dukes classification and CEA.

group at the Mallory Institute. With an average survival of 13 months, CEA as a parameter indicated that a value >10 ng/ml at randomization demonstrated prognostic significance. For example, 5 of 134 patients with Dukes B_2 lesions and a CEA of <10 ng/ml at randomization had recurrences, but 4 of 6 patients with a CEA >10 ng/ml following definitive surgery had recurrences over the same period of time. Similarly, for patients with more than four lymph nodes, designated by this group as C_2 patients, 18 of 52 with a CEA <10 ng/ml had recurrences, and 6 of 7 or 80% with more than four positive nodes and a CEA >10 ng/ml had recurrences after an average of 13 months of follow-up. These are preliminary data, but the staging usefulness of this parameter appears verified in spite of the difficulties of group measurement and evaluation of data.

Based on these considerations, we believe strongly that CEA is useful for staging and has some predictive value. It may not be a great deal more predictive than the Dukes classification, and may be no more predictive

TABLE 4 CEA Versus Dukes Classification in Colon Cancer

Dukes classification	Pre-CEA	Number of patients	Number recurred	Percent
B	0-2.5	31	2	<10
	2.6+	16	4	25
C	0-2.5	18	6	33
	2.6+	16	8	50

Source: From Ref. 46.

than an accurate TNM classification, but it does give a number, and it may be additive. It should, for example, provide a reasonable basis for staging in patients in whom preoperative therapy is being considered. We also feel that prospective randomized studies for adjuvant effect should be stratified by using CEA as well as by careful histopathological staging of the extent of disease to assure optimum comparability between individuals in different treatment groups.

4. CEA and Second Look

Herrera reported our earlier data on patients with colon and rectal cancer who were followed after definitive surgery with periodic examination, routine chemistry studies, x-rays, and scans as well as CEA determinations [38]. We found, as have many others, that serial CEA does generally indicate disease course. If surgery is definitive CEA falls to normal levels over 30 to 45 days, depending on how elevated it was initially. However, a fall in CEA to normal levels does not guarantee freedom from disease, and many of those whose values fall to normal after surgery will all develop recurrent disease [48].

As these patients are followed, 80-85% of those who develop clinical recurrences show an elevation of CEA either before or simultaneously with other evidence of disease progression. Nevertheless, in 15-20% of patients other clinical evidence of progression occurs without a rise in CEA. Operationally, this is not too harmful since these patients would not have been diagnosed in any event, and some reassurance may be gained in knowing that a patient at considerable risk whose CEA remains low is 80% safe from having clinically significant disease.

Most monitoring data following surgery for "cure" confirms work reported by Herrera [38]. As we have suggested, false negatives do not put us at too much of a disadvantage as long as we are aware that they may occur. False positives are a problem. If we require a definite elevation of CEA above 2.5 ng/ml in our laboratory, and if we require two consecutive values above this level on two occasions some weeks apart, and even if we insist that the second value be equal to or greater than the first, we still find between 10 and 20% false positive determinations when we review our data retrospectively.

Many of the problems of CEA follow-up have been stressed by Mach and others [43,44]. One problem that has occurred in trying to assess the possible usefulness of CEA as a monitor of disease after surgery is that the lead time between an elevated CEA and clinical detection of disease may often be one clinic visit. When clinic visits for follow-up occur every month, CEA will confer only a 30-day lead time. Similarly, a considerable percentage of patients appear to gain 90 days of lead time, but these are patients who are seen every 3 months. Nevertheless, as we reported, in a balanced review of literature, CEA appears to be elevated in a good

3 / Carcinoembryonic Antigen

one-third of patients as subclinical cancer of the colon or rectum proceeds before it is detected by other means. And in a large percentage, perhaps half of these, there is a good lead time of 3 or more months [49].

The basic question is whether or not the information gained above allows a meaningful therapeutic intervention that will affect survival. This question reaches into the colon second-look problem [50, 51]. If we follow our patients at Roswell Park after definitive surgery for cancer of the colon or rectum, are we able to salvage more than we can otherwise by CEA based second-look surgery and resection? The answer is not known. No prospective randomized study of "CEA second look" versus second look based on history, physical examination, and other biochemical parameters of possible tumor growth is underway at present. Several studies are evaluating second look subsequent to finding an elevated CEA, including those of Minton, Wilson, and our own at Roswell Park carried on by Mittelman and Evans. So far we have encountered 47 recurrences in 161 patients followed [46].

On careful review 6 of 47 patients with recurrence, only 15%, had a CEA abnormality as the sole evidence of recurrent disease proved by histological diagnosis at exploratory laparotomy—this in spite of a full diagnostic workup in the face of the elevated CEA in an effort to confirm and locate disease. Three additional patients were followed by CEA, and on the basis of similar elevations of CEA were explored, but no tumor was found. Still, two of three patients explored did have tumor. This 6 of 47 patients represented about 15% or about the percentage that we anticipated would show an elevated CEA significantly early. Of these six patients, two had definitive surgery and one had surgery and subsequent directed radiotherapy of a moderately sized pelvic wall residual. Both have remained free of disease to date.

This experience spells out our review of the problem to date. We are not sure whether this salvage is worth the vexation of facing false positive elevations and negative laparotomies. We are particularly not sure how much we gain in overall salvage over waiting for symptoms and then operating.

Minton and his group feel the answer is definitive and affirms the value of CEA [51]. Trying to eliminate technical variation by establishing a nomogram and calling significant any confirmed CEA more than 2 standard deviations greater than a postsurgical low, they have attempted to increase the sensitivity of the assay for follow-up. Minton's group now reports that of the last 14 patients in their series in whom a rise in CEA was confirmed at greater than 2 standard deviations above the postsurgical low, 11 presented with localized tumor, and complete resection was technically feasible in all 11.

No one else has figures like these, and, as Minton believes, they may be due to early laparotomy at about 1 month after the first rise in CEA in this series. But our experience to date, based on what we know about the histopathology of the spread of colon cancer, indicates that this series is different from ours and most of the others with which we are familiar in detail.

On the positive side, we have learned from following our own patients that many of the transient rises in CEA which we did not understand before may be due to chemotherapy, particularly in our experience with MeCCNU. What can be accomplished using CEA as an indicator in second look awaits more study. Our original estimate that we can perhaps salvage 10% of all cancer patients with this technique still appears possible. How many of these will be above and beyond what we can salvage by operating after other clinical evidence of recurrence appears is moot, as is the question of how many individuals rendered technically free of disease at second look will remain so. In our experience at Roswell Park Memorial Institute, patients with recurrent disease have remained free of disease following definitive resection in about the same manner and for the same period as our patients with primary disease (Fig. 3). While we may remain skeptical about resection of recurrent colon cancer, and in particular, of multiple liver metastases, the answer is not to be found in what can or cannot be accomplished using CEA as a means to identify recurrence and to move to earlier

FIGURE 3   Recurrence time in months.

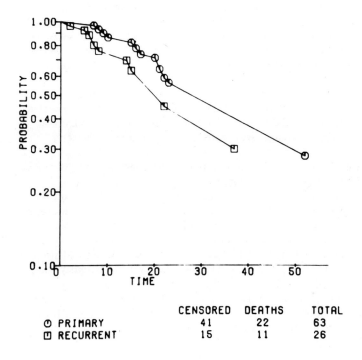

	CENSORED	DEATHS	TOTAL
⊙ PRIMARY	41	22	63
▢ RECURRENT	15	11	26

intervention. Long-term follow-up of treated recurrent disease is needed.

Summing up with regard to following patients with CEA assay after definitive surgery using CEA elevation as a basis for laparotomy, our preliminary evidence is simply that, properly employed, this assay can save patients. And we believe that the advantages of monitoring patients, particularly those with a reasonable risk of recurrence following definitive surgery for primary disease, will outweigh the disadvantages, particularly since the disadvantages can be reduced with sound clinical judgment and an increasing knowledge of the natural history of this disease and the vicissitudes of the CEA assay. In our hands two of three individuals with an elevated CEA as the only sign of tumor confirmed over 1 month did have tumor, and in half of these disease has been controllable.

5. Monitoring Advanced Disease

We can use CEA to monitor the response of metastastic colorectal cancer treated with radiation or chemotherapy. But, partly because so few chemotherapeutic agents cause a meaningful response, and partly because the assay does not always represent tumor burden and may indicate changing levels of plasma CEA for a variety of changes associated with therapy not related to tumor status, we agree with the Mayo Clinic study that at present CEA does not have a role to play outside of a research setting for these patients. However, it should be emphasized that this problem is completely different from that of monitoring patients following definitive surgery for primary disease.

6. CEA in other Malignancies

Carcinoembryonic antigen, or at least a group of carcinoembryonic antigens with an apparently identical reactive site, have been identified in the serum and plasma of patients with mammary cancer, cancer of the lung, and gynecologic malignancy, and in some patients with a variety of urological and other soft-tissue tumors.

7. Mammary Cancer

Several studies have clearly indicated that assay for CEA in plasma is simply not a reasonable marker to consider for screening or an an adjuvant in diagnosis. Searle and Bagshaw record a 3% positivity for patients with primary mammary cancer in Stages I, II, or III, and a 27% positivity for Stage IV or metastatic tumor [52]. Such low preoperative values obviously imply a restriction on the usefulness of staging or prognosticating primary disease [52]. The question of monitoring disease has been reviewed by Zamcheck's group, who felt it might be feasible [53].

At Roswell Park, Chu and Nemoto reported an overall positivity of 68% in 83 patients with metastatic disease [54]. However, in their postmastectomy patients the CEA level did not reflect the presence or absence of early metastases. This was not unexpected in light of the observation that overall only 41% of patients with soft-tissue metastases showed an elevated CEA.

In this series there were also quite a few instances of unexplained transient rises in CEA not related to metastases. In patients treated hormonally or chemotherapeutically, Chu and Nemoto attempted to correlate tumor response to CEA levels in serum. With more pronounced progression or regression, better correlation was seen. However, in the overall opinion of these investigators, it was not good enough to be clinically useful in monitoring the course of disease.

One approach to the problem of monitoring breast cancer has been reported by the group in Boston as well as by Tormey et al. [55,56]. Essentially, it uses multiple markers, nonspecific compared to CEA, but additive. Tormey found that if he measured CEA, N^2, N^2-dimethylguanosine and HCG, at least one of these assays was positive in 97% of patients with metastases. While data are cited in this study that these other markers, the nucleoside and HCG, are not usually elevated in control populations, the report suffers from not having simultaneously tested a normal symptomatic population or even a normal young and healthy population.
On the other hand published data in this area are enough to justify continuation of this type of study, especially with sequential follow-up.

8. Lung Cancer

CEA is elevated in patients with cancer of the lung [13]. However, it is also elevated, although to a lesser extent, in individuals who are heavy smokers or who have emphysema. For diagnosis, either through population screening or as a diagnostic adjunct, it does not seem that CEA is useful for patients with this malignancy. There is some difference of opinion in that Vincent and Chu have indicated that it may be useful as a prognostic parameter in patients with pulmonary cancer [57,58]. However, Concannon has reported data which indicate that CEA does not correlate well with stage of disease or patient survival [59]. Furthermore, the assay failed to show an elevation of CEA in a considerable proportion of patients with extensive metastatic disease.

We feel at present that the assay is really not a great deal of help in treating patients with cancer of the lung. Of course it should be pointed out that poor survival and poor treatability are partly responsible for this shortcoming. In the event of better chemotherapy, an appropriate marker assay should be relatively more important.

9. Gynecological Malignancies

Several studies have demonstrated that gynecological malignancies produce CEA [12,60,61]. Since in most studies serum levels of patients with such neoplasma apparently correlate with extent of disease, several empirical clinical evaluations of the possible usefulness of CEA for monitoring have been undertaken in patients with gynecological malignancy [14, 15]. DeSaia et al. have reported that plasma elevation of CEA occurs in almost 50% of patients with invasive gynecological cancer, and of patients with recurrent squamous cell cancer of the cervix, 84% had a positive CEA [14]. Van Nagell et al. reported that 81 of 100 patients with gynecologic metastases and 17 of 95 patients with benign disease had a CEA greater than 2.5 ng/ml [61]. However, Seppälä felt that a cutoff of 5 ng/ml was more meaningful, in which case he encountered a much lower incidence of positivity with known tumor, but fewer false positives [62]. Van Nagell felt that serial CEAs might indeed turn out to be useful in monitoring patients with gynecological malignancy, as does Makin.

At present there are not enough properly collected data to tell us whether the CEA assay is specific or sensitive enough in patients with cancer of the cervix or ovaries to allow its real use in the treatment of these tumors. The situation is similar to that of colon cancer in that significant proportion of these patients demonstrate an elevated serum plasma level with progression, and that occult recurrence cannot be detected in some patients with both intrapelvic and intra-abdominal disease. More prospective work needs to be done in this area.

We should also note at this point that a major effort is needed toward the identification of other markers which might be elevated in relationship to gynecological malignancy. Haines recently reported on cervical carcinoma antigen or CCA [63]. This may be of use in tissue or exfoliative cells at least. Chiang et al. reported on the CCA (CA-58) with an RaI assay indicating sensitivity and specificity enough to warrant more study [64]. Several other possible squamous cell related antigens have similarly been reported, including TAA, TA-4, and herpes simplex antigen [65-68]. The situation at present is the same for cystadenocarcinoma-associated antigens (OCAA) and Ov C-antigen, as well as others, for ovarian cancer. These need to be carefully studied not in massive numbers, but in moderately sized studies in which a prospective attempt is made to see if they are predictive, i.e., useful for staging and stratification, and if they are useful as monitors to the point that they can affect care advantageously. The caution must be entertained that such study requires sophisticated follow-up and statistical analyses, and that with the means of therapy available to us today for progressive disease, even if such markers are moderately accurate, it may not be possible to demonstrate any effect on survival except in small select groups which may require quite a long time to accumulate.

10. Other Tumors

For patients with gastric and pancreatic tumors, CEA does tell us some information about tumor burden and outlook [69, 70]. But these are not usually very meaningful data. The Mayo Clinic has indicated that performance status does as well as CEA for staging advanced gastrointestinal cancer, if not better [71]. In addition, particularly for patients with pancreatic cancer, we are discussing a predicted survival difference based on a CEA of 3 to 6 weeks. The limitation of such a predictive capability needs no comment.

An initial hope that urinary CEA might be useful in patients with bladder tumor has not been realized [72, 73]. The problems encountered are irregular levels of antigen in infection and benign disease, and the fact that urine contains a variety of related glycoproteins that squelch and interfere with RaI for CEA.

Other uses of CEA have been suggested but need proof of efficacy in a clinical care setting: to prepare antibody for scan, to assay pleural or ascitic fluid, to determine whether the bulk of a tumor has been located within a field of radiation during preoperative therapy for rectal cancer, and to determine whether pancreatic juice can be analyzed for CEA in a way that will help in the diagnosis of pancreatic cancer [74-77]. All of these possible uses have been partially explored and all have advocates, but none have really proved useful to date.

III. CONCLUSION

After almost 10 years of clinical study, CEA has become a prototype tumor marker, particularly a tumor antigen marker. Its use is still being explored in the numerous areas described above. At present we believe its use in general care should be confined to patients with colon or rectal cancer who should have a preoperative baseline value to assist in staging. We feel that following definitive surgery for this disease for 2 years, especially in patients with B_2 or C_2 lesions, the assay should be carried out at least quarterly, and that a rise that persists is reason to consider a second-look approach. This approach is not yet of proven efficacy, but it should be labeled "possibly effective" at this juncture. We strongly advocate that the assay be used in stratifying adjunct study treatment programs for colorectal cancer. In following patients it is necessary to observe other causes for a rise in CEA, in particular chemotherapy, as patients are monitored.

At this point while it is possible that CEA may prove useful in following a number of patients with gynecological tumors, most of the other tumor type data reviewed in this article do not indicate a great deal of opportunity for use. In part this is because of the paucity of studies adequately designed to test whether the CEA assay is truly useful as a diagnostic adjunct or as a monitor. What are needed now are careful clinical studies carried on long enough to answer these questions, not in thousands of patients, but in hundreds. Selected second-look studies should be encouraged and prospective decisions made. We need a new quality and standard of clinical work to study this and other markers if we are to make progress.

ACKNOWLEDGMENTS

Supported by U.S.P.H.S. Contracts NO1-CM-43782, NO1-CM-47894, NO1-CM-57034, NO1-CB-33858, and Grants CA-18410 and CA-15263-04.

REFERENCES

1. E. J. Foley, Cancer Res. 13:835-837 (1953).
2. G. Klein, Cancer Res. 19:343-358 (1959).
3. T. A. Korosteleva, Problems of Oncology 3:641-652 (1957).
4. G. I. Abelev, S. D. Perova, N. J. Khramkova, Z. A. Postnikova, and I. S. Irlin, Transplantation 1:174 (1963).
5. Y. S. Tartarinov, Vopr. Med. Khim. 10:90 (1964).
6. P. Gold, S. O. Freedman, J. Exp. Med. 121:439-459 (1965).
7. S. A. Berson, R. S. Yalow, Clin. Chim. Acta 22:51-69 (1968).
8. D. M. P. Thompson, J. Krupey, S. O. Freedman, et al., Proc. Natl. Acad. Sci. USA 64:161-167 (1969).
9. R. S. Farr, J. Infect. Dis. 103:239-362 (1958).
10. M. K. Schwartz, Antibiotics Chemother. 22:176-184 (1978).
11. W. M. Hunter, F. C. Greenwood, Nature 194:495-496 (1962).
12. G. Reynoso, T. M. Chu, E. D. Holyoke, JAMA 220:261-365 (1972).
13. D. J. R. Laurence, U. Stevens, R. Bettelheim, D. Darcy, C. Leese, C. Turberville, P. Alexander, E. W. Johns, and A. Monroe Neville, Brit. Mech. J. 3 (1972).
14. P. DeSaia, B. J. Haverback, B. J. Dyce, and M. Morrow, Surg. Gynecol. Obstet. 138:542-544 (1974).
15. S. K. Khoo, E. V. Mackay, Aust. N.Z. J. Obstet. Gynecol. 13:1-17 (1973).
16. H. J. Hansen, J. J. Snyder, E. Miller, et al., J. Hum. Pathol. 5:139-147 (1974).
17. T. M. Chu and E. D. Holyoke, Proc. XI Int. Cancer Congress, Florence, Italy, 1:351 (1974).
18. P. W. Dykes and J. King, Progress report: Carcinoembryonic antigen. Gut 13:1000-1013 (1974).
19. P. Burtin, E. Martin, M. Sabme, S. Von Kleist, J. Natl. Cancer Inst. 48:25-29 (1972).
20. H. Denk, G. Tappeiner, R. Eckerstofer, and J. H. Holznar, Int. J. Cancer 10:262-272 (1972).
21. M. Bordes, R. Michiels, and F. Martin, Digestion 9:106 (1973).
22. T. L. Moore, H. Z. Kupchik, N. Marcon, and N. Zamcheck, Am. J. Dig. Dis. 16:1-7 (1971).

23. P. LoGerfo, J. Krupey, and H. J. Hansen, N. Engl. J. Med. 285:138-141 (1971).
24. N. Zamcheck, T. L. Moore, P. Dhar, and H. Kupchik, N. Engl. J. Med. 286:83-86 (1972).
25. T. M. Chu, Unpublished data.
26. R. B. Herberman, Am. J. Clin. Pathol. 68:688-698 (1977)
27. P. B. Chretien, Centennial Conference on Laryngeal Cancer, New York, Appleton-Century-Crofts, 1976, pp. 339-349.
28. Joint National Cancer Institute of Canada/American Cancer Scoeity Investigation, Can. Med. Assoc. J. 107:25-33 (1972).
29. T. M. Chu and G. P. Murphy, Evaluation of carcinoembryonic antigen as a screening assay in non-cancerous clinics, N.Y. State J. Med. 78:879-882 (1978).
30. R. S. Grinnell and N. Lane, Inst. Abst. Surg. 106:519-538 (1958).
31. M. Kancho, Mt. Sinai J. Med. 39:103-111 (1972).
32. T. L. Moore, P. A. Kantrowitz, and N. Zamcheck, JAMA 222:944-947 (1972).
33. N. Zamcheck, Adv. Intern. Med. 19:413-432 (1974).
34. W. B. Doos, W. I. Wolff, H. Shinya, A. De Chabon, R. J. Stenges, L. S. Gottlieb, N. Zamcheck, Cancer 36:1996-2003 (1975).
35. M. E. Costanza, S. Das, L. Nathenson, A. Rule, and R. S. Schwartz, Cancer 33:583-590 (1974).
36. S. K. Khoo, P. S. Hunt, and I. R. Mackay, Gut 14:545-548 (1973).
37. S. K. Khoo and I. R. Mackay, J. Clin. Pathol. 26:470-475 (1973).
38. M. A. Herrera, T. M. Chu, and E. D. Holyoke, Am. Surg. 183:5-9 (1970).
39. E. D. Holyoke, Oncofetal Antigen and Metastasis Monitoring in Fundamental Aspects of Metastasis. L. Weiss, North Holland Press, Amsterdam, 1970, pp. 407-423.
40. E. D. Holyoke, T. M. Chu, and G. P. Murphy, Cancer 35:830-836 (1975).
41. W. H. McCartney, P. B. Hoffer, Radiology 110:325-328 (1974).
42. E. D. Holyoke, G. Reynoso, and T. M. Chu, Am. Surg. 176:559-564 (1972).
43. N. Zamcheck, Cancer 36:2460-2470 (1975).
44. J. P. Mach, P. H. Jaeger, M. M. Bertholet, C. H. Ruegsegger, R. M. Loosli, and J. Pettard, Lancet 2:535 (1974).
45. T. M. Chu and E. D. Holyoke, Proc. Third International Meeting on Cancer Diagnosis and Prevention, Marcel Dekker, New York, 1978, pp. 606-615.
46. J. T. Evans and A. Mittelman, Unpublished data.
47. K. P. Ramming, N. Zamcheck, and E. D. Holyoke, Abstract, ASCO (May 1979).
48. P. LoGerfo, J. Krupey, and H. J. Hansen, N. Engl. J. Med. 285:138-141 (1972).

49. J. T. Evans, A. Mittelman, T. M. Chu, and E. D. Holyoke, Cancer 42:23-25 (1978).
50. E. W. Martin, W. E. Kibbey, L. DeVecchio, G. Anderson, and P. Catalano, Cancer 37:62-81 (1976).
51. J. P. Minton, K. K. James, P. E. Hurtubise, L. Rinker, S. Joyce, and E. W. Martin, Surg. Gynecol. Obstet. 147:208-210 (1978).
52. F. Searle, A. C. Louesay, B. A. Roberts, G. T. Rodgers, and K. D. Bagshaw, J. Immunol. Methods 4:113-119 (1974).
53. A. M. Steward, D. Nixon, N. Zamcheck, and A. Aisenberg, Cancer 33:1246-1252 (1974).
54. T. M. Chu and T. Nemoto, J. Natl. Cancer Inst. 51:1119-1122 (1973).
55. D. C. Tormey, D. Ahmann, C. W. Gehrke, R. W. Zumwatt, J. Snyder, and H. Hansen, Cancer 35:1095-1100 (1975).
56. D. C. Tormey, T. P. Waalkes, J. J. Snyder, and R. M. Simon, Cancer 39:2397-2404 (1977).
57. R. Vincent and T. M. Chu, J. Thorac. Cardiovasc. Surg. 66:320-328 (1973).
58. R. Vincent, T. M. Chu, and W. W. Lane, Cancer 44:685-691 (1979).
59. J. P. Concannon, M. H. Dalbow, G. H. Lieber, K. E. Blake, C. S. Weil, and J. W. Cooper, Cancer 34:184-192 (1974).
60. S. A. Gall, J. Walling, and J. Pearl, Am. J. Obstet. Gynecol. 115:387-393 (1974).
61. J. R. van Nagell, Jr., E. S. Donaldson, E. C. Wood, and D. M. Goldenberg, Cancer 42:1527-1532 (1978).
62. M. Seppälä, H. Pihko, E. Rusolanti, Cancer 35:1377-1381 (1975).
63. H. Haines, S. Nordquist, A. B. Ng, and R. Leif, Compendium of Assays for Immunodiagnosis of Human Cancer (R. B. Herberman, ed.), Elsevier North-Holland, Amsterdam, 1979, pp. 485-487.
64. W. T. Chiang, K. C. Chen, and E. R. Alexander, Compendium of Assays for Immunodiagnosis of Human Cancer (R. B. Herberman, ed.), Elsevier North Holland, Amsterdam, 1979, pp. 481-484.
65. A. Hollingshead, Compendium of Assays for Immunodiagnosis of Human Cancer (R. B. Herberman, ed), Elsevier North Holland, Amsterdam, 1979, pp. 335-343.
66. A. N. Ibraham and A. Nahmas, Compendium of Assays for Immunodiagnosis of Human Cancer (R. B. Herberman, ed.), Elsevier North Holland, Amsterdam, 1979, pp. 285-291.
67. H. Kato, Compendium of Assays for Immunodiagnosis of Human Cancer (R. B. Herberman, ed.), Elsevier North Holland, Amsterdam, 1979, pp. 507-514.
68. S. Sprecher-Goldberger, Compendium of Assays for Immunodiagnosis of Human Cancer (R. B. Herberman, ed.), Elsevier North Holland, Amsterdam, 1979, pp. 515-526.
69. M. Ravry, K. R. McIntire, C. G. Moertel, T. A. Waldmann, A. J. Schutt, and V. L. W. Go, J. Natl. Cancer Inst. 52:1019 (1974).

70. F. Ona, N. Zamcheck, P. Dhar, T. L. Moore, and H. Z. Kupchik, Cancer 31:324-327 (1973).
71. A. D. T. Shani, M. J. O'Connell, C. G. Moertel, A. J. Schutt, A. Silvers, and L. W. Go, Ann. Intern. Med. 88:627-630 (1978).
72. P. Guinan, R. J. Ablin, N. Sadoughi, and T. M. Bush, J. Surg. Oncol. 6:127-131 (1974).
73. Z. Wajsman, C. E. Merrin, T. M. Chu, R. H. Moore, and G. P. Murphy, J. Urol. 114:879-893 (1975).
74. D. M. Goldenberg, F. DeLend, and E. Kim, N. Engl. J. Med. 208:1384-1386 (1978).
75. A. Bhargava, J. Chen, E. D. Holyoke, and T. M. Chu, Clin. Res. 23:714 (1974)
76. P. H. Sugarbaker, W. D. Bloomer, E. D. Corbett, and J. T. Chaffey, Am. J. Roentgenol Ther. Nucl. Med. 127:641-644 (1976).
77. M. P. Sharma, J. A. Gregg, M. S. Loewenstein, R. P. McCabe, and N. Zamcheck, Cancer 38:2457-2461 (1976).

4

ENZYMES AS TUMOR MARKERS

MORTON K. SCHWARTZ / Department of Biochemistry, Memorial Sloan-Kettering Cancer Center, New York, New York

    I. Introduction 81
   II. Lysozyme 82
  III. γ-Glutamyl Transpeptidase 83
  IV. Ribonuclease 84
   V. Glycosyltransferases 85
      A. Sialytransferase 85
      B. Galactosyltransferase 86
      C. Fucosyltransferase 88
  VI. Lactic Dehydrogenase 88
 VII. Aryl Sulfatase 89
VIII. Histaminase 89
  IX. Hexosaminidase 90
   X. Terminal Deoxynucleotidyl Transferase 90
      References 90

I. INTRODUCTION

Differences in enzyme activity between normal and cancerous tissues have been sought to explain the phenomenon of uncontrolled malignant growth. While basic research has focused on enzymatic and isoenzymatic differences between normal and cancerous tissue, clinical studies have attempted to exploit these findings to develop specific and sensitive diagnostic tests. The diagnostic use of enzyme assays has traditionally been to determine activity in serum. However, in recent years enzymes have been measured from the diagnostic point of view in urine, effusions, cyst or spinal fluid, body cavity washings, biopsy material, and tissue scrapings. Efforts to use enzyme assays have extended from initial attempts to establish the presence of cancer, to define the involved organ, to describe the extent and stage of the cancer, to permit the clinician to evaluate the progression or regression of the disease by following changes in activity, to define high risk groups by enzyme configurations, and to predict whether a particular chemotherapeutic agent will be effective based on the presence or absence of an enzyme required to convert a drug to an active metabolite [1-5].

II. LYSOZYME

Lysozyme is primarily found in neutrophils and monocytes. Granulocyte destruction usually leads to increased serum lysozyme concentrations. Since lysozyme is eliminated primarily by the kidney there is a relationship between urine and serum lysozyme levels in patients with normal renal function. In a survey of the literature it was found that serum lysozyme was elevated in 54 of 121 patients with acute granulocytic leukemia (AGL), but in the urine of only 4 of 65 of these patients [6].

In 73 cases of acute myelomonocytic leukemia (AMML) 56 had increased serum lysozyme activity and 31 of 52 had lysozuria. In patients with acute myeloblastic leukemia (AML) there were serum elevations in 64 of 72 patients and supranormal urinary excretions in 58 of 60 patients. However, in acute lymphoblastic leukemia (ALL), elevations were seen in the serum of only 1 of 85 patients and in the urine of none of 46 patients. It can be concluded that elevated concentrations of serum and/or urinary lysozyme rule out the presence of ALL.

The enzyme can also be very useful in following the course of the disease and in evaluating chemotherapy. Serum lysozyme reflects bone marrow status better than the peripheral leukocyte count does. In one study of patients in remission there was a rise in serum and urinary lysozyme at a time when the peripheral blood still indicated remission. The patients experienced a hematological relapse some time later. Some investigators have reported that in AML and AMML serial lysozyme values provide as much information as serial bone marrow examinations. It has also been reported that patients with initially high serum lysozyme activities had fewer bacterial infections and in general had longer survival and a better prognosis.

The methods of measuring lysozyme are of interest. These have been a turbidimetric catalytic assay using Micrococcus lysodeikticus or the cell wall polysaccharide, an immunochemical diffusion method, and the Laurell method of electroimmunoassay. In the catalytic method egg white lysozyme is used as the standard, but in the other methods antibody of necessity is prepared to human lysozyme and human lysozyme is used as standard. The values obtained by these methods all correlate well, and clinical data obtained by them are similar. Values with the catalytic method are about one-third to one-fourth the values obtained with either the lysoplate method or the immunoelectrophoretic technique [6].

III. γ-GLUTAMYL TRANSPEPTIDASE

γ-Glutamyl transpeptidase (γ-GTP) is not found in adult rat liver but is present in hepatomas and preneoplastic liver nodules of these animals [7]. In cultured liver cells it has been observed in 25-90% of malignant epithelial liver cell lines, but in none of fibroblastic or nonmalignant epithelial cells [7]. It has been suggested that the enzyme can be a marker in cultured liver cells for those cells that have been spontaneously or chemically induced to a malignant state.

Although γ-GTP has been of specific value in monitoring liver disease, it has not been extensively evaluated as a monitor of primary liver carcinoma. Its extreme sensitivity may predicate against its use in the evaluation of cancer metastatic to the liver. Elevations have been observed in alcoholism, malabsorption, biliary obstruction, hepatitis, myocardial infarction, and neurological diseases, and in patients undergoing anticonvulsant therapy [1]. The ingestion of 80 g of alcohol within 18 hours of blood collection may elevate γ-GTP as much as 40% above normal [1]. Recently, γ-GTP in serum has been separated into 12 bands by polyacrylamide gradient gel electrophoresis. Band II was detected in serum of 22 of 63 patients (35%) with hepatoma, but not in serum from patients with other diseases. The authors concluded that patients with hepatoma may exhibit a characteristic γ-GTP isoenzyme pattern [8].

Kim and his associates evaluated alkaline phosphatase, 5'-nucleotidase, γ-GTP, and glutamate dehydrogenase in serum of 116 patients in diagnosing metastases to the liver [9]. They found that 5'-nucleotidase had the greatest utility for this purpose. It showed the lowest false positive results (7.4%) with the highest positive predictive value (85.7%). Although γ-GTP had the lowest false negative values (2.8%), it was least specific and had a false positive rate of 35%.

Roberts and his associates evaluated 5'-nucleotidase, alkaline phosphatase, and γ-GTP in 276 patients with malignant hematological diseases and concluded that the γ-GTP levels added little to the clinical data base in following patients with leukemia [10]. In Hodgkin's disease and non-Hodgkin's lymphoma elevations of γ-GTP were associated with active and widespread disease; elevations were observed in 15 of 50 patients with Hodgkin's disease and 14 of 26 patients with non-Hodgkin's lymphoma. It was suggested that in lymphoma elevations are not necessarily related to liver involvement and that γ-GTP could be used to monitor the disease. In myeloma elevations were observed in 7 of 12 patients and presumably were related to myeloma infiltration of the liver.

An important consideration in the use of γ-GTP in cancer patients may be its combination with the assay of other constituents. The simultaneous assay of carcinoembryonic antigen (CEA) and either phosphohexose isomerase of γ-GTP has been used to achieve a better prediction of the presence of hepatic metastases. The serum enzyme indicates that the liver is abnormal, and the CEA discriminates between hepatomegaly due to benign liver disease and cancer metastatic to the liver. With the combination of serum enzyme and tumor-associated antigen assays, the presence of hepatic metastases may be predicted 3-9 months before clinical evidence of recurrence [3].

IV. RIBONUCLEASE

Ribonuclease (RNase) has been studied by Reddi and Holland using poly(C) as substrate [11]. They found that serum and pancreatic RNase have similar properties. An important consideration in this method is that RNase from leukocytes, which has been reported to be the majority of serum RNase, utilizes poly(U) as a substrate and is almost completely inactive with poly(C) as a substrate. In their study 52 normal persons were found to have mean serum activity of 104 ± 24.3 (SD) units/ml. In 30 patients with pancreatic carcinoma the mean serum value was 384 ± 145 (SD) units/ml. Two of these patients had values of 99 and 150 units, while the other 28 activities ranged from 247 to 714 units/ml of serum. Only 1 of 10 patients with pancreatitis had an elevated value (206 units), and the mean value (120 ± 39.4 units/ml) was not different from that of normal persons.

Elevations were observed in small numbers of patients with other cancers. In 28 patients with breast cancer, 3 had elevated values of 244, 250, and 275 units/ml; in 20 colon cancer patients, 3 had values of 252, 295, and 300 units; and in 11 patients with lung cancer, 2 had values of 244 and 272 units. None of 6 patients with gastric cancer or of the 4 with liver cancer had elevations, but each of 4 patients with kidney cancer had elevated values even higher than those observed in pancreatic cancer (628 ± 89.4 units) with a range of 545 to 732 units. Elevations were also found in each of 2 patients with gallbladder cancer, one individual with cancer of the ampulla, and one with ovarian cancer. Six patients in renal failure who did not have cancer also had marked elevations with values between 1730 and 3500 units/ml. If a threshold value of 250 units is used, 90% of patients with pancreatic cancer are included, and all normal persons and patients with pancreatitis as well as 90% of patients with other forms of cancer are excluded. In this study no information was provided concerning the extent of disease, staging, or presence of metastases.

Fitzgerald and his associates found that with the procedure proposed by Reddi and Holland and a normal-abnormal cutoff of 250 units/ml, elevations were observed in 8 of 16 patients (50%) with cancer in the region of the head of the pancreas, in 3 of 10 (30%) with other cancers, and in 4 of 11 (36%) with noncancerous pancreatic disease [12].

In another study tRNA was used as substrate, and enzymatically released oligonucleotides were determined by measuring their absorbance at

260 nm [13]. Similar plasma activities were observed in healthy women and women with benign gynecological tumors. However, increased activity was observed in 21 of 22 patients with ovarian carcinomas. The 22 patients included 2 patients with Stage 1A disease and 2 with 1C ovarian cancer. The remainder had advanced disease. The plasma enzyme activity returned to normal levels in most patients following surgical removal of the tumor.

Maor and her associates measured serum ribonuclease by following the nucleotide formation at 260 nm when polyuridylic acid, polycytidylic acid, and a combination of polyadenylic and polyuridylic acid were used as substrates [14]. In this study 40 of 53 patients (75%) with lung cancer and 49 of 74 smokers without cancer (55%) had elevated activity compared to only 13 of 179 nonsmokers (13%). Elevated values were observed in 95% of the patients with epidermoid cancer (20 of 21 patients) compared to 13 of 19 patients (68%) with adenocarcinoma, 4 of 8 patients with large cell carcinomas, and 3 of 5 patients (60%) with oat cell carcinomas. Further studies are required to establish the true role of ribonuclease in pancreatic, ovarian, and other forms of cancer.

V. GLYCOSYLTRANSFERASES

The glycostyltransferases are involved in the biosynthesis of glycoproteins. They are a family of enzymes which catalyze the sequential addition of monosaccharides to specific protein and glycoprotein acceptors. The monosaccharides include among others sialic acid (sialyltransferase); galactose (galactosyltransferase); fucose (fucosyltransferase); and N-acetylglucosamine (N-acetylglucosaminyltransferase).

A. Sialyltransferase

The most widely studied of the glycosyltransferases is sialyltransferase, which catalyzes the transfer of sialic acid from the nucleotide cytidine monophosphate sialic acid to the galactose residue of desialated fetuin. Henderson and Kessel reported their findings in plasma of 134 randomly selected patients who had pathologically proven primary or metastatic cancer [15]. In 20 normal persons the values ranged from 147 to 330 units with a mean of 240 ± 55 units. The upper limit (±2 SD) was 340 units. In the 134 cancer patients, 340 values were obtained and 293 (86%) were elevated. In sequential studies carried out in 57 patients, it was found that decreasing plasma sialyltransferase accompanied tumor response in 17 of 20 patients (85%) in whom clinical improvement was observed. In 29 of 37 (78%) of the patients without a clinical response, enzyme levels did not change or increase. Overall the plasma enzyme activity responded to the clinical course in 46 of 51 patients. Responses to the clinical state were more pronounced in patients with massive metastatic tumors (18 of 20 patients) than in patients whose tumor burden was not considered massive (27 of 41). Correlation was observed in 20 of 26 patients with liver metastases, 11 of 14 with lung metastases, and in 36 of 50 with other forms of metastatic cancer. Plasma sialyltransferase levels rose about 18% within 1 week of surgery but fell to 80% of their original values in 30-45 days in those patients in whom successful surgery had been carried out. Immunlogical skin testing and drug-induced hepatotoxicity resulted in transient elevations. Plasma elevations have also been observed in individuals with acute and chornic nonneoplastic liver disease, rheumatoid arthritis, and a variety of other non-cancer-related diseases. It seems that the probable role of plasma sialyltransferase in cancer patients is to monitor the response to therapy in association with other more conventional indicators of change of tumor mass.

In another study sialyltransferase was assayed in the serum of more than 500 patients [16]. Abnormal activity was observed in 80% of them compared to elevations of CEA in only 43%. Elevations were observed in all forms of cancers. Breast cancer patients with T-1-2, NO, MO disease had elevated activity, and patients without metastases who had elevated values showed decreases in activity following surgery or radiation therapy. The activities began to rise when recurrence developed.

B. Galactosyltransferase

Bhattacharya measured a UDP-galactose: glycoprotein galactosyl transferase in serum and tissue of patients with ovarian cancer [17]. In ovarian tissue the enzyme level was three to five times higher than in normal tissue in the presence or absence of fetuin. In 10 normal controls the serum activities ranged from 147 to 179 pmol of galactose transferase per 10 μl of serum per 60 min (165 ± 10.6) and did not seem to vary with blood type. Significant elevations were seen in patients with ovarian cancer. In 11 patients the activity ranged from 227 to 519 units. In each case in which there was a postoperative specimen a fall of 13-20% was observed. In one patient who was declared cured of disease the serum level returned to normal. In serial studies increasing levels preceded relapses that were subsequently confirmed clinically.

In a later study by this group various glycosyltransferases were studied and it was concluded that galactosyltransferase was the most useful marker in ovarian cancer [18]. In 30 patients with ovarian cancer (25 with advanced disease, 4 with Stage I or II cancer, and 1 with a recurrent borderline malignancy), values ranged from 43 to 78 pmol hr^{-1} ml^{-1} serum (mean ± SD, 51 ± 8) compared to a range of 19 to 39 (29 ± 6) in a control group of healthy donors who were 22 to 27 years old. Values between 22 and 33 units (28 ± 4) were observed in 10 patients who had no clinical evidence of disease at least 6 to 12 months before the assay.

UDP-Galactose N-acetylglucosamine galactosyl transferase was studied in patients with breast cancer [19]. Mean serum activities were higher in the cancer group than in either healthy women or women with noncancerous diseases. In specimens obtained before therapy, significant elevations were observed in 14.3% of Stage I patients, 66.7% of those with Stage II disease, 78.6% of Stage III patients, and 96.5% of women with Stage IV breast cancer. In 84.6% of the patients serum activity returned to normal following modified radical mastectomy. The control group included six patients who had active liver disease not related to cancer. Elevated levels were observed in these patients.

In another study it was found that galactosyltransferase (UDP-galactose: glycoprotein galactosyltransferase) activity was elevated in 40 tissue specimens in bladder transitional cell sarcoma of both an invasive and noninvasive nature (24.4 to 180.0 cpm/μg protein) compared to activity in normal or inflamed mucosa of 0.8-46.1 cpm/μg protein. All normal tissues were below 22.6 cpm/μg protein and all cancerous tissues above 24.4 cpm/μg protein. Increased activity appeared independent of tumor stage or grade. Galactosyltransferase activity was not measurable in urine or bladder washings [20].

As is clear from the above reports, total galactosyltransferase activity is elevated in serum of patients with a variety of cancers. Most of this enzyme activity has been found as a single broad band on polyacrylamide gel electrophoresis and has been termed galactosyltransferase isoenzyme I. A second band, galactosyltransferase II (GT II) is found in lesser amounts but is apparently primarily observed in serum of persons with cancer [21].

In a study of 232 patients with cancer, elevated amounts of GT II were observed in serum of 165 patients (71%), including 85 of 117 patients (73%) with colorectal cancer, 15 of 18 patients (83%) with pancreatic cancer, 12 of 16 persons (75%) with gastric cancer, 18 of 23 women (78%) with breast cancer, and 13 of 20 patients (65%) with lung cancer [22]. The enzyme was not found in serum of 58 healthy control patients, nor in patients with ulcerative colitis, Crohn's ileocolitis, pancreatitis, viral hepatitis or biliary tract disease. The enzyme was observed in 3 of 15 patients (20%) with severe alcoholic hepatitis and in 18 of 20 patients with celiac disease.

Although the enzyme activity was not related to the site of the colon cancer, it was related to the extent of colon cancer at operation. Enzyme was not detectable in any of 3 patients with Dukes A disease, in 7 of 9 patients with Dukes B cancer with a mean activity less than 3 units and a maximum activity less than 5 units, and in 23 of 31 patients with Dukes C colon cancer. In the patients with extended disease the highest activity was somewhat less than 8 units and the mean less than 5 units. Activity elevated to even higher levels was observed in 14 of 19 patients with distant metastases. Activity became undetectable in the Dukes B patients following "curative" surgery and was not detectable in serum of 14 persons who were disease-free 5 or more years after surgery. In 20 patients followed sequentially a correlation was observed between the disease status and the serum activity, and increasing activity in 3 patients preceded clinically apparent recurrent disease by 3 to 7 months.

C. Fucosyltransferase

Bauer and his associates studied the glycosyltransferases which add L-fucose or N-acetylneuraminic acid to acceptors [23]. In patients with colon cancer there was a drop in enzyme activity 4 to 6 weeks after surgery, and activity reached normal values in 14 days. In patients with residual disease the serum enzyme levels fell only slightly after surgery and then tended to increase. In 16 patients with breast cancer, elevated serum activity before therapy reflected successful treatment (surgery or chemotherapy), and serum levels fell or rose as the clinical condition changed. Fucosyltransferase (GPP-fucose: galactosyl fucosyltransferase) was elevated in patients with acute myelogenous leukemia and with nonresponding non-Hodgkin's lymphoma. In acute myelogenous leukemia the enzyme activity correlated with the percentage of marrow myeloblasts and returned to normal during remission. There was no correlation between disease state and either sialyltransferase or galactosyltransferase. Another fucosyltransferase (GDP fucose: N-acetylglucosaminide fucosyltransferase) was elevated in all these patients but did not reflect the clinical status.

VI. LACTIC DEHYDROGENASE

Lactic dehydrogenase is involved in a wide variety of metabolic pathways and is elevated in the serum of a large percentage of patients with cancer metastatic to the liver. Elevations are observed in about 70% of such patients. Elevations ranged from 33% in patients with metastatic cervical cancer to 76% of patients with metastatic colorectal cancer [3, 24]. At least one group has concluded that total LDH assays do not add to the clinical impression for diagnosis [25]. Assays of isoenzymes of LDH may be a more important use in the cancer patient. In cancer tissue there is usually a predominance of the LDH_5 (muscle-liver) isoenzyme. Such observations have been made for human breast, uterus, prostate, brain, lung, stomach, and kidney. Elevation of the LDH_5 fraction has been reported in serum of some patients. Whereas total serum LDH was elevated in only 24% of patients with colorectal cancer, LDH_5 was abnormal in 52% of the patients [24].

The most impressive data were observed in tissue. The ratio of LDH_5 to LDH_1 was six times greater in cancerous breast tissue than in breast tissue from patients with fibroadenoma; the LDH_5 proportion of the total activity rose from 4% in normal uterine tissue to 42% in cervical carcinoma [3]; the ratio of LDH_5 to LDH_1 was greater than 1 in 90 of 116 specimens of prostatic carcinoma and less than 1 in 473 of 503 specimens of benign prostatic hypertrophy [25]. In bladder tissue the LDH_5 to LDH_1 ratio was 0.27 in normal tissue and 4.74 in cancerous tissue [26].

In colorectal carcinoma Langvad observed a mean LDH_4 to LDH_2 ratio of 1.67 in 420 tissue samples from 36 cancerous surgical specimens from patients with colorectal cancer, and of 0.72 in 317 tumor negative specimens. However, in a number of cases increased ratios were observed in morphologically uninvolved tissues. When the clinical status of these patients was evaluated 5-7 years later, it was observed that a significant number of patients in whom the LDH_4 to LDH_2 ratio of the resection edge of the original tissue was high (mean ratio 0.92) succumbed with local recurrence, whereas those patients in whom the ratio was low (mean ratio 0.65) remained clinically without evidence of disease [27, 28].

VII. ARYL SULFATASE

Aryl sufatase is found in a variety of tissues. In colon tissue, the enzyme exists in three variant forms, A, B, and C, which may be differentiated by their substrate specificities. Aryl sulfatase B has been observed in higher concentrations in colon cancer tissue than in the homologous normal tissue, as well as in the urine of patients with colon carcinoma. Increased urinary activity has been correlated with the extent of disease. Elevations have been observed in only 28% of patients with Dukes A disease, 55% of those with Dukes B, and in more than 75% of patients with Dukes C and D lesions [29].

VIII. HISTAMINASE

Histaminase catalyzes the deamination of aliphatic and aromatic amines. Very high concentrations are observed in placenta, and increased activity is observed in serum of women during pregnancy. Elevated concentrations have been observed in thyroid tumor tissue, and elevations have been reported in serum of patients with medullary thyroid cancer [30, 31]. Seventy percent of patients with metastatic medullary cancer are reported to have elevated serum activities. Elevated histaminase activity has been observed in effusion fluids from patients with cancer of the ovary, breast, stomach, or lung. Such elevations were seen in 44% of ascitic fluids from women with with cancer of the ovary [32]. This enzyme has also been observed in patients with small cell carcinoma of lung [33] and has been observed to be a part of the APUD system of endocrine cells (A = amines; PU = amine precursor uptake; D = L-aromatic amino acid decarboxylase). Histaminase was increased in each of six primary oat cell tumors to levels 3 to 14,000 times that in surrounding normal lung tissue [34].

IX. HEXOSAMINIDASE

This enzyme (N-acetyl-β-D-glucosaminidase) catalyzes the hydrolysis of hexosamines and exists as an acidic isoenzyme (hexosamidase A) and a basic isoenzyme (hexosamidase B). Elevated amounts of total enzyme activity have been reported in serum of patients with leukemia or solid tumors [35], and recently the B form of the enzyme has been observed as the predominant isoenzyme in human colon carcinoma, compared to the predominance of the A variant in normal colon tissue [36]. The B form has also been reported to be present as the predominant isoenzyme in the serum of patients with solid tumors, although the activity of both variants was increased above that observed in serum of healthy controls or patients without cancer [37]. Elevated activity has also been observed in children with acute lymphoblastic leukemia, but not in those with acute myelocytic leukemia [38].

X. TERMINAL DEOXYNUCLEOTIDYL TRANSFERASE

Terminal deoxynucleotidyl transferase (TdT) is a unique DNA-polymerizing enzyme which transfers deoxyribonucleotides onto a primer molecule without the presence of a template polynucleotide. This enzyme may be present in large amounts in cells from patients with certain forms of leukemia. The activity in lymphocytes from normal peripheral blood is low and is generally at the lower limit of the sensitivity of the assay. In adult patients the activity is high in the cells of all patients with acute lymphoblastic leukemia, of 10% of patients with acute myelocytic leukemia, of 30% of patients with chronic myelogenous leukemia, and of a small number of patients with non-Hodgkin's leukemia [39]. The enzyme activity is not elevated in peripheral lymphocytes or marrow cells from patients with acute myelomonocytic or monocytic leukemia or in individuals with Hodgkin's disease, solid tumors, infectious diseases, or a variety of other diseases. The enzyme activity is not detectable when the leukemia is in remission, but it may, by increased activity, reflect an exacerbation of the disease [39].

Adenosine deaminase and TdT can be used in the classification of acute lymphoblastic leukemia. T-cell disease is quite distinct biochemically from B-cell acute lymphoblastic leukemia. In B-cell leukemia both TdT activity and adenosine activity are very low. In T-cell disease the deaminase activity is very high.

REFERENCES

1. M. K. Schwartz, Clin. Chem. 19:10 (1973).
2. M. K. Schwartz, in Biological Markers of Neoplasia (R. Ruddon, ed.), Elsevier, New York, 1978, p. 503.

3. M. K. Schwartz, Cancer 37:542 (1976).
4. M. K. Schwartz, Ann. Clin. Lab. Sci. 7:99 (1977).
5. M. K. Schwartz, Cancer 36:2334 (1975).
6. P. E. Perille and S. L. Finch, in Lysozyme (E. F. Osserman, R. E. Canfield, and S. Beychock, eds.), Academic Press, New York, 1974, p. 359.
7. E. Huberman, R. Montesano, C. Drevon, T. Kuroki, L. St. Vincent, T. D. Pugh, and S. Goldfarb, Cancer Res. 39:269 (1979).
8. N. Sawabu, M. Nakagen, M. Yoneda, H. Makino, S. Kameda, K. Kobayashi, N. Hattori, and M. Ishii, Gann 69:601 (1978).
9. N. K. Kim, W. G. Yasmineh, E. F. Freier, A. I. Goldman, and A. Theologides, Clin. Chem. 23:2034 (1977).
10. B. E. Roberts, J. A. Child, E. H. Cooper, R. Turner, and J. Stone, Acta Hematologica 59:65 (1978).
11. K. K. Reddi and J. F. Holland, Proc. Natl. Acad. Sci. USA 73:2308 (1976).
12. P. J. Fitzgerald, J. G. Fortner, R. C. Watson, M. K. Schwartz, P. Sherlock, R. S. Benua, A. L. Cubilla, D. Schottenfeld, D. Miller, S. J. Winawer, C. J. Lightdale, S. P. Leidner, J. S. Nisselbaum, C. J. Menendez-Botet, and M. H. Poleski, Cancer 41:868 (1978).
13. B. Shied, T. Lu, L. Pedrinan, and J. H. Nelson, Cancer 39:2204 (1977).
14. D. Maor, M. E. Klein, D. E. Kenady, P. P. Cretien, and M. R. Mardiney, Jr., JAMA 239:2766 (1978).
15. M. Henderson and D. Kessel, Cancer 39:1129 (1977).
16. U. Ganzinger, Wien Klin. Wochenschr. 89:594 (1977).
17. M. Bhattacharya, S. K. Chatterjee, and J. J. Barlow, Cancer Res. 36:2096 (1976).
18. S. K. Chatterjee, M. Bhattacharya, and J. J. Barlow, Cancer Res. 39:1943 (1979).
19. J. F. Paone, T. P. Waalkes, R. R. Baker, and J. H. Shaper, Surg. Forum 29:158 (1978).
20. K. Hagen-Cooke, G. R. Prout, G. M. Plotkin, S. L. Gilbert, and G. Wolf, Surg. Forum 29:627 (1978).
21. M. M. Weiser, D. K. Podolsky, and K. J. Isselbacher, Proc. Natl. Acad. Sci. USA 73:1319 (1976).
22. D. K. Podolsky, M. M. Weiser, K. J. Isselbacher, and A. M. Cohen, N. Engl. J. Med. 299:703 (1978).
23. C. H. Bauer, W. G. Reutter, K. P. Erhart, E. Kottgen, and W. Gerok, Science 201:1232 (1978).
24. D. C. Wood, V. Varela, M. Palmquist, and F. Weber, J. Surg. Oncol. 5:251 (1977).
25. S. Clark and V. Srinivasan, J. Urol. 109:444 (1973).
26. H. C. Bredin, J. J. Daly, and G. R. Prout, J. Urol. 113:487 (1977).
27. E. Langvad, Int. J. Cancer 3:17 (1968).

28. E. Langvad and B. Jemec, Br. J. Cancer 31:661 (1975).
29. L. R. Morgan, M. S. Samuels, W. Thomas, E. T. Krementz, and W. Meeker, Cancer 36:2332 (1975).
30. S. B. Baylin, M. A. Beaven, K. Engelman, and A. Sjoerdsma, N. Engl. J. Med. 283:1239 (1970).
31. S. B. Baylin, M. A. Beaven, H. R. Keiser, A. H. Tashijian, Jr., and K. E. W. Melvin, Lancet 1:455 (1972).
32. C. Lin, M. K. Orcutt, L. L. Stollbach, Cancer Res. 35:2762 (1975).
33. S. B. Baylin, M. D. Alcloff, K. C. Wierman, J. W. Tomford, and D. S. Ettinger, N. Engl. J. Med. 293:1286 (1975).
34. S. B. Baylin, W. R. Weisburger, J. C. Eggleston, G. Mendelsohn, M. A. Beaven, M. D. Abeloff, and D. S. Ettinger, N. Engl. J. Med. 299:105 (1978).
35. J. W. Woollen and P. Turner, Clin. Chim. Acta 12:671 (1965).
36. M. G. Brattain, P. M. Kimball, and T. G. Pretlow, Cancer Res. 37:731 (1977).
37. C. H. Lo and D. Kritchevsky, J. Med. Clin. Exp. Theor. 9:313 (1978).
38. R. B. Ellis, N. T. Rapson, A. D. Patrick, and M. F. Greaves, N. Engl. J. Med. 298:476 (1978).
39. J. J. Hutton and F. F. Bollum, in Biological Markers of Neoplasia (R. Ruddon, ed.), Elsevier, New York, 1978, p. 569.

5

DEVELOPMENTAL ALKALINE PHOSPHATASES AS BIOCHEMICAL TUMOR MARKERS

RANDALL H. KOTTEL[*] and WILLIAM H. FISHMAN / La Jolla Cancer Research Foundation, La Jolla, California

I. Introduction 93
   A. Background 93
   B. Methods of Isoenzyme Analysis 94
   C. Alkaline Phosphatase Isoenzymes of Normal Tissue 96
II. Developmental (Placental) Alkaline Phosphatases 98
   A. Term Placental Isoenzyme and Variants 98
   B. Early Placental Isoenzyme 99
III. Developmental Alkaline Phosphatase Isoenzymes in Human Cancer 100
   A. Regan Isoenzyme 100
   B. Nagao Isoenzyme 103
   C. Regan or Hepatoma Variant (Kasahara Isoenzyme) 104
   D. Non-Regan Isoenzymes 105
   E. Clinical Application 106
IV. Summary 108
   References 109

I. INTRODUCTION

A. Background

The measurement of alkaline phosphatase (E.C. 3.1.3.1) activity in serum has been extremely valuable in the clinical diagnosis of hepatobiliary and bone disease for over 40 years. With the introduction of techniques of isoenzyme analysis, it became evident that the enzyme existed in several distinct forms, and that the resolution and measurement of these various forms may provide clinical information relevant to a number of specific disease states.

   The importance of these advances with respect to developmental alkaline phosphatases as potential biochemical markers in cancer was

---

[*]Present affiliation: Department of Biology, Mesa College, San Diego, California.

first appreciated in 1968 when Fishman et al. discovered an isoenzyme (Regan isoenzyme) in the sera and tissue of a patient with lung carcinoma that was apparently identical to that expressed by the term placenta [1]. Since that time an increasing number of investigations has revealed that human tumors of diverse tissue types also express this placental alkaline phosphatase phenotype. More recent work has demonstrated that the isoenzyme in term placenta is preceded in development by isoenzymes found in first trimester placenta which have physical biochemical and immunological properties distinct from those of term placenta [2]. It is interesting that a wide variety of tumor types has been observed to express an alkaline phosphatase activity which most closely resembles this early placental (non-Regan) phenotype.

These findings appear to indicate that the different enzyme phenotypes of alkaline phosphatase which occur during normal trophoblast development may be re-expressed in the neoplastic cell (Table 1). Such evidence supports the notion that cancer is a disease of gene regulation involving in part certain genes or gene sets which are characteristically active only during normal fetal or trophoblast development. It is as yet unknown whether the expression of such genes is coincidental in the tumor cell or if the activity of these genes is required to establish and/or maintain the neoplastic state.

More important in the context of the present discussion is the fact that such developmental proteins, by virtue of their unique characteristics, may serve as reliable biochemical markers, the measurement of which could prove to be important for detection of neoplasia in humans. It should be kept in mind that at the present time no biochemical marker exists which is considered to be a universally reliable indicator of the presence of the variety of different malignant diseases in man. Rather it is to be expected that a battery of biochemical tumor markers will evolve and that the simultaneous analysis of all of these will provide a reliable laboratory index useful in the clinical diagnoses of the future.

B. Methods of Isoenzyme Analysis

Description of the more recently discovered "developmental" alkaline phosphatases which appear to be re-expressed in malignant disease has relied on criteria and conditions adopted over the past 20 years. Therefore some of the same methods that were previously used to define biochemically the isoenzymes of different tissue of origin in normal individuals have also been applied to characterize the activities found not only in the early stages of trophoblast development, but also in sera and/or tissues of patients with certain malignant diseases. It is important to realize that these methods are valuable in that they have aided in the characterization of these recently described isoenzymes but that, with one exception, there is still no direct evidence to indicate a common genetic origin.

TABLE 1 Oncodevelopmental Alkaline Phosphatases

Year	Neoplastic	Developmental
1968	Regan	
1968	Non-Regan	
1969	Hepatoma variant	
1970	Nagao	
1971		
1972		Placental F, FS, S [a]
1973		Placental D variant [b]
1974		Fast (FL) amnion
1975		Chorionic phase 1 < 10 weeks

[a] Common electrophoretic phenotypes of alkaline phosphatase.
[b] Rare electrophoretic phenotype.
Source: Fishman et al. [64].

The most frequently used criteria for isoenzyme characterization include heat inactivation, uncompetitive inhibition by organ-specific amino acid inhibitors, and electrophoretic analysis. The last method is often coupled with use of the same uncompetitive inhibitors and/or monospecific antisera to determine biochemical and immunological activity of individual isoenzyme bands. Heat sensitivity has been used clinically to differentiate liver and bone type isoenzymes [3-6]. Extreme resistance to heating serves as the major distinguishing feature of placental isoenzyme and its counterparts associated with neoplasia [1, 7]. Moss et al. pointed out the variability which exists in clinical samples with respect to heat stability [8]. In this regard, Fishman and Ghosh emphasized the need to combine use of all distinguishing methods for analysis of the isoenzymes [9].

The discovery of differential uncompetitive inhibition of alkaline phosphatase isoenzymes by L-phenylalanine and L-homoarginine has proved to be extremely valuable in such analysis. Thus L-phonylalanine inhibits human intestinal and placental phosphatases [10, 11], while L-homoarginine selectivity inhibits the liver and bone type activities [12, 13]. The conditions of assay and inhibitor concentrations routinely utilized in such analysis have been described in detail [14].

With the introduction of electrophoretic techniques, the problem of physically separating the various alkaline phosphatase isoenzymes was solved. Although starch gel electrophoresis was introduced initially and is still considered useful in discriminating the genetic polymorphism of placental alkaline phosphatase [15], polyacrylamide gel electrophoresis is more rapid and excellent resolution has been achieved. Smith et al. demonstrated separation using 5% gels with an alkaline buffering system [16]. Human liver, bone, and intestinal enzyme activities were readily identified on the basis of mobility alone. Green and his associates utilized similar techniques coupled with use of the organ-specific inhibitors L-phenylalanine and L-homoarginine to identify the source of isoenzyme in sera from 56 normal subjects [17]. More recently L. Fishman introduced the use of the detergent Triton X-100 in the acrylamide gel matrix to facilitate entry and migration of enzyme which previously failed to resolve in this system minus detergent [18]. Both fast- and slow-migrating protein fractions of human liver, placenta, intestine, and bone were well separated.

Another technique which has been adopted for routine analysis, especially of very small samples, is microzone electrophoresis [19]. The method employs cellulose acetate membrane electrophoresis to separate isoenzymes and the fluorogenic substrate napthol ASMX-phosphate to visualize activity. The technique offers both rapid and extremely sensitive means for identifying alkaline phosphatase isoenzymes. Additional methods which have been used often to distinguish isoenzymes include inactivation by urea [20, 21], alteration of electrophoretic mobility by neuraminidase treatment [1, 16], and inhibition of activity by EDTA and L-leucine [22, 23].

Some of the same distinguishing techniques described above have also been adapted for use in histochemical characterization of the isoenzymes in thin sections of tumor tissue samples. For example, Sasaki and Fishman incorporated the uncompetitive inhibitors L-phenylalanine and L-homoarginine in histochemical staining studies of alkaline phosphatase isoenzymes of human ovarian cancer cells [24]. More recently the availability of nonspecific antisera developed toward particular isoenzymes has made possible development of direct immunoperoxidase staining techniques. Miyayama et al. presented evidence that tumor cells which were found to express L-phenylalanine-sensitive alkaline phosphatase activity also exhibited a high level of immunoperoxidase staining when antisera toward term placental isozyme was used [25]. Such studies have also confirmed the plasma membrane location of both placental and tumor enzyme.

C. Alkaline Phosphatase Isoenzymes of Normal Tissue

In a recent review Moss presented evidence that there are now accepted three major antigenic categories of tissue-specific alkaline phosphatase in humans which appear to be expressed by at least three corresponding genetic loci [26]. These include alkaline phosphatase of term placental, intestinal, and nonintestinal origins (the last group characteristically includes liver and bone activities, which share very similar properties).

TABLE 2 Distinguishing Properties of Alkaline Phosphatase Isoenzymes

Criteria	Placental	Intestinal	Nonintestinal Liver	Nonintestinal Bone
Inhibition by L-phenylalanine (%)	75	75	0-10	0-10
Inhibition by L-homoarginine (%)	5	5	78	78
Heat inactivation	0	50-60	50-70	90-100
Alteration of electrophoretic mobility by neuraminidase	+	0	+	+
Antigenic determinants of placental isoenzyme	+	0	+	0
Antigenic determinants of liver isoenzyme	0	0	+	+
Antigenic determinants of intestinal isoenzyme	0	+	0	0

Source: Fishman [9].

The major criteria developed over the last two decades for distinguishing these major classes are shown in Table 2 and have been described in detail in a recent review by Fishman [9].

The term placental enzyme is identified clinically by its extreme resistance to denaturation by heat (65°C), significant inhibition by L-phenylalanine (75%), and cross-reaction with antisera directed toward the purified isoenzyme. That the placental isoenzyme derives from a genetic origin distinct from the other classes is supported by the finding of Donald and Robson of 44 electrophoretic phenotypes which arise from 18 alleles at the single gene locus [27]. Earlier studies by Beckman and his co-workers of variants of placental alkaline phosphatase in a variety of ethnic groups are also consistent with this statement [28-31]. One of these variants, the so-called rare D variant, is of interest because of the appearance at high frequency of a similar activity (Nagao isoenzyme) in certain malignant diseases. This activity is characterized by its inhibition by L-leucine and EDTA and by its slow-migrating characteristics on starch gel. Although the D-variant has been found in less than 1.0% of placental extracts [32, 33], it is important because it represents a second normal placental phenotype which has a counterpart expressed in certain malignant diseases.

Alkaline phosphatase isoenzyme of intestinal origin has been shown to share some of the properties of the term placental isoenzyme, including uncompetitive inhibition by L-phenylalanine and a comparable rate of hydrolysis with various substrates [34]. In addition, purified intestinal phosphatase cross-reacts with antisera directed toward placental isoenzyme [35], indicating at least partial structural similarity. However, intestinal isoenzyme fails to demonstrate either the resistance to heat denaturation or the allelic polymorphism of the placental counterpart. Neuraminidase treatment fails to alter the net molecular charge of intestinal isozyme, whereas all other alkaline phosphatase isoenzymes treated in similar fashion reveal reduced migration rates after isozyme electrophoresis.

Recent studies by Hirano et al. show that purified intestinal isozyme differs from that of term placenta in both dimer and subunit molecular weights and in amino acid composition [36]. Such evidence is consistent with the existence of a separate structural gene for this isoenzyme.

The third major category of alkaline phosphatases, the nonintestinal isozymes, includes activities found in bone, liver, and kidney. Enzyme from these sources is typically more sensitive to inactivation by heat, is inhibited to a large extent (75%) by L-homoarginine, and is insensitive to inhibition by L-phenylalanine (<10%). Antisera directed toward liver isoenzyme will react with these isozymes, whereas antisera directed toward intestinal or placental isoenzymes is immunologically nonreactive, within the group. Moss recently reviewed the indirect evidence that this group of biochemically similar isozymes may arise from a common genetic origin [26], and pointed out the contradictory immunochemical results discussed by Fishman [9]. It is possible that the minor differences observed result from postgenetic modification of the carbohydrate moieties of the protein and/or peptides and lipids associated with the isozyme [26].

II. DEVELOPMENTAL (PLACENTAL) ALKALINE PHOSPHATASES

A. Term Placental Isoenzyme and Variants

There is abundant evidence that the hyperphosphatasemia observed first in sera of pregnant women is the result of the presence of the placental isoenzyme of alkaline phosphatase [10, 37]. Because of the appearance in neoplasia of an activity similar in most respects to this isoenzyme, much work has been done to purify and characterize the enzyme present in the normal term placenta. This information has proved invaluable in comparative studies of this protein and in those alkaline phosphatases produced ectopically by malignant tissues or cell lines growing in vitro [38].

5 / Developmental Alkaline Phosphatases

The distinctive properties of this enzyme, as mentioned above, include its resistance to heat inactivation (5 min, 65°C), high pH optimum (pH 10.7), inhibition by L-phenylalanine, and immunological specificity with rabbit antisera directed toward the purified protein. More detailed analysis of term placental enzyme by Ghosh and Fishman [39] revealed two major molecular weight variants (A and B) after Sephadex G-200 chromatography. The latter form was found to be very heterogeneous with a molecular weight greater than 200,000 daltons. It is likely that this electrophoretically slow form represents enzyme that is incompletely solubilized from the original plasma membrane site. A molecular weight of 70,000 daltons was obtained for the homogeneous A variant. Identical subunit molecular weight was reported by Doellgast and Fishman in similar studies [40].

Holmgren and Stigbrand presented evidence of comparable values for isoenzymes from two genetically classified placenta [41]. Dimer molecular weights for the SS or FF isozyme type as determined by SDS gel electrophoresis were 122,000 and 119,000 daltons, respectively. The subunit (monomer) molecular weights of 65,000 and 63,000 daltons, respectively, were somewhat lower than those reported by others. This difference can be explained by the fact that previous workers utilized enzyme derived from pooled placental tissue, and preparations therefore contained a mixture of several isozyme variants. Additional physicochemical properties such as sedimentation velocity, isoelectric point, and partial specific volume were consistently similar and generally in agreement with the results of the previous studies.

B. Early Placental Isoenzyme

Until very recently the term placental isozyme was the only developmental alkaline phosphatase isozyme with counterparts recognized in human cancer. Fishman et al. designed experiments to elucidate the time during trophoblast development when the genes that produce this isozyme are first active [2]. When very early (8-10 week) placentas were examined, however, they were found to contain significant amounts of alkaline phosphatase activity which was apparently different from that of term placenta. An activity with similar properties had previously been observed in ultrastructural studies of human ovarian cancers [24] in which it was classified as non-Regan alkaline phosphatase.

As a result of the combined use of biochemical, immunological, and electrophoretic analyses, these workers were able to define three phases of alkaline phosphatase expression in the developing trophoblast. Phase 1 or chorionic phase (6-10 week placenta) was characterized by the presence of two bands of activity after microzone electrophoresis on cellulose acetate. Both components were heat-sensitive and inhibited to a large extent by L-homoarginine but not by L-phenylalanine. The more slowly migrating B band was found to be immunologically reactive with antisera directed toward liver isoenzyme and therefore possessed antigenic determinants of liver or bone type enzyme.

More interesting was the finding that the fast-migrating species (chorionic A band) failed to react with individual antisera specific for placental, intestinal, or liver-bone type determinants. At times between 10 and 13 weeks of trophoblast development, a pattern was observed which indicated a mixture of Phase 1 and term placental isozymes. During this time (Phase 2), the immunologically distinct fast A band and liver-bone B band, as well as a band characterized histochemically as term placental isoenzyme, were evident. Phase 3 (greater than 13 weeks) was represented solely by heat stable, L-phenylalanine-sensitive term placental type isoenzyme.

Parallel studies involving histochemical localization and inhibitor analysis of enzyme activities in the various placental tissue sections confirmed the biochemical data. These studies also revealed that alkaline phosphatase of early placenta is associated exclusively with the microvilli, a location which is shared by the alkaline phosphatase of term placenta.

The current information indicates that the temporal change in alkaline phosphatase may represent alternate expression of two separate trophoblast genes. One is active early in trophoblast development and is turned off by approximately 12 weeks. The other is not expressed until after 10 weeks of gestation, but continues to be active until parturition. It should be noted, however, that until direct evidence is available this change could also be attributed to the postgenetic modification of the product of a single trophoblast gene. Additional studies will be required to characterize thoroughly the interesting early placental (Phase 1) isoenzymes, but such studies have been hampered by the limited availability of material from this early stage of development.

Chou has recently reported the establishment of several clonal human early placental cell lines which synthesize human chorionicgonadotropin [42]. It is possible that such cell lines may be exploited as an excellent source of material for purification and characterization of these isoenzymes. It will then be possible to carry out comparative studies of physiochemical properties of the early trophoblast isozyme(s) and the similar non-Regan enzymes observed in human tumors. Such studies could lead to the identification of a second developmental alkaline phosphatase gene which is re-expressed in malignant disease and which is distinct from the gene that codes for the term placental isoenzyme.

III. DEVELOPMENTAL ALKALINE PHOSPHATASE ISOENZYMES IN HUMAN CANCER

A. Regan Isoenzyme

The first report of the expression of developmental alkaline phosphatases was published in 1968 by Fishman et al. [1]. In studies of alkaline phosphatase isoenzymes in serum of cancer patients, an individual with metastatic bronchogenic carcinoma was found to possess in his serum high levels of an isoenzyme which most closely resembled the pregnancy-associated term placental isoenzyme. Enzyme activity of the same phenotype was also demonstrated at autopsy in the neoplasm and its metastases by using sensitive histochemical enzyme staining techniques. In serum it was observed that approximately 50% of the total activity was inhibited by L-phenylalanine, and only 1-15% of total activity was sensitive to heat treatment (5 min, 65°C). The activity associated with tumor tissue of the lung and subsequent metastases to the lymph nodes, adrenal gland, and spleen was found to be inhibited 70-74% by L-phenylalanine and completely resistant to heat inactivation. Electrophoretic analysis revealed that the heat-stable activity of both serum and cancer-involved tissue samples co-migrated with the heat-stable L-phenylalanine-sensitive alkaline phosphatase of pregnancy serum. This activity was also found to precipitate in the presence of antiserum to human term placental isoenzyme [43]. Using the biochemical, electrophoretic, and histochemical criteria available at the time, Fishman et al. concluded that the enzyme expressed in this case of lung cancer was indistinguishable from placental alkaline phosphatase [1]. Alkaline phosphatase activity of this type has been termed the Regan isoenzyme after the name of the patient in which it was first discovered.

Since the initial observations of tumor production of term placental isoenzyme, much attention has been focused on the examination of many different tumor types to determine the incidence of this developmental phosphatase in neoplasia. Initial studies by Stolbach et al. detected the Regan isoenzyme in 4.6% of the sera of 590 patients with a variety of malignancies [44]. Of the 27 patients with Regan-positive sera, tissue samples from seven were obtained at autopsy and butanol extracts were prepared to determine if tumor production of enzyme occurs. In all cases such extracts contained heat-stable L-phenylalanine-sensitive alkaline phosphatase which was immunologically reactive with placental antiserum. Moreover, similar extracts of tumor cancer from 23 other patients with no detectable Regan isoenzyme serum failed to exhibit activity.

It was apparent from this study that the Regan isoenzyme which was detected initially in sera was the result of ectopic production by tumor cells. In contrast to the rather limited tumor types which express such fetal protein as carcinoembryonic antigen or alpha fetoprotein, the authors pointed out the variety of organs and histological characteristics with which Regan isoenzyme may be associated.

In further studies Nathanson and Fishman examined the sera of over 300 patients with cancers at various sites for the presence of Regan isoenzyme [45]. It was observed that sera of approximately 14% of the untreated cancer patients contained detectable levels of the term placental isoenzyme. The highest percentage of positive values occurred in sera of patients with carcinoma of the ovary, pancreas, stomach, lung, or breast. The authors pointed out that ovarian cancer and testicular teratocarcinoma displayed the greatest percentage of sera positive for Regan isoenzyme.

Results of a more recent series of investigations of sera and effusion fluids from carcinoma of the breast, lung, and ovary appear to substantiate this finding [46]. Thus 10% and 4% of sera from active cases of breast and lung carcinoma, respectively, were found to be positive for Regan isoenzymes, while 39% of sera from patients with ovarian cancer were positive. Moreover, 58% of malignant effusion fluids from patients with ovarian cancer also revealed the Regan isoenzyme. It was suggested that the increased incidence of developmental phosphatase in these germ cell tumor types strengthened the emerging concept that a relationship existed between undifferentiated embryonic cells and those cells that are associated with the neoplastic state.

An additional finding of this study was the relatively high incidence of Regan-positive sera from individuals with conditions such as ulcerative colitis, familial polyposis, and cirrhosis [46]. All of these nonmalignant conditions were associated with a predisposition to the development of neoplastic disease. Thus it was apparent that particular genetic information may be expressed in certain precancerous states and monitored before any histological manifestations become apparent. Such a possibility was encouraging from the standpoint of the early clinical detection of malignant disease in humans.

Other investigations have substantiated the fact that human tumor cells produce an isoenzyme which is apparently identical to that of term placenta. Kang et al. reported production of placental type isoenzyme in a case of lung cancer [47]. Cell lines derived from human tumor tissue have also been reported to express an isoenzyme with term placental characteristics. Different HeLa cell lines, derived from a case of adenocarcinoma of the cervix, have been shown to express such activity by several groups [48, 49]. Recently, Herz et al. reported similar activity in the C41 cell line (squamous carcinoma of the cervix) [50]. Singer et al. observed the production of Regan isoenzyme in the HCT-8 cell line derived from cancer of the lower gastrointestinal tract [51]. In all of the above studies the identification of alkaline phosphatase isoenzyme as term placental type was based on biochemical and immunological properties.

Of more interest, however, was the finding by Greene and Sussman that purified Regan isozyme from human cancer shared physical and chemical properties with the purified enzyme of term placental [52]. The alkaline phosphatase with which placental protein was compared was derived from liver metastases of a giant cell carcinoma of the lung. Comparative studies of enzymes reveal identical subunit molecular weights and very similar isoelectric points. In addition, both enzymes showed homologous N-terminal amino acid sequences over the last four residues. (Parallel studies of purified liver isozyme revealed a different amino acid sequence.) Furthermore, after partially digested tumor and placental preparations were subjected to two-dimensional peptide mapping, it was found that only insignificant differences existed.

These findings were subsequently supported by Luduena and Sussman who characterized the enzyme produced in vitro by the KB cell line, derived from a nasopharyngeal tumor [53]. Native enzyme from cells proved to be identical to that purified from human term placenta with respect to pH optimum, amino acid inhibition profile, isoelectric point, and peptide maps. Analysis of subunits on SDS polyacrylamide gels, however, revealed nonidentical molecular weights. The 69,000 dalton molecular weight monomer was identical to that observed for the placental isoenzyme. The 72,000 dalton molecular weight monomer was found to vary only in the N-terminal amino acid sequence and behaved in all other respects like the placental protein. It was suggested that such minor differences probably represent allelic variation, which is characteristic of the placental alkaline phosphatase gene and is consistent with the high degree of polymorphism reported previously. Current information from both in vivo and in vitro source material supports the fact that the Regan isoenzyme found in human cancer is produced by the same structural gene that is expressed normally in placental development (Table 1).

B. Nagao Isoenzyme

Several years after the discovery of the Regan isoenzyme by Fishman, an activity of alkaline phosphatase was identified which shared some but not all of the characteristics of the placental isoenzyme. Nakayama et al. described an activity derived from a patient (Mr. Nagao) with pleural carcinomatosis which was identical to that of placental isoenzyme in certain immunological and biochemical characteristics, including heat stability and inhibition by L-phenylalanine [22]. Such activity was evident in both the serum and pleural fluid of this individual. Further characterization showed that the activity was also inhibited by L-leucine, L-isoleucine, and L-valine, as well as by EDTA. These properties had not been reported previously for any of the alkaline phosphatase isoenzymes. It was noted, however, that with respect to L-leucine inhibition this activity most closely resembled the rare D-variant isoenzyme of term placental alkaline phosphatase.

Within the next year two additional studies revealed the presence of this isoenzyme in other malignant disorders. Jacoby and Bagshawe found that the activity in both serum and tumor tissue of adenocarcinoma of the bile duct was similar in some respects to term placental alkaline phosphatase [54]. Activity was found to elute at a position identical to that of purified isozyme after G-200 chromatography, and it migrated to the same position after electrophoresis. In addition, the activity was precipitated by antisera directed toward the term placental isoenzyme. Increased sensitivity of activity to inhibition by both L-leucine and EDTA, however, was consistent with that characteristic of the Nagao isoenzyme. Alkaline phosphatase with these same characteristics was also described by Nakayama et al. in a patient with adenocarcinoma of the tail of the pancreas [23].

Additional evidence for the apparent similarity of the malignancy associated Nagao enzyme and the normal placental D-variant was presented by Inglis et al. [15]. The activities of 18 of 29 sera which were identified as Regan-positive were also sensitive to L-leucine inhibition and were found to co-migrate with D-variant enzyme after starch gel electrophoresis. More recently, Doellgast and Fishman have demonstrated that differences were discerned in inhibition assays by utilizing various peptides which contain L-leucine. Quantitation of Nagao isoenzyme presently involves the use of L-leucine inhibition [15]; however, it is possible that peptides containing this amino acid may serve as more critical discriminants in the future.

It is not apparent from the existing evidence whether the gene coding for the D-variant isoenzyme of normal placenta is re-expressed in the malignant cell to produce the very similar Nagao isoenzyme. Moss has pointed out that the placental type alkaline phosphatases produced by neoplastic tissues in vivo are frequently observed to be structurally and/or functionally altered [26]. A recent study by Nakayama and Kitamura revealed that Nagao isozyme was observed in almost the same frequency as Regan isozyme in a large number of sera, effusion fluids, and body fluids [56]. These workers observed, however, that electrophoresis of samples of Nagao isozyme indicated fast-moving components unlike those that characterize the D-variant [57, 58].

C. Regan or Hepatoma Variant (Kasahara Isoenzyme)

A third alkaline phosphatase isoenzyme which also possesses some of the properties of placental isoenzyme was first observed by Warnock and Reisman in extracts of 8 of 10 patients with hepatocellular carcinoma [59]. This activity, also termed the Kasahara isoenzyme, was found to be inhibited by L-phenylalanine, was notably more susceptible to heat inactivation at 65°C than placental isoenzyme but more stable than liver isozyme, and migrated anodally ahead of normal liver isoenzyme.

Higashino et al. observed a similar activity in 3 of 10 hepatoma patients [60, 61]. More detailed investigations by Higashino's group have indicated that this activity exhibits properties of both the placental D-variant and the liver isoenzymes. However, in some respects the isoenzyme is unique and possesses properties unlike either of these activities. For example, enzyme purified over 300-fold from a patient with hepatocellular carcinoma was inhibited by L-phenylalanine but not by L-homoarginine, was inactivated by urea, and was cleaved by neuraminidase. It shared both the molecular weight and the antigenic determinants of placental isoenzyme. In addition, it was observed that the activity was inhibited by L-leucine, was more heat-labile than the term placental isoenzyme, and displayed a much lower Michaelis constant. These properties indicate that the activity also possesses characteristics of the placental D-variant isozyme. This same activity was similar to that of liver isoenzyme with respect to pH optimum (10.0) and resistance to inhibition by 5 mM phosphate (placental isozyme is inhibited 50%).

Fishman suggested that the available evidence supported the fact that a significant number of hepatomas produce an isoenzyme which most closely resembles a hybrid composed of subunits of both the liver and placental (D-variant) alkaline phosphatases [9]. Continued studies by Honda et al. [62] and Higashino et al. [63] have indicated, however, that the Kasahara isozyme shares many of the same biochemical and immunological properties of fast-moving isozymes of the FL cell line which is derived from human amniotic membrane. Fishman et al. designated this hepatoma variant activity as Fast (FL) Amnion type alkaline phosphatase and noted that it shared antigenic determinants with the intestinal isoenzyme [64].

Other studies by Otani et al. revealed that Kasahara isoenzyme with these same properties occurred in 11 of 30 hepatoma specimens [65]. In addition, it was observed that a fetal intestinal type alkaline phosphatase activity appeared concomitantly with Kasahara isozyme in these tissues. Detailed studies revealed the similarity of both molecules to fetal and adult intestinal alkaline phosphatases as well as to the fast components of FL amnion cells. These workers have stressed the need for complete analysis of molecular structure to determine the relationship of these two activities to each other and to their developmental counterpart (i.e., amnion-intestinal alkaline phosphatase).

D. Non-Regan Isoenzymes

Before the discovery by Fishman et al. that the early placenta expresses alkaline phosphatase isoenzyme activities distinct from those of term placenta [1], many investigators reported enzyme activities in cancer patients which were not inhibited by L-phenylalanine and which were heat-sensitive. The so-called non-Regan alkaline phosphatases were described in a variety of malignancies including cancer of the pancreas [66], craniopharyngiomas [67], lung cancers [68], meningioma [69], and ovarian cancers [70]. Activity of this nature was characteristic of liver (or bone) isoenzymes and thus was not considered in a developmental context. More recently Ehrmeyer et al. detected an electrophoretic form of alkaline phosphatase in a high percentage of the sera of patients with cancers in a variety of tissue types [71]. This activity was found to be fast-migrating after cellulose acetate electrophoresis, inhibited (70%) by L-homoarginine, minimally inhibited (20%) by L-phenylalanine, and extremely heat-sensitive (85% loss of activity). In blind analysis of 200 specimens these workers observed an incidence of the fast homoarginine-sensitive alkaline phosphatase (FHAP) in 3% of normal controls, 35% of patients with gastrointestinal cancer, 14% of those with gynecological benign disease, and 53% of those with gynecological cancer.

This newly observed activity was examined in detail after separation from other alkaline phosphatase isoenzymes on DEAE-Sephadex A-50. The FHAP was found to differ from Regan, Nagao, and Kasahara isoenzymes and most closely resembled liver alkaline phosphatase. Results of inhibitor studies showed similarity between this FHAP and the unique fast band of early placenta described by Fishman [2]. Others have also observed this unique activity in a high proportion of cancer patients who do not have any complicating hepatic involvement [72]. It is also important that this FHAP can be distinguished from both the Regan and liver isoenzymes of alkaline phosphatase by immunochemical methods, and that it shares the same electrophoretic mobility as the fash (chorionic A) band of alkaline phosphatase from early (8-10 week) placentae. It is not known at present if the activity of the early placental chorionic A band is identical to the FHAP described in serum by Ehrmeyer et al. [71], but such a possibility should be considered. In fact, a preliminary syudy by Fishman et al. [64] demonstrated activity that was indistinguishable from this early placental activity in extracts of various neoplasms including testicular teratomas and lung tumors.

It is possible that the development of specific immunological reagents will make it possible to design tests similar to those currently used for Regan isozyme to detect early placental isoenzymes as well. The importance of such reagents is substantiated by the fact that expression of alkaline phosphatase activity by neoplastic tissue is not limited to a particular developmental (i.e., term placental) isoenzyme type. For example, Fishman et al. showed that testicular teratocarcinoma exhibits activities characteristic of all phases of development [64]. The same study revealed both chorionic A and B bands in 5 of 14 homogenates of lung cancer tissue. The remaining samples exhibited only the B band. Previous studies of ovarian cancer showed that both Regan and non-Regan isoenzymes may be co-expressed in certain cases [70]. Indeed, the ability to monitor carefully for the range of developmental alkaline phosphatases described thus far may greatly increase the percentage of neoplasms detectable by these methods.

E. Clinical Application

At the present time only the reagents and techniques for quantitation of Regan isoenzyme are available for clinical assays for developmental alkaline phosphatases. Both immunochemical and radioimmunoassay techniques have been developed, and as little as 1 ng of protein corresponding to Regan isozyme can be detected [73-75]. In addition, the sensitive histochemical and immunohistochemical techniques described earlier for detection of Regan isozyme in tissues have also found clinical application [24, 25].

As stated earlier, the incidence of this isoenzyme in patients with a variety of neoplasms varies from 3 to 15%. Despite this somewhat low incidence, it should be pointed out that increased levels of Regan isoenzyme in serum occur in a variety of different neoplastic diseases. This multiplicity contrasts with the more limited types of malignancies in which one encounters the other oncodevelopmental proteins such as carcinoembryonic antigen (CEA) and alpha fetoprotein (AFP).

Several investigators have reported a concomitant decrease in measured Regan activity as a result of therapy, suggesting potential use of this assay in cancer management programs. Stolbach et al. measured serum levels of Regan isozyme in a patient with carcinoma of the colon which had metastasized to the liver and who was undergoing treatment with 5-fluorouracil [44]. They observed a gradual decrease of Regan isozyme to undetectable levels over a period of approximately 4 months, during which time liver size also decreased. Continued treatment was accompanied by a slow increase in liver size and a return of serum Regan isozyme to levels comparable to those observed initially. These workers proposed that measurement of Regan isozyme in serum may be a valuable index of patient response to different therapeutic and/or surgical regimens.

Such a proposal was supported by the additional findings of Nathanson and Fishman [45]. They found that levels of Regan isoenzyme in two patients with breast cancer who underwent adrenalectomy dropped sharply within a very short time after surgery. Clinical improvement was indicated in both cases. The levels of placental type isoenzyme also decreased abruptly in two other patients who were surgically treated and presumed cured for carcinoma of the colon. Additional patients undergoing chemotherapy, and whose serum enzyme levels correlated with response, included patients with carcinomas of the pancreas and ovary. Similar correlation between levels of Regan isoenzyme in serum and changes resulting from therapy were reported by Charles [76] and Belliveau [77] in patients with lung carcinoma and mediastinal choriocarcinoma, respectively. Taken together these results indicate that the monitoring of Regan isoenzyme levels in selected patients can serve as a useful measure of the effectiveness of treatment of the disease.

With respect to developmental gene activation as a generalized phenomenon in neoplasia, the detection or measurement of placental alkaline phosphatase as one of several established markers has proved valuable. There have been many reports of the co-expression of markers, including human chorionic gonadotropin (HCG), alpha fetoprotein (AFP), and carcinoembryonic antigen (CEA), as well as alkaline phosphatase. Portugal et al. observed simultaneous production of AFP and placental alkaline phosphatase variant (Kasahara) isoenzyme in six cases of hepatoma [78]. Belliveau et al. detected substantial levels of CEA, HCG, and Regan isoenzyme in the blood of a patient with mediastinal choriocarcinoma [77]. Co-expression of alkaline phosphatases and other developmental proteins has also been demonstrated in lung carcinoma [76], gastric carcinoma [79, 80], and in cell lines derived from adenocarcinoma of the lower gastrointestinal tract [51].

Studies by Stolbach et al. [46] indicate the advantages of using such a battery of tests. These workers measured levels of the tumor-associated markers HCG, CEA, and Regan isoenzyme in sera of 215 patients with cancer of the breast, ovary, or lung. They found that at least one of these markers appeared in 82% of sera of breast cancer patients. Comparable values of 81 and 89% from patients with lung and ovarian cancers, respectively, were also reported. When malignant effusion fluids were monitored, it was observed that 100% of the patients with ovarian cancer expressed at least one marker.

Perhaps the very recent description of the presence of additional developmental alkaline phosphatase isoenzymes offers the most potential for future clinical application. That these activities are also apparent in many neoplastic diseases supports the notion that developmental phosphatase isozymes may be expressed at high frequency in cancer, and indicates the need for further studies that can provide the required reagents. In fact, because of the more widespread occurrence of non-Regan alkaline phosphatase isozymes in neoplasia, it is likely that the introduction of a battery of tests for such activities which possess developmentally unique properties will prove to be a useful addition to the current analysis for Regan isoenzyme alone. As pointed out by Fishman et al. [64], it is possible that the type(s) of phosphatase(s) expressed may prove valuable in determining the extent of embryonic gene activation in transformation and/or neoplasia. Such information may allow clinical interpretation and evaluation of both the stage of neoplastic development and subsequent metastatic potential.

IV. SUMMARY

The expression of a developmental alkaline phosphatase in a case of neoplastic disease was first described in 1968. Since that time it has been established that the isoenzyme found in term placenta and the Regan isoenzyme expressed in a variety of different cancers represent the product of the same structural gene. This finding has allowed antisera to be directed toward purified placental isozyme in sensitive radioimmunoassay techniques for detection of the cancer-associated placental proteins. While the primary usefulness of such assays may involve cases of gynecological cancer in which the incidence of expression has been shown to be greatest, it should be pointed out that this isoenzyme is much more widespread in distribution than other oncodevelopmental proteins such as AFP and CEA. Current trends suggest the incorporation of assays for all established oncodevelopmental products into a battery of tests, since it has become apparent that the presence of one or more of these products is indicated in a very high percentage of patients with a variety of cancers. As additional assay methods are added it is expected that the prognostic and diagnostic value of detection of these unique gene products will increase.

It is certain that investigations of oncodevelopmental alkaline phosphatases in the future will concentrate on elucidating the properties of the more recently described isoenzymes which have counterparts in neoplasia. Of particular interest, because of its unique immunological characteristics, is the fast-moving chorionic A band of early placenta. The presence of a similar activity in sera of patients with various malignant diseases provides the impetus to generate the necessary reagents for use in clinical and histological analyses of serum and tissue specimens, respectively. When this is accomplished it may then be possible to extend the number of malignant diseases in humans which are detectable by critical analysis of these developmental isoenzymes.

ACKNOWLEDGMENTS

The experimental work from this laboratory described in this review has been supported in part by grants-in-aid (RO1 21967) from the National Cancer Institute awarded to William H. Fishman.

REFERENCES

1. W. H. Fishman, N. R. Inglis, L. L. Stolbach, and M. J. Krant, A serum alkaline phosphatase isoenzyme of human neoplastic cell origin. Cancer Res. 28:150 (1968).
2. L. Fishman, H. Miyayama, S. G. Driscoll, and W. H. Fishman, Developmental phase-specific alkaline phosphatase isoenzymes of human placenta and their occurrence in human cancer. Cancer Res. 36:2268 (1976).
3. J. F. Kerkhoff, A rapid serum screening test for increased osteoblastic activity. Clin. Chim. Acta 22:231 (1968).
4. M. X. Fitzgerald, J. J. Fennelly, and K. McGeeney, The value of differential alkaline phosphatase thermostability in clinical diagnosis. Am. J. Clin. Pathol. 51:194 (1969).
5. I. K. Tan, L. F. Chio, and L. Teow-Suah, Heat stability of human serum alkaline phosphatase in bone and liver diseases. Clin. Chim. Acta 41:329 (1972).
6. B. J. Cadeau and A. Malkin, A relative heat stability test for the identification of serum alkaline phosphatase isoenzymes. Clin. Chim. Acta 45:235 (1973).
7. N. G. Beratis and K. Hirschhorn, Properties of placental alkaline phosphatase. III. Thermostability and urea inhibition of isolated components of the three phenotypes. Biochem. Genet. 6:1 (1972).

8. D. W. Moss, M. J. Shakespeare, and D. M. Thomas, Observations on the heat stability of alkaline phosphatase isoenzymes in serum. Clin. Chim. Acta 40:35 (1972).
9. W. H. Fishman, Perspectives on alkaline phosphatase isoenzymes. Am. J. Med. 56:617 (1974).
10. W. H. Fishman and N. K. Ghosh, Isoenzymes of human alkaline phosphatase. Adv. Clin. Chem. 10:255 (1967).
11. D. A. Byers, H. N. Fernley, and P. G. Walker, Studies on alkaline phosphatase. Inhibition of human placental phosphoryl phosphatase by L-phenylalanine. Eur. J. Biochem. 29:197 (1972).
12. C. W. Lin and W. H. Fishman, L-homoarginine: An organ specific uncompetitive inhibitor of human liver and bone alkaline phosphohydrolases. J. Biol. Chem. 247:3082 (1972).
13. S. Green, C. L. Anstiss, and W. H. Fishman, Automated differential isoenzyme analysis. II. The fractionation of serum alkaline phosphatases into "liver," "intestinal" and "other" components. Enzymologia 41:9 (1971).
14. L. Fishman, N. R. Inglis, and W. H. Fishman, Preparation of two antigens of human liver isoenzymes of alkaline phosphatase. Clin. Chim. Acta 34:393 (1971).
15. N. R. Inglis, S. Kirley, L. L. Stolbach, and W. H. Fishman, Phenotypes of the Regan isoenzyme and identity between the placental D-variant and the Nagao isoenzyme. Cancer Res. 33:1657 (1973).
16. I. Smith, J. D. Perry, and P. J. Lightstone, Disc electrophoresis of alkaline phosphatase: Mobility changes caused by neuraminidase. Clin. Chim. Acta 25:17 (1969).
17. S. Green, F. Cantor, N. Inglis, and W. H. Fishman, Normal serum alkaline phosphatase isoenzymes examined by acrylamide and starch gel electrophoresis and by isoenzyme analysis using organ-specific inhibitors. Am. J. Clin. Pathol. 57:52 (1972).
18. L. Fishman, Acrylamide disc gel electrophoresis of alkaline phosphatase of human tissues, serum and ascites fluid using Triton X-100 in the sample and the gel matrix. Biochem. Med. 9:309 (1974).
19. N. R. Inglis, D. L. Guzek, S. Kirley, S. Green, and W. H. Fishman, Rapid electrophoretic microzone membrane techniques for Regan isoenzyme (placental type alkaline phosphatase) using a fluorogenic substrate. Clin. Chim. Acta 33:287 (1971).
20. J. J. Fennelly, J. Dunne, K. McGeeney, L. Chong, and M. Fitzgerald, The importance of varying molecular size, differential heat and urea inactivation of phosphatase in the identification of disease patterns. Ann. N.Y. Acad. Sci. 166:794 (1969).
21. J. J. Fennelly, M. X. Fitzgerald, and K. McGeeney, Value of differential thermostability, urea inhibition and gel filtration of alkaline phosphatase in the identification of disease states. Gut 10:45 (1969).

22. T. Nakayama, M. Yoshida, and M. Kitamura, L-leucine sensitive heat-stable alkaline phosphatase isoenzyme detected in a patient with pleuritis carcinomatosa. Clin. Chim. Acta 30:546 (1970).
23. T. Nakayama, M. Yoshida, M. Kitamura, and K. Saito, Heat stable alkaline phosphatase isoenzyme (Nagao isoenzyme) detected in adenocarcinoma of pancreas tail. Rynsho-Kagaku 11:80 (1971).
24. M. Sasaki and W. H. Fishman, Ultrastructural studies on Regan and non-Regan isoenzymes of alkaline phosphatase in human ovarian cancer cells. Cancer Res. 33:3008 (1973).
25. H. Miyayama, G. J. Doellgast, V. Memoli, L. Gandbhir, and W. H. Fishman, Direct immunoperoxidase staining for Regan isoenzyme of alkaline phosphatase in human tumor tissues. Cancer 38:1237 (1976).
26. D. W. Moss, Isoenzymes in medicine. Molec. Aspects Med. 1:477 (1977).
27. L. J. Donald and E. B. Robson, Rare variants of placental alkaline phosphatase. Ann. Hum. Genet. 37:303 (1973).
28. L. Beckman, G. Bjorling, and C. Christodoulou, Pregnancy enzymes and placental polymorphism. I. Alkaline phosphatase. Acta Genet. 17:406 (1967).
29. L. Beckman, G. Beckman, C. Christodoulou, and A. Ifekwumiqive, Variations in human placental alkaline phosphatase. Acta Genet. 17:413 (1967).
30. G. Beckman and E. O. Johannson, Distribution of placental alkaline phosphatase types in the Icelandic population. Acta Genet. 17:413 (1967).
31. G. Beckman and L. Beckman, The placental alkaline phosphatase polymorphism, variation in Hawaiian subpopulations. Hum. Hered. 19:524 (1969).
32. S. H. Boyer, Alkaline phosphatases in human sera and placentae. Science 134:1002 (1961).
33. L. Beckman and G. Beckman, A genetic variant of placental alkaline phosphatase with unusual electrophoretic properties. Acta Genet. 18:543 (1968).
34. W. H. Fishman and H. G. Sie, Organ-specific inhibition of human alkaline phosphatase isoenzyme of liver, bone, intestine and placenta: L-phenylalanine, L-tryptophan, and L-homoarginine. Enzymologia 41:141 (1971).
35. F. G. Lehmann, Immunological relationship between human placental and intestinal alkaline phosphatase. Clin. Chim. Acta 65:257 (1975).
36. K. Hirano, M. Sugiura, Y. Miki, S. Iino, H. Suzuki, and T. Oda, Characterization of tissue-specific isozyme of alkaline phosphatase from human placenta and intestine. Chem. Pharm. Bull. (Tokyo) 25:2524 (1977).
37. S. Posen, Alkaline phosphatase. Ann. Intern. Med. 67:183 (1967).

38. H. H. Sussman, Oncoplacental phosphatase isoenzyme, in Chemistry of Tumor Associated Antigens (E. Ruoslahti and E. Engvall, eds.), Scand. J. Immunol. Suppl. 6:126 (1978).
39. N. K. Ghosh and W. H. Fishman, Purification and properties of molecular weight variants of human placental alkaline phosphatase. Biochem. J. 108:779 (1968).
40. G. J. Doellgast and W. H. Fishman, Purification of human placental alkaline phosphatase. Biochem. J. 141:103 (1974).
41. P. A. Holmgren and T. Stigbrand, Purification and partial characterization of two genetic variants of placental alkaline phosphatase. Biochem. Genet. 14:777 (1976).
42. J. Y. Chou, Establishment of clonal human placental cells synthesizing human choriogonadotropin. Proc. Natl. Acad. Sci. USA 75:1854 (1978).
43. W. H. Fishman, N. R. Inglis, S. Green, C. L. Anstiss, N. K. Ghosh, R. E. Reif, R. Rustigian, M. J. Krant, and L. L. Stolbach, Immunology and biochemistry of Regan isoenzyme of alkaline phosphatase in human cancer. Nature 219:697 (1968).
44. L. L. Stolbach, M. J. Krant, and W. H. Fishman, Ectopic production of an alkaline phosphatase isoenzyme in patients with cancer. N. Engl. J. Med. 281:457 (1969).
45. L. Nathanson and W. H. Fishman, New observations on the Regan isoenzyme of alkaline phosphatase in cancer patients. Cancer 27:1388 (1971).
46. L. Stolbach, N. Inglis, C. Lin, R. Turksoy, W. Fishman, D. Marchant, and A. Rule, Measurement of Regan isoenzyme, HCG, CEA and histaminase in the serum and effusion fluids of patients with cancer of the breast, ovary and lung, in Onco-Developmental Gene Expression (W. H. Fishman and S. Sell, eds.), Academic Press, New York, 1976, pp. 433-443.
47. K-Y Kang, K. Higashino, M. Hashinotsume, Y. Takahashi, T. Aoki, E. Tsubura, and Y. Yamamura, Production of the placental type alkaline phosphatase isoenzyme by lung cancer tissue. Gann 63:217 (1972).
48. N. A. Elson and R. P. Cox, Production of fetal-like alkaline phosphatase by Hela cells. Biochem. Genet. 3:549 (1969).
49. B. Beckman, L. Beckman, and E. Lundgren, Isoenzyme variations in human cells grown in vitro. IV. Identity between alkaline phosphatase from Hela cells and placenta. Hum. Hered. 20:1 (1970).
50. F. Herz, O. J. Miller, D. A. Miller, N. Auersperg, and L. G. Koss, Chromosome analysis and alkaline phosphatase of C41, a cell line of human cervical origin, distinct from Hela. Cancer Res. 37:3209 (1977).

51. R. M. Singer, W. A. F. Tompkins, L. J. White, and J. E. Perry, Coproduction of Regan isoenzyme and carcinoembryonic antigen in HCT-8 cells. J. Natl. Cancer Inst. 56:175-178 (1976).
52. P. J. Greene and H. H. Sussman, Structural comparison of ectopic and normal placental alkaline phosphatase. Proc. Natl. Acad. Sci. USA 70:2936 (1973).
53. E. A. Ludeva and H. N. Sussman, Characterization of KB cell alkaline phosphatase: Evidence of similarity to placental alkaline phosphatase. J. Biol. Chem. 251:2620 (1976).
54. B. Jacoby and K. D. Bagshawe, Placental-type alkaline phosphatase from human tumor tissue. Clin. Chim. Acta 35:473 (1971).
55. G. J. Doellgast and W. H. Fishman, Inhibition of human placental-type alkaline phosphatase variants by peptides containing L-leucine. Clin. Chim. Acta 75:449 (1977).
56. T. Nakayama and M. Kitamura, L-leucine sensitive alkaline phosphatase isozyme from cancer tissue. Ann. N.Y. Acad. Sci. 259:325 (1975).
57. S. H. Boyer, Alkaline phosphatases in human sera and placentae. Science 134:1002 (1961).
58. L. Beckman and G. Beckman, A genetic variant of placental alkaline phosphatase with unusual electrophoretic properties. Acta Genet. 18:543 (1968).
59. M. L. Warnock and R. Reisman, Variant alkaline phosphatase in human hepatocellular cancer. Clin. Chim. Acta 24:5 (1969).
60. L. Higashino, M. Hashinotsume, K. Y. Kang, Y. Takahashi, and Y. Yamamura, Studies on a variant alkaline phosphatase in sera of patients with hepatocellular carcinoma. Clin. Chim. Acta 40:67 (1972).
61. K. Higashino, S. Kudo, and Y. Yamamura, Further investigation of a variant of the placental alkaline phosphatase in human hepatic carcinoma. Cancer Res. 34:3347 (1974).
62. T. Honda, T. Kurabori, S. Ishigami, and J. Sakurai, An alkaline phosphatase of FL cells. Iqaku-no-Ayumi (Japan) 86:313 (1973).
63. K. Higashino, S. Kudo, R. Ohtani, Y. Yamamura, T. Honda, and J. Sakurai, A hepatoma-associated alkaline phosphatase, the Kasahara isozyme, compared with one of the isozymes of FL amnion cells. Ann. N.Y. Acad. Sci. 259:337 (1975).
64. W. H. Fishman, T. Nishiyama, A. Rule, S. Green, N. R. Inglis, and L. Fishman, Onco-developmental alkaline phosphatase isozymes, in Onco-Developmental Gene Expression (W. H. Fishman and S. Sell, eds.), Academic Press, New York, 1976, p. 165.
65. R. Otani, S. Kudo, K. Higashino, and Y. Yamamura, A group of oncofetal alkaline phosphatases, in Onco-Developmental Gene

Expression (W. H. Fishman and S. Sell, eds.), Academic Press, New York, 1976, p. 727.
66. T. W. Warnes, W. R. Timperley, P. Hine, and G. Kay, Pancreatic alkaline phosphatase and tumor variant. Gut 13:513 (1972).
67. W. R. Timperley, P. Turner, and S. Davies, Alkaline phosphatase in craniopharyngiomas. J. Pathol. 103:257 (1971).
68. W. R. Timperley, Alkaline phosphatase secreting tumor of the lung. Lancet 2:356 (1968).
69. W. R. Timperley and T. Warnes, Alkaline phosphatase in meningiomas. Cancer 26:100 (1970).
70. W. H. Fishman, N. R. Inglis, J. Vaitukaitis, and L. L. Stolbach, Regan isoenzyme and human chorionic gonadotrophin in ovarian cancer. Natl. Cancer Inst. Monograph 42:63 (1975).
71. S. L. Ehrmeyer, B. L. Joiner, L. Kahan, F. C. Larson, and R. L. Metzenberg, A cancer-associated fast, homoarginine-sensitive electrophoretic form of serum alkaline phosphatase. Cancer Res. 38:599 (1978).
72. B. Nordentoft-Jensen, The fast moving fraction of serum alkaline phosphatase in patients with various diseases. Danish Med. Bull. 13:175 (1966).
73. S. Iino, K. Abe, T. Oda, and H. Suzuki, A new method of radioimmunoassay for human placental alkaline phosphatase. Clin. Chim. Acta 42:161 (1972).
74. M. Usategui-Gomez, F. M. Yeager, and A. F. de Castro, A sensitive immunochemical method for the determination of the Regan isoenzyme in serum. Cancer Res. 33:1574 (1973).
75. C. H. Chang, S. Raam, D. Angellis, G. J. Doellgast, and W. H. Fishman, A simple radioimmunoassay of human placental alkaline phosphatase (Regan isoenzyme) using specific antibody polymers. Cancer Res. 35:1706 (1975).
76. M. A. Charles, R. Claypool, M. Schaaf, W. E. Rosen, and B. D. Weintraub, Lung carcinoma associated with the production of three placental proteins: Ectopic human chorionic gonadotrophin, human chorionic somatomammotropin and placental alkaline phosphatase. Arch. Intern. Med. 132:427 (1973).
77. R. E. Belliveau, P. N. Wiernik, and E. A. Sickles, Blood carcinoembryonic antigen, Regan isoenzyme and human chorionic gonadotropin in a man with primary mediastinal choriocarcinoma. Lancet 1:22 (1973).
78. M. L. Portugal, M. S. Azevedo, and C. Manso, Serum alphafetoprotein and variant alkaline phosphatase in human hepatocellular carcinoma. Int. J. Cancer 6:383 (1970).

79. M. Nishimura, K. Nawata, Y. Makisaka, and T. Fujita, Variant alkaline phosphatase in human gastric cancer. Jap. J. Gastroenterol. 71:651 (1974).
80. M. Masuzawa, P-K Lee, T. Kamada, T. Akeyama, H. Abe, T. Shimano, T. Mori, H. Morino, and S. Ishiguro, Carcinoembryonic antigen alpha-fetoprotein and carcinoplacental alkaline phosphatase in gastric carcinoma metastatic to the liver. Cancer 39:1175 (1977).

6

PROSTATIC ACID PHOSPHATASE IN HUMAN
PROSTATE CANCER

T. MING CHU, MING C. WANG, CHING-LI LEE, CARL S. KILLIAN, and GERALD P. MURPHY / Department of Diagnostic Immunology Research and Biochemistry, Roswell Park Memorial Institute, Buffalo, New York

    I. Introduction   117
   II. Biochemical Nature of Prostatic Acid Phosphatase   118
  III. Immunological Specificity of Prostatic Acid Phosphatase   120
  IV. Radioimmunoassay for Prostatic Acid Phosphatase   120
   V. Counterimmunoelectrophoresis for Prostatic Acid Phosphatase   122
  VI. Fluorescent Immunoassay for Prostatic Acid Phosphatase   126
 VII. Solid-Phase Immunoadsorbent Assay   126
VIII. Clinical Evaluation of Immunoassays for Prostatic Acid Phosphatase   127
  IX. Bone Marrow Acid Phosphatase   129
   X. Summary   131
       References   132

I. INTRODUCTION

Acid phosphatases are the enzymes that hydrolyze phosphoric monoesters at an acidic pH, therefore the International Union of Biochemistry classifies them as the orthophosphoric monoester phosphohydrolases (EC 3.1.3.2) [1]. Acid phosphatase is one of the oldest tumor markers: The assay has been used as an aid in the diagnosis and treatment of prostate cancer since Gutman and Gutman published their report in 1938 [2] indicating the association of elevated acid phosphatase activity with metastasis of prostate cancer.

    In addition to prostate, which has 100-1000 times the specific enzyme activity of other tissues, other sources of acid phosphatases are liver, kidney, spleen, bone, intestine, and virtually all other tissues, as well as erythrocytes, leukocytes, and platelets. Hence the acid phosphatase in circulation is a combination of enzyme activity from all tissues and blood cellular components [3].

Unlike most of the other enzymes, which are highly substrate-specific, acid phosphatase hydrolyzes a wide variety of phosphate monoesters. Current chemical assays for serum acid phosphatase employ many different substrates, such as α-naphthyl phosphate, β-naphthyl phosphate, α-glycerol phosphate, β-glycerol phosphate, phenyl phosphate, p-nitrophenyl phosphate, adenosine 5'-monophosphate, phenylphthalein monophosphate, and sodium thymolphthalein monophosphate [4]. Although "specific" substrates and various chemical inhibitors have been reported in the measurement of prostate-specific acid phosphatase, as differentiated from acid phosphatases of other origins, the validity of these assays has always been questioned [5-8]. This is primarily due to the fact that similar catalytic activities toward different substrates are exhibited by different acid phosphatases, although they may have different relative rates of hydrolysis upon different substrates [4]. Consequently, the conventional spectrophotometric methods, or so-called chemical or biochemical methods, are unable to distinguish between acid phosphatase of prostate origin and those of other sources.

In recent years, several new and specific techniques, which primarily utilize immunological approaches—namely radioimmunoassay, counterimmunoelectrophoresis, and immunofluoroassay—have been developed to measure prostatic acid phosphatase. The common principle of these techniques is the immunological specificity of the prostatic acid phosphatase antiserum. These new assays of acid phosphatase and their role as diagnostic tools for prostate cancer are the subjects of this communication. Excellent general reviews of the clinical significance of acid phosphatase are available elsewhere [3, 4, 9-11], and will not be discussed in this report.

II. BIOCHEMICAL NATURE OF PROSTATIC ACID PHOSPHATASE

Acid phosphatase of human prostate has been purified to homogeneity by several investigators [12-16]. Biochemical analysis showed that the prostatic acid phosphatase is of glycoprotein in nature [17]. It consists of approximately 95% protein and 5% carbohydrate with a molecular weight of about 100,000 (Table 1). The amino acid and carbohydrate moieties contain 18 amino acid residues and 6 monosaccharides, respectively.

By isoelectric focusing technique, a homogeneous prostatic acid phosphatase preparation can be separated into at least eight isoenzymes [18, 19]. The pIs of these isoenzymes were between 4.2 to 5.5, although they may vary from preparation to preparation. Under identical experimental conditions, sera from patients with prostate cancer were found to exhibit similar isoenzyme patterns. In addition, as the enzyme activities increased, the pIs of isoenzymes shifted toward more acidic ranges. Acid phosphatase from normal human erythrocytes produced a single enzyme band at a pI of about 6.0. Similarly, serum specimens drawn from patients with

TABLE 1 Biochemical Composition of Prostatic Acid Phosphatase

Carbohydrate (mol/molecular of enzyme, 100,000)	
N-acetylglucosamine	12.7
Mannose	9.5
Sialic acid	6.4
Galactose	3.5
Fucose	2.1
Total (g/mol)	7000
Amino acid (residues/molecule)	
Glutamic acid	100
Leucine	93
Aspartic acid	54
Serine	54
Proline	50
Threonine	50
Lysine	45
Tyrosine	43
Glycine	42
Valine	34
Arginine	33
Phenylalanine	32
Alanine	27
Isoleucine	27
Histidine	26
Methionine	20
Tryptophan	18
Cysteine	16
Total (g/mol)	93,000

Gaucher's disease were also found to demonstrate a single enzyme band at a pI around 5.0. These data therefore suggest that, at least in some aspects, prostatic acid phosphatase is different biochemically from acid phosphatase of other origins.

III. IMMUNOLOGICAL SPECIFICITY OF PROSTATIC ACID PHOSPHATASE

Since prostatic acid phosphatase is a protein of molecular weight 100,000, it is a fairly strong immunogen, and antiserum can be produced easily in rabbits or goats. It has been shown that the reaction of antiprostatic acid phosphatase serum and enzyme resulted in a single immunoprecipitin line upon staining for both protein and acid phosphatase activity [13, 20]. The antiserum showed immunological reactivity only with purified acid phosphatase or prostate tissue extract, and exhibited no reactivity with extracts prepared from other tissues [13, 20-23].

Prostatic acid phosphatase thus seems to exhibit immunological specificity and to have a different site for its antibody binding than for its enzyme activity. Furthermore, the antiserum appears to have the ability to stabilize the enzyme for about 3 days even at room temperature [13, 24]. These properties are the basis for the development of counterimmunoelectrophoresis and immunofluoroassay for the specific measurement of serum prostatic acid phosphatase.

IV. RADIOIMMUNOASSAY FOR PROSTATIC ACID PHOSPHATASE

The radioimmunoassay for prostatic acid phosphatase was first reported in 1974 by Cooper and Foti [25]. This method was similar to any other radioimmunoassay that combined the sensitivity of radioisotopic technique with the specificity of immunological reagents. Acid phosphatase was isolated from normal prostatic fluid by a simple gel filtration on Sephadex G-100. Although it is questionable whether this single step was sufficient to purify the acid phosphatase, the authors showed by polyacrylamide gel electrophoresis that their purified enzyme preparation exhibited an acid phosphatase enzyme activity band. This acid phosphatase preparation was used without further purification for immunizing rabbits to produce antiserum to the enzyme.

A double antibody radioimmunoassay technique was developed using the rabbit antiserum and the radioactive ^{125}iodine-labeled acid phosphatase. A double incubation procedure, each requiring 48 hours, was needed to complete the assay. By this method a normal range of 15-65 ng of acid phosphatase protein per milliliter of serum was obtained from 107 healthy adult males ranging from 16 to 50 years of age. This procedure was reliable for measuring prostatic acid phosphatase in serum from 10 to 300 ng/ml. The sensitivity of this method was 10 ng/ml of serum.

A year later the double antibody radioimmunoassay was improved, which resulted in a solid-phase radioimmunoassay [26]. Polypropylene tubes coated with antiserum to prostatic acid phosphatase were used as the solid support in the assay. Double incubation was no longer required, and only a 48-hr incubation was employed in this improved technique. By means of this solid-phase radioimmunoassay, a similar normal range, 11.2-37.2 ng/ml, of serum prostatic acid phosphatase protein was reported. A preliminary clinical evaluation with limited numbers of prostatic cancer patients indicated that serum prostatic acid phosphatase levels were elevated in patients with advanced carcinoma of the prostate, and that patients with other malignancies were shown to have normal ranges of serum prostatic acid phosphatase.

The sensitivity of this solid-phase radioimmunoassay for human prostatic acid phosphatase was shown to be superior to that of a chemical method using p-phenyl phosphate as the substrate [27, 28]. In a group of 109 patients with prostate cancer, the radioimmunoassay detected 73%, versus 31% for the chemical method. According to the stages of prostatic carcinoma the data were as follows: In 44 patients with Stages I and II prostatic cancer, 43% were detected to have an elevated serum acid phosphatase by radioimmunoassay, and only 9% were so detected by the chemical assay; in the rest of the 65 patients with Stages III and IV prostatic cancer, the radioimmunoassay detected an elevation of enzyme in 94% of patients, and the chemical assay in only 46%. Results from patients with benign prostatic hypertrophy indicated that there were no false positive results recorded for either the radioimmunoassay or chemical assay. However, a recent report has indicated a significant false positive rate, as high as 13%, in patients with other malignancies and in normal individuals [29].

Three other groups of investigators have also reported the development of double antibody radioimmunoassay for prostatic acid phosphatase. In 1978 Choe et al. reported a radioimmunoassay using acid phosphatase purified from ejaculate of normal young males [21]. In addition to gel filtration, ion exchange chromatography, and Affi-Gel Blue affinity chromatography, Con A-Sepharose were used in the purification of acid phosphatase. Antiserum to the purified prostatic acid phosphatase preparation was raised in rabbit, which was then used in a double antibody radioimmunoassay system. A double incubation, 24 hours each, was employed in this assay. Normal males, 162 in total, ages 23 to 40, were found to have serum prostatic acid phosphatase levels up to 32 ng/ml. From a series of a limited number of cancer patients and patients with other diseases, serum and bone marrow aspirate from patients with metastases of prostatic carcinoma exhibited an elevated level of prostatic acid phosphatase.

Serum acid phosphatase of nonprostatic carcinoma patients and patients with benign prostatic hypertrophy were shown to be the same as those of normal males.

Belville et al., using antiserum raised against acid phosphatase prepared also from ejaculated human seminal fluid plasma, developed another double-antibody immunoassay technique [30]. Although no results concerning its clinical application in serum have been reported, they have used this assay to study the acid phosphatase in bone marrow aspirates. Results indicated that prostatic acid phosphatase in bone marrow aspirate as measured by this radioimmunoassay were found to correlate well with increased clinical stages of the disease in patients with prostate carcinoma. In contrast to radioimmunoassay, chemical assay using α-naphthyl phosphate as the substrate showed a poor correlation with the presence of prostatic adenocarcinoma.

More recently Vihko et al. [22] reported a solid-phase radioimmunoassay similar to that of Foti et al. [24] using polypropylene tubes. Vihko used a purified homogeneous acid phosphatase preparation, which was isolated from benign hypertrophic prostate by powerful affinity chromatography and gel filtration, as well as by a preparative isoelectric focusing technique. This procedure resulted in a highly purified acid phosphatase, claimed to be the most pure acid phosphatase with the highest enzyme activity [15]. Similarly, specific antiserum was raised in rabbits against this highly purified acid phosphatase. By means of a double incubation, again 48 hr each, sera from 53 normal subjects, 30 to 60 years of age, were found to contain acid phosphatase from 1 to 10 ng/ml of serum. Eleven patients with benign hypertrophy were also shown to have a range similar to that of healthy males, but 12 patients with prostatic carcinoma were shown to have a markedly elevated serum acid phosphatase, up to several hundred-fold higher than that of normal controls.

Although the radioimmunoassay technique of Foti et al. has been evaluated in a large-scale trial of up to 10,000 patients [31], no clinical information from such an extensive evaluation has been reported. The methods of Belville, Vihko, and Choe have not been extensively evaluated clinically. Undoubtedly, the radioimmunoassay technique is basically a sound procedure which provides good sensitivity. However, it also has some disadvantages, such as the instability of the labeled enzyme, the requirement of purified enzyme as the primary standard, the possible contamination of radioactive material, the rather long time needed to perform the assay, and the high cost.

V. COUNTERIMMUNOELECTROPHORESIS FOR PROSTATIC ACID PHOSPHATASE

Another immunoassay for serum prostatic acid phosphatase, perhaps a more practical one, is the so-called counterimmunoelectrophoretic (CIEP) assay. As described first by Chu et al. in 1976 [32], a specific rabbit antiserum was raised against purified prostatic acid phosphatase isolated from human malignant prostate by a series of gel filtrations, and ion exchange column and Con A-Sepharose chromatographies. Antiserum to prostatic acid phosphatase was shown to react with purified prostatic acid phosphatase and extract of prostate tissue, and did not react with acid phosphatase preparations from normal kidney, bladder, bone, liver, spleen, brain, intestine, or erythrocytes. In addition to the method of Chu et al. [13, 32], three other reports are available describing the CIEP assay for the measurement of circulating prostatic acid phosphatase [33-35].

The essence of this technique is movement of acid phosphatase and its antiserum in opposite directions under an electric field to form an immune precipitin line. Although protein staining of the enzyme-antibody precipitin complex is not sensitive enough to detect circulating acid phosphatase at the nanogram level, an enzyme activity staining technique can readily detect prostatic acid phosphatase with high sensitivity.

The sensitivity of the CIEP assays for serum prostatic acid phosphatase varied when the specific antiserum and the enzyme activities staining technique were used. As reported, it ranged from 20 ng/ml of prostatic acid phosphatase protein by Chu et al. [13] to 200 ng/ml by Foti et al. [26]. The following variables may explain these differing results: The avidity of antisera produced by various investigators, the purities of acid phosphatase employed as the reference, the pH of electrophoresis, and the different staining reagents used. Nevertheless, this simple and specific immunoassay has been shown by many laboratories to be rapid, reproducible, and sensitive for determination of serum prostatic acid phosphatase.

The CIEP technique of Chu et al. has been evaluated most extensively. The result of their preliminary study was quite encouraging [13]. In one patient with Stage A prostate cancer (confined to the prostate and not palpable), no serum prostatic acid phosphatase could be detected by this assay [36]. Certainly, one really could not draw any conclusion from a single patient. However, among 30 patients with Stage B disease (confined to the prostate but palpable, and with no metastasis), 6 or 20% gave positive results. Fifty-four percent of patients with Stage C disease (locally invasive beyond the capsule of the prostate) and 85% of those with Stage D (distant metastases) tested positive.

To determine the rate of possible false positives, serum specimens from 19 patients with benign hypertrophy, from 107 normal healthy volunteers, and from 50 normal age-matched older men were examined. All gave negative results. Furthermore, 87 patients with other carcinomas such as cancers of the breast, colon, lung, pancreas, and stomach were tested. Only one of these patients gave a positive result for serum prostatic acid phosphatase; this patient had a primary lung adenocarcinoma and a prostate cancer. Twelve patients with Gaucher's disease also tested negative.

Their CIEP assay was further evaluated nationally during 1977-1978 by more than 25 medical centers and hospitals [37]. In this expanded study, a total of 962 serum specimens from patients with prostate cancer were analyzed. Serum acid phosphatase assays were simultaneously performed by both the CIEP of Chu et al. and the chemical methods used by individual hospitals. Of 64 patients with Stage A prostate cancer, 38% had positive results with the CIEP assay while only 12% had elevated results when tested by conventional chemical methods. Thirty-five percent of 178 samples from Stage B patients were positive by the CIEP technique compared with only 17% by the conventional assays. Among 485 samples from patients with Stage D cancer, 69% were positive by the CIEP and 51% by the chemical methods.

In the national trial study the CIEP assay was also tested in patients without a confirmed diagnosis of prostate cancer. A total of 598 samples were obtained from the participating centers from patients who were thought to be free of prostate cancer. The test results indicated that 85% of these patients had a negative CIEP result and a negative rectal examination. However, 84 individuals or 14% of the patients showed a positive CIEP result. In 50 of these 84 patients a biopsy was performed and no carcinoma of the prostate could be diagnosed. A possible explanation for the false positive rate may be that the specimens were collected immediately after rectal examination of the prostate, which could cause an elevation of circulating prostatic acid phosphatase. In three patients with a positive CIEP assay, subsequent biopsy revealed evidence of prostatic carcinoma. No biopsy information was available on the other 31 patients.

Using the same reagents used for their radioimmunoassay, McDonald et al. employed a CIEP technique for rapid identification of circulating prostatic acid phosphatase [21, 33]. They reported that the CIEP was sensitive enough (25 ng/ml) to detect the circulating serum prostatic acid phosphatase, and they studied the levels as determined by a radioimmunoassay and by CIEP in the serum of normal males and females and in serum specimens from patients with prostatic and nonprostatic tumors. An excellent correlation was observed between these two assay results. They concluded that the CIEP assay was a simple and reliable screening method for the circulating prostatic acid phosphatase in the clinical diagnosis of prostatic cancer.

Romas et al. also developed a CIEP assay to detect prostatic acid phosphatase [35]. A crude acid phosphatase preparation was prepared from benign human prostatic tissues, and the antiserum was produced in rabbit. After absorption with normal female serum and normal lung tissue extract, the absorbed antiserum, shown to be specific to prostatic acid phosphatase, was used in their CIEP technique. The sensitivity of their CIEP technique was 40 ng/ml of prostatic acid phosphatase protein. Clinical evaluation, although with a smaller number of patients, produced results as promising as those reported by Chu et al. [13] and McDonald et al. [33]. They also concluded that CIEP test for prostatic acid phosphatase represented a useful clinical method for the study of patients with hyperacid phosphatasemia.

Despite these promising clinical data reported by investigators using three CIEP techniques (different acid phosphatase preparations, different antiserum reagents, and different conditions for electrophoresis and staining reagents), Foti et al. reported a CIEP method with a greatly decreased sensitivity, 200 ng/ml, although the reagents used were the same as those in their radioimmunoassay [34]. Consequently, they could not detect circulating prostatic acid phosphatase in patients with early stages of prostatic cancer. Only those patients with moderately increased enzyme activity showed positive CIEP results. In contrast, they reported that they were able to obtain better results using the radioimmunoassay. They indicated that in their method some of the serum proteins migrated with the acid phosphatase and interfered with the staining for acid phosphatase. It also appeared that their electrophoresis system was conducted at pH 8.6. At such an alkaline pH the acid phosphatase may be denatured during electrophoresis, which could explain the decreased sensitivity of their enzyme activity staining system.

One of the advantages of the CIEP assay, from the theoretical point of view, is that it measures the acid phosphatase protein by both the antigenicity and hydrolytic activity of the enzyme, and may be superior to the activity assay by conventional chemical methods and the protein assay by radioimmunoassay. Due to the built-in technical handicap, this is basically a semi-quantitative technique. Most recently, Killian et al. by means of densitometry and fixed-point kinetic spectrophotometry, achieved the quantitation of prostatic acid phosphatase by the CIEP method [38]. After electrophoresis, washing, and staining, the agarose gel was scanned and integrated by a densitometer. Alternatively, before staining the enzyme activity, the gel between the wells of antigen (serum or referenced prostatic acid phosphatase) and antibody (antiprostatic acid phosphase serum) can be punched out, and the enzyme activity in antibody-enzyme precipitin complex can be measured spectrophotometrically. These two improvements on the CIEP technique produced sensitivities of 1 ng/ml and 16 ng/ml, respectively, which are comparable to the current sensitivity of radioimmunoassay.

A linear standard curve up to 250 ng/ml of prostatic acid phosphatase protein, and recoveries of more than 95% were also realized with these two modifications. Furthermore, a primary reference for the acid phosphatase, such as α-naphthol, can be incorporated into the system, which represented a further refinement of the CIEP technique.

When the improved CIEP was used, the levels of serum prostatic acid phosphatase in normal, apparently healthy controls and in patients with nonprostatic tumors were shown to be similar, and those in patients with early stages of prostatic cancer were significantly elevated. These data thus indicated that the original CIEP can be modified to become a quantitative immunoassay for serum prostatic acid phosphatase and is suitable for use as a screening method for clinical prostatic neoplasm.

VI. FLUORESCENT IMMUNOASSAY FOR PROSTATIC ACID PHOSPHATASE

A new and even more sensitive immunoassay known as solid-phase fluorescent immunoassay for human prostatic acid phosphatase (PAP) was developed by Lee et al. [20]. In addition to the immunological specificity of the anti-PAP, this new assay utilized the fluorescent property of α-naphthol, the enzyme hydrolysis product of α-naphthyl phosphate. IgG antibody to prostatic acid phosphatase was isolated from antiserum and conjugated to a solid-phase support, Sepharose 4B beads, which were then used to bind and separate prostatic acid phosphatase from serum specimens. The bound enzyme then reacted with α-naphthyl phosphate to produce a fluorogenic compound, α-naphthol, which permits quantitation of enzyme activity with highly sensitive spectrophotofluorometry.

The sensitivity of this new immunoassay for prostatic acid phosphatase has been reported to be 20 pg/ml, several hundred-fold more sensitive than any previously reported immunoassays. Further, this technique requires neither isotope nor a long incubation period, as does the radioimmunoassay system. The normal range of serum prostatic acid phosphatase from a group of 30 normal healthy males is 1.399 to 9.838 ng/ml. The results obtained from a preliminary clinical evaluation indicated that the enzyme was elevated in almost all untreated patients with Stage A prostatic cancer and in about 60-70% of patients with Stage B and C who were receiving standard therapy at the time when blood was drawn for tests, and in all patients with an advanced stage of prostatic cancer. Patients with cancer other than prostatic tumor were found to have normal ranges of circulating prostatic acid phosphatase. Therefore this assay, by retaining its immunological specificity, has produced a much greater sensitivity. Large-scale testing is being used to evaluate its potential role in the early detection of prostate cancer.

VII. SOLID-PHASE IMMUNOADSORBENT ASSAY

More recently, Lee et al. further developed a new immune-colorimetric assay [39] for prostatic acid phosphatase, modified from their previously reported immunofluoroassay and also utilizing specific anti-PAP antibodies conjugated to CNBr-activated Sepharose 4B. The serum prostatic acid phosphatase was bound, and separated from other acid phosphatases and serum proteins, by the solid-phase anti-PAP IgG Sepharose 4B. The enzyme activity in the conjugate was quantitated by measuring, spectrophotometrically instead of spectrophotofluorometrically, the enzyme hydrolytic product, α-naphthol, from a primary standard solution.

The sensitivity of this method was 0.22 IU/liter of enzyme activity, or 0.88 ng of prostatic acid phosphatase protein per milliliter of serum. The normal range of serum PAP was the same as that of the solid-phase immunofluoroassay, or, as expressed in enzyme units, 0.2-2.4 IU/liter. Initial clinical evaluation showed that 19 of 25 patients with clinically staged A, B, and C prostatic cancer, and 12 of 14 patients with metastatic cancer, demonstrated an elevated PAP enzyme level (overall 79%), compared with only 6 and 8 patients, respectively (overall 36%), by a conventional chemical method (α-naphthyl phosphate).

One major difference of this assay from the previous fluoroassay is employment of a spectrophotometer, which is available in most clinical laboratories, rather than a spectrofluorometer. Among 48 patients with cancer other than prostate cancer, one lung cancer patient and two gastric cancer patients also registered positive results, as did one of eight patients with benign prostatic hypertrophy. Whether these three patients also had prostate cancer remained to be verified. In addition, of 39 patients with histologically confirmed adenocarcinoma of the prostate, 32 were shown to have a higher value of PAP by the immunoassay than was measured by the chemical assay, which was a consistent observation when PAP was quantitated by the immunoassays. Although no definitive factors that may cause these discrepancies have been identified, it is likely that PAP inhibitors, including some proteins and inorganic phosphate in the serum, could be removed in the immunoassay system and thus render the immunoassay more sensitive.

In considering all pros and cons, counterimmunoelectrophoresis is as good as radioimmunoassay; and without question, at this stage of development the new solid-phase immunofluorescent and immunoadsorbent assays appear to be superior.

VIII. CLINICAL EVALUATION OF IMMUNOASSAYS FOR PROSTATIC ACID PHOSPHATASE

Table 2 compares the diagnostic results of PAP in prostatic cancer among various immunoassays. Results from two radioimmunoassays [40, 41] and the original CIEP of Chu were quite similar, but different from the report of Foti's radioimmunoassay. Data from Lee's two new immunoassays appeared to agree with each other, although in expanded evaluation both clinical and surgical staged patients were included [42]. It should be noted, though, that the practice of clinical staging in prostate cancer may vary from group to group, and that some patients may receive treatment at the time of the blood test.

TABLE 2 Elevation of Serum Prostatic Acid Phosphatase Measured by Immunoassays in Various Clinical Stages of Prostate Cancer

Stage	Percentage positive in various stages					
	RIA Foti [27]	RIA Mahan [40]	RIA Griffiths [41]	CIEP Chu [13]	SPIF[a] Lee [42]	SPIA Lee [39]
A	33	13	12	–	73	75
B	79	26	32	20	56	75
C	71	30	47	55	82	77
D	92	94	86	79	86	86

[a] From a total of 133 patients with stages determined clinically and surgically. In their original report [20] from a limited number of patients staged clinically, the results were A, 4/4; B, 2/5; C, 7/11; and D, 4/4.

So far, only a handful of reports have shown data from patients with surgically staged prostate cancer and their PAP levels. By means of a solid-phase immunofluoroassay [42], 53 patients with surgically staged prostate cancer were evaluated for PAP level. The positive/total results were: A_1, 0/1; A_2, 2/2; B_1, 6/12; B_2, 11/19; C_1, 3/4; C_2, 3/4; D_1, 2/6; D_2, 4/5. A few patients with metastatic disease were under treatment and others were not at the time of assay. Therefore this study showed that elevated PAP levels were obtained in patients with localized, nonmetastatic disease, as confirmed by surgical staging and careful inspection of all pelvic lymph node materials. An elevated serum PAP can occur in stage A_2 patients; although the number of patients is still very small, this may alter current concepts of the staging of prostatic cancer.

Most recently, the original qualitative CIEP of Chu et al. has been modified and improved by Killian et al. to become a quantitative assay [43]. Also, a clinical evaluation using sera from early stages (A, B, and C) of prostate cancer patients, which initially demonstrated normal acid phosphatase activity by conventional chemical methods, has resulted in the detection of elevated prostatic acid phosphatase by the newly developed quantitative counterimmunoelectrophoresis in a substantial number of surgically staged patients: A, 2 of 7; B, 11 of 23; and C, 8 of 18.

In addition, PAP levels as measured by immunoassay procedures have been shown to be valuable in monitoring the efficacy of treatment for patients with Stage D prostate cancer, and in correlating patients' clinical statuses [44]. This is so primarily because secondary and metastatic prostatic tumors do express the prostate-specific acid phosphatase, as shown recently by Nadji et al. with a sensitive immunoperoxidase technique using specific anti-PAP serum [45]. Furthermore, serial monitoring of serum PAP levels has shown encouraging results in the early detection of recurrent disease in prostatectomy patients during the clinical trial of the National Prostatic Cancer Project [44].

IX. BONE MARROW ACID PHOSPHATASE

In addition to the immunological determinations of prostatic acid phosphatase, the use of bone marrow acid phosphatase in the clinical management of prostate cancer has been a major development in recent years. In 1973 Reynolds et al. reported that bone marrow acid phosphatase determination produced the highest accuracy in detecting metastatic disease for prostatic cancer [46]. They reported that when comparing serum acid phosphatase, histological determinations of the bone scan, skeletal roentgenograms, and ^{18}fluorine total body bone scan, the bone marrow acid phosphatase determination was best in detecting bony metastases. Therefore they advocated using the bone marrow acid phosphatase determination to stage all patients with carcinoma of the prostate.

Subsequently, at least three other groups supported the determination of bone marrow acid phosphatase as a more sensitive parameter for bony metastases in patients with prostate cancer [47-49]. In 1976 Yarrison et al. further showed that the bone marrow acid phosphatase assay could be a very good tool for detecting early osseous metastasis from any primary site of cancer, including the prostate [50].

A common weakness of these reports was the lack of control patients. Although some investigators used patients with benign prostatic hypertrophy as controls, no patients with other neoplastic and nonneoplastic diseases were employed in their studies.

In 1977 Khan et al. were among the first to point out that false positive results may be common for the measurement of bone marrow acid phosphatase [51], especially in patients with primary hematological disorders. They cautioned that a careful interpretation of the bone marrow acid phosphatase determination should be given so that some patients would not be denied appropriate therapy, and that the definitive role of bone marrow acid phosphatase should be studied in a long-term follow-up of patients. Pontes et al. found as many as 61% false positive bone marrow acid phosphatase determinations in patients with various hematological diseases [52].

Some recent reports went even further to state that bone marrow acid phosphatase determination has no clinical use at all. For instance, Sadlowski studied 47 patients with early stages of prostatic cancer and concluded that pelvic lymphadenectomy appeared to have a well-defined role in the diagnostic study of early stages of prostatic cancer, and that bone marrow acid phosphatase determination did not have any clinical value [53].

Boehme et al. reported that bone marrow acid phosphatase determination did not differ significantly between those patients with or without bone metastases, and that the patients with prostatic cancer did not have higher levels of prostatic acid phosphatase than subjects with other nonmalignant conditions [54].

Fossa et al. also observed significant levels of bone marrow acid phosphatase only in advanced Stage D patients with significantly increased serum enzyme levels [55]. Therefore the levels of bone marrow acid phosphatase gave no supplementary diagnostic information in patients with prostatic cancer.

All these authors expressed doubt concerning the hypothesis that increased activity of bone marrow acid phosphatase is diagnostic for early metastases in patients with prostatic cancer. It should be noted that the determinations of bone marrow acid phosphatase in the above reports were performed by the conventional chemical or enzymatic methods.

With the availability of immunochemical assays for acid phosphatase, the controversial role of bone marrow acid phosphatase determination has recently been extensively reinvestigated. As early as 1972 Moncure and his associates, using a simple gel diffusion method in a series of bone marrow aspirates, were able to demonstrate a positive reaction of prostatic acid phosphatase in two cases in which cytological examinations were negative [56]. But in their more recent report [57], with a larger series of patients, no false positives were found in patients with benign prostatic hypertrophy or various metabolic and nonprostatic neoplastic disorders. False positive results with chemical methods for the detection of bone marrow acid phosphatase were reported by Pontes et al. in 61% of 22 patients with nonprostatic malignancy, and only one false positive occurred with either counterimmunoelectrophoresis or radioimmunoassay [52].

By means of solid-phase radioimmunoassay, Cooper et al. also studied the acid phosphatase in bone marrow aspirates from patients with prostatic cancer and with benign prostatic hypertrophy [58]. Although no definite conclusion was made in their study, they did indicate that radioimmunoassay appeared to be technically superior to the enzymatic methods in measuring bone marrow acid phosphatase.

Similarly, using a radioimmunoassay technique, Belville et al. studied 118 patients with prostate cancer and 50 patients with benign prostatic hypertrophy [30]. About 10% of their patients with benign prostatic hypertrophy showed an elevated bone marrow acid phosphatase; in the majority of patients with prostatic cancer, levels of prostatic acid phosphatase in bone marrow aspirate correlated well with the increasing clinical stage of the disease. Although no other control patients were used in their study, they concluded that the determination of bone marrow prostatic acid phosphatase by radioimmunoassay can be a valuable adjunct to the clinical pathological staging of prostate cancer.

By utilizing the counterimmunoelectrophoresis technique of Chu et al. [13], Catane et al. studied acid phosphatase simultaneously in bone marrow and in serum from patients with prostatic cancer, and in patients with hematological disorders and other cancers [59]. They noticed a high rate of false positive results (more than 50%) in the determination of bone marrow acid phosphatase in control patients when the conventional chemical method (using α-naphthyl phosphate as the substrate) was applied. With the CIEP technique, no false positives were observed. Furthermore, the positive rates in serum and bone marrow acid phosphatase determinations in their patients were similar. It appeared that immunoassay may eliminate the need for bone marrow acid phosphatase determination in most cancer patients with prostatic cancer. Certainly, considerably more clinical application is necessary for complete evaluation of this problem.

X. SUMMARY

Acid phosphatase is one of the oldest tumor markers; the assay has been used as an aid in diagnosing and treating prostate cancer since the 1930s. In addition to the prostate, virtually all tissues and blood cellular components contain acid phosphatase. Therefore the acid phosphatase in circulation is a combination of enzyme activity from all these sources. Conventional chemical or biochemical methods for determining acid phosphatase have been unsatisfactory in distinguishing between the acid phosphatase of prostate origin and that of other sources.

Recent advances in immunological technology have resulted in several new immunochemical assays, namely, radioimmunoassay, counterimmunoelectrophoresis, and immunofluoroassay for prostatic acid phosphatase. These newly developed techniques are based on the immunological specificity of prostatic acid phosphatase. Both radioimmunoassay and counterimmunoelectrophoresis have been shown to be more specific and sensitive than the conventional spectrophotometric techniques in detecting serum-prostatic-specific acid phosphatase; and the fluorescent immunoassay, at least at present, seems to be the superior technique. It is reasonable to expect that early detection of prostatic cancer can soon be realized through these immunodiagnostic tests.

ACKNOWLEDGMENTS

Our work reviewed in this report was supported in part by research grants CA-15126, CA-15437, and CA-23990, awarded by the National Cancer Institute, DHEW. Ms. J. Ogledzinski's and Ms. B. A. Richards-Smith's assistance in preparation of this report is acknowledged.

REFERENCES

1. Recommendations of the International Unions of Pure and Applied Chemistry and the International Union of Biochemistry, in <u>Enzyme Nomenclature</u>, Elsvier Scientific Publishing Company, Amsterdam and New York, 1972, p. 198.
2. A. B. Gutman and E. N. Gutman, An "acid" phosphatase occurring in the serum of patients with metastasizing carcinoma of the prostate gland. J. Clin. Invest. 17:473-478 (1938).
3. H. A. Woodard, Acid and alkaline glycerophosphatase in tissues and serum. Cancer Res. 2:497-508 (1942).
4. O. Bodansky, Enzymes in cancer: The phosphohydrolases, in <u>Biochemistry of Human Cancer</u>, Academic Press, New York, 1975, pp. 61-71.
5. E. J. King and K. A. Jegatheesan, A method for the determination of tartrate-labile prostatic acid phosphatase in serum. J. Clin. Pathol. 12:85-89 (1959).
6. A. V. Roy, M. R. Brower, and J. E. Hayden, Sodium thymolphthalein monophosphate. A new acid phosphatase substrate with greater specificity for the prostatic enzyme in serum. Clin. Chem. 17:1093-1096 (1971).
7. G. P. Murphy, G. Reynoso, G. M. Kenny, and J. F. Gaeta, Comparison of total and prostatic fraction serum acid phosphatase levels in patients with differentiated and undifferentiated prostate carcinoma. Cancer 23:1309-1312 (1969).
8. L. M. Ewen and R. W. Spitzer, Improved determination of prostatic acid phosphatase (sodium thymolphthalein monophosphate) substrate. Clin. Chem. 22:627-629 (1976).
9. A. B. Gutman, The development of the acid phosphatase test for prostatic carcinoma. Bull. N.Y. Acad. Med. 44:63-76 (1968).
10. M. K. Schwartz, M. Fleisher, and O. Bodansky, Clinical application of phosphohydrolase measurement in cancer. Ann. N.Y. Acad. Sci. 166:775-793 (1969).
11. L. T. Yam, Clinical significance of the human acid phosphatase. A review. Am. J. Med. 56:604-616 (1974).
12. W. Ostrowski, Further characterization of acid phosphomonoesterase of human prostate. Acta Biochem. Polonica 15:213-225 (1968).
13. T. M. Chu, M. C. Wang, W. W. Scott, R. P. Gibbons, D. E. Johnson, J. D. Schmidt, S. A. Loening, G. R. Prout, and G. P. Murphy, Immunochemical detection of serum prostatic acid phosphatase. Methodology and clinical evaluation. Invest. Urol. 15:319-323 (1978).
14. B. K. Choe, E. J. Pontes, S. Bloink, and N. R. Rose, Human prostatic acid phosphatases. I. Isolation. Arch. Andrology 1:221-226 (1978).

15. P. Vihko, M. Kontturi, and L. K. Korhonen, Purification of human prostatic acid phosphatase by affinity chromatography and isoelectric focusing. Part 1. Clin. Chem. 24:466-470 (1978).
16. R. L. Van Etten and M. S. Saini, Selective purification of tartrate-inhibitable acid phosphatases: Rapid and efficient purification (to homogeneity) of human and canine prostatic acid phosphatases. Clin. Chem. 24:1525-1530 (1978).
17. T. M. Chu, A. K. Bhargava, E. A. Barnard, W. Ostrowski, M. J. Varkarakis, C. Merrin, and G. P. Murphy, The tumor antigen and acid phosphatase isoenzyme in prostatic cancer. Cancer Chemother. Rep. 59:97-103 (1975).
18. T. M. Chu, M. C. Wang, P. Kuciel, L. Valenzuela, and G. P. Murphy, Enzyme markers in human prostatic carcinoma. Cancer Chemother. Rep. 61:193-200 (1977).
19. T. M. Chu, M. C. Wang, L. Valenzuela, C. Merrin, and G. P. Murphy, Isoenzymes of human prostatic acid phosphatase. Oncology 35:198-200 (1978).
20. C. L. Lee, M. C. Wang, G. P. Murphy, and T. M. Chu, A solid-phase fluorescent immunoassay for human prostatic acid phosphatase. Cancer Res. 38:2871-2878 (1978).
21. B. K. Choe, E. J. Pontes, M. K. Morrison, and N. R. Rose, Human prostatic acid phosphatase. II. A double-antibody radioimmunoassay. Arch. Andrology 1:227-233 (1978).
22. P. Vihko, E. Sanjati, O. Janne, L. Peltonen, and R. Vihko, Serum prostate-specific acid phosphatase. Development and validation of a specific radioimmunoassay. Clin. Chem. 25:1915-1919 (1978).
23. S. Shulman, L. Mamrod, M. J. Gonder, and W. A. Soanes, The detection of prostatic acid phosphatase by antibody reactions in gel diffusion. J. Immunol. 93:474-480 (1964).
24. A. G. Foti, H. Herschman, J. F. Cooper, and H. Imfeld, The effect of antibody on human prostatic acid phosphatase. Substrate utilization by enzyme or enzyme-antibody complex. Arch. Biochem. Biophys. 176:154-158 (1976).
25. J. F. Cooper and A. G. Foti, A radioimmunoassay for prostatic acid phosphatases. I. Methodology and range of normal male serum values. Invest. Urol. 12:98-102 (1974).
26. A. G. Foti, H. Herschman, and J. F. Cooper, A solid phase radioimmunoassay for human prostatic acid phosphatase. Cancer Res. 35:2446-2452 (1975).
27. A. G. Foti, J. F. Cooper, H. Herschman, and R. R. Malbaez, Detection of prostatic cancer by solid phase radioimmunoassay of serum prostatic acid phosphatase. N. Engl. J. Med. 297:1357-1361 (1977).

28. J. F. Cooper, A. G. Foti, H. Herschman, and W. Finkle, A solid phase radioimmunoassay for prostatic acid phosphatase. J. Urol. 119:388-391 (1978).
29. A. G. Foti, H. Herschman, and J. F. Cooper, Comparison of human prostatic acid phosphatase by measurement of enzymatic activity and by radioimmunoassay. Clin. Chem. 23:95-99 (1977).
30. W. D. Belville, H. D. Cox, D. E. Mehan, J. P. Olmert, B. T. Mittenmeyer, and A. W. Bruce, Bone marrow acid phosphatase by radioimmunoassay. Cancer 41:2286-2291 (1978).
31. A. G. Foti, J. F. Cooper, H. Herschman, and S. R. Sapon, The detection of prostatic cancer by radioimmunoassay: A review. Hum. Pathol. 9:618-620 (1978).
32. T. M. Chu, M. C. Wang, R. Kajdasz, E. A. Barnard, P. Kucil, and G. P. Murphy, Prostate-specific acid phosphohydrolase in the diagnosis of prostate cancer. Proc. Am. Assoc. Cancer Res. 17:191 (1976) (abstract).
33. I. McDonald, N. R. Rose, E. J. Pontes, and B. K. Choe, Human prostatic acid phosphatase. III. Counterimmunoelectrophoresis for rapid detection. Arch. Andrology 1:225-239 (1978).
34. A. G. Foti, J. F. Cooper, and H. Herschman, Counterimmuno-electrophoresis in determination of prostatic acid phosphatase in human serum. Clin. Chem. 24:140-142 (1978).
35. A. N. Romas, K. C. Hsu, P. Tomashefsky, and M. Tannebaum, Counterimmunoelectrophoresis for detection of human prostatic acid phosphatase. Urology 12:79-83 (1978).
36. W. F. Whitmore, The rationale and results of ablative surgery for prostatic cancer. Cancer 16:1119-1121 (1963).
37. Z. Wajsman, T. M. Chu, J. Saroff, N. Slack, and G. P. Murphy, Two new, direct and specific methods of acid phosphatase determinations. A national field trial. Urology 13:8-11 (1979).
38. C. S. Killian, F. P. Vargas, C. L. Lee, M. C. Wang, G. P. Murphy, and T. M. Chu, A quantitative counterimmunoelectrophoresis assay for prostatic acid phosphatase. Clin. Chem. 25:1084 (1979).
39. C. L. Lee, C. S. Killian, G. P. Murphy, and T. M. Chu, A solid-phase immunoadsorbent assay for serum prostatic acid phosphatase. Clin. Chim. Acta 101:209-216 (1980).
40. D. E. Mahan and D. P. Doctor, A radioimmune assay for human prostatic acid phosphatase levels in prostatic disease. Clin. Biochem. 12:10-17 (1979).
41. J. C. Griffiths, Prostate-specific acid phosphatase: Re-evaluation of radioimmunoassay in diagnosing prostatic disease. Clin. Chem. 26:433-436 (1980).
42. C. L. Lee, T. M. Chu, Z. Wajsman, N. H. Slack, and G. P. Murphy, Value of a new fluorescent immunoassay for human prostatic acid phosphatase in prostate cancer. Urology 15:338-341 (1980).

43. C. S. Killian, F. P. Vargas, C. L. Lee, M. C. Wang, G. P. Murphy, and T. M. Chu, Quantitative counterimmunoelectrophoresis assay for prostatic acid phosphatase. Invest. Urol. 18:219-224 (1980).
44. C. S. Killian, F. P. Vargas, N. H. Slack, G. P. Murphy, and T. M. Chu, Prostatic specific acid phosphatase versus acid phosphatase in monitoring patients with prostate cancer. Clin. Chem. 27:1064 (1981) (Abstract).
45. M. Nadji, S. Z. Tabei, A. Castro, T. M. Chu, and A. R. Morales, Prostatic origin of tumors. Am. J. Clin. Pathol. 73:735-739 (1980).
46. R. D. Reynolds, B. R. Greenberg, C. D. Martin, R. N. Lucas, C. N. Gaffney, and L. Hawn, Usefulness of bone marrow serum acid phosphatase in staging carcinoma of the prostate. Cancer 32:181-184 (1973).
47. E. O. Gursel, M. Rezvan, F. A. Sy, and R. J. Veenema, Comparative evaluation of bone marrow acid phosphatase and bone scanning in staging of prostate cancer. J. Urol. 111:53-57 (1974).
48. S. Marshall, R. P. Lyon, and M. P. Scott, Prostatic acid phosphatase levels. Significance in serum and bone marrow. Urology 4:435-438 (1974).
49. J. D. Pontes, S. W. Alcorn, A. J. Thomas, and J. M. Pearce, Jr., Bone marrow acid phosphatase in staging prostatic carcinoma. J. Urol. 114:422-424 (1975).
50. G. Yarrison, B. F. Mertens, and J. C. Mathies, New diagnostic use of bone marrow acid and alkaline phosphatase. Am. J. Clin. Pathol. 66:667-671 (1976).
51. R. Khan, B. Turner, M. Edson, and M. Dolan, Bone marrow acid phosphatase: Another look. J. Urol. 117:79-80 (1977).
52. J. E. Pontes, B. K. Choe, N. R. Rose, and J. M. Pearce, Jr., Bone marrow acid phosphatase in staging of prostatic cancer: How reliable is it? J. Urol. 119:772-776 (1978).
53. R. W. Sadlowski, Early stage prostatic cancer investigated by pelvic lymph node biopsy and bone marrow acid phosphatase. J. Urol. 119:89-93 (1978).
54. W. M. Boehme, R. R. Augspurger, S. F. Wallner, and R. E. Donahue, Lack of usefulness of bone marrow enzymes and calcium in staging patients with prostatic cancers. Cancer 41:1433-1439 (1978).
55. S. D. Foosa, J. Sokolowski, and L. Theodorsen, The significance of bone marrow acid phosphatase in patients with prostatic carcinoma. Br. J. Urol. 50:185-189 (1978).
56. C. W. Moncure, C. L. Johnston, Jr., M. J. V. Smith, and W. W. Kootz, Jr., Immunological and histochemical evaluation of marrow aspirates in patients with prostatic carcinoma. J. Urol. 108:609-610 (1972).
57. J. R. Drucker, C. W. Moncure, C. L. Johnson, M. J. V. Smith, and W. W. Kootz, Jr., Immunological staging of prostatic carcinoma: Three years of experience. J. Urol. 119:94-98 (1978).

58. J. F. Cooper, A. G. Foti, and P. W. Shank, Radioimmunochemical measurement of bone marrow acid phosphatase. J. Urol. 119:392-395 (1978).
59. R. Catane, S. Madajewicz, Z. Wajsman, T. M. Chu, A. O. Mittelman, and G. P. Murphy, Prostatic cancer: Immunochemical detection of prostatic acid phosphatase in serum and bone marrow. N.Y. State J. Med. 78:1060-1061 (1978).

7

BIOCHEMICAL MARKERS FOR THE DIFFERENTIAL DIAGNOSIS OF LEUKEMIAS

B. I. SAHAI SRIVASTAVA / Department of Experimental Therapeutics, Grace Cancer Drug Center, Roswell Park Memorial Institute, Buffalo, New York

- I. Introduction 137
- II. Purine and Pyrimidine Metabolizing Enzymes 139
- III. DNA Polymerases 140
- IV. Cell Membrane Enzymes 141
- V. Acid Hydrolases 142
- VI. Esterases 144
- VII. Drug and Radiation Sensitivity of Leukemic Cells 145
- VIII. Miscellaneous Markers 145
- IX. Terminal Deoxynucleotidyl Transferase (TdT) 147
    - A. Properties of TdT 147
    - B. TdT Assay and Characterization 149
    - C. Animal Studies with TdT 154
    - D. Clinical Studies with TdT 158
- X. Protease and DNase Activity 162
- XI. Conclusion 162
  - References 163

I. INTRODUCTION

At present, the classification and treatment of patients with leukemias and malignant lymphomas relies largely on the subjective interpretation of morphological findings (Hayhoe and Cawley, 1972; Nelson, 1976; Tan and Lamberg, 1977). Although this approach has been generally useful, the range of morphologic variation in acute lymphoblastic leukemia (ALL), acute myeloblastic leukemia (AML), and the blastic phase of chronic myeloid leukemia (CML), as well as in other leukemias, and variable responses to treatment within each group have been evident for some time. Thus, in recent years, specialized methods such as cytochemistry, cytogenetics, electron microscopy, in vitro growth pattern on soft agar, and immunological and biochemical markers have been used to define subgroups and to ascertain if there are any correlations between these subgroups and clinical and laboratory findings, response to treatment, and prognosis.

The usefulness and limitations of cytochemical staining techniques in differentiating the various subtypes of acute leukemia have been recently reviewed (Shaw, 1976), and a French-American-British cooperative group (Bennett et al., 1976; Gralnick et al., 1977) has recently proposed the evaluation of both blast cell morphology and certain cytochemical staining to differentiate between different types of acute leukemias. Similarly, the chromosome analysis by recent banding techniques in leukemic patients has been helpful in diagnosis and in evaluating prognosis (Rowley and DeLa Chapelle, 1978). For example, the recognition of Ph^1 chromosome is important for the diagnosis of CML and for identifying patients with Ph^1 chromosome positive acute leukemia who have no preceding chronic phase and may present with ALL or AML but have a poorer prognosis (Rowley, 1978).

Examination of the surface of blast cells by scanning electron microscopy has been claimed to distinguish between lymphoblastic and nonlymphoblastic leukemias (Polliack et al., 1976), although it is now realized that the cell surface, as seen by scanning electron microscopy, may show variations depending on the functional state and the environment of the cell and its place in differentiation (Roath et al., 1978).

Evaluation of the in vitro growth pattern of colony-forming cells obtained from bone marrow and peripheral blood has also been used as an aid in classifying leukemias, monitoring clinical status, and predicting the patient's response to therapy and the subsequent remission rate (Moore, 1975; Moore, 1976; Robinson and Stonington, Jr., 1976; Mangalik et al., 1977; Bakkeren et al., 1976; Moore, 1977; Spitzer et al., 1977). Although this procedure appears to be useful for the early detection of relapse or remission in AML, and for prediction of the onset of blastic transformation in CML, it has limitations in differentiating between acute leukemias (Vincent et al., 1977; Elias and Greenberg, 1977; Morris et al., 1977).

Cell surface phenotyping of leukemia and lymphoma cells (Brouet et al., 1976; Greaves, 1975; Greaves, 1977; Winchester et al., 1977; Coccia, 1977; Roberts et al., 1978; Filippa et al., 1978; Koziner et al., 1977; Minowada et al., 1978; Mathé, 1977; Janossy et al., 1978a; Minowada et al., 1977; Bloomfield et al., 1976; Berard et al., 1978; Brouet and Seligmann, 1978) has led to the recognition of T-,* B-,† and non-T/non-B leukemias and lymphomas which differ in prognosis and response to therapy. For example, differences in prognosis or response to therapy have been found

---

*Thymus-derived cell.
†Bursa-equivalent-derived cell.

7 / Markers for Diagnosing Leukemias

between T-lymphoblastic and non-T/non-B-lymphoblastic or B-lymphocytic lymphomas (Mathé, 1977), between B- and non-T/non-B diffuse lymphoma (Bloomfield et al., 1976), between B-ALL, T-ALL and non-T/non-B ALL (Chessells et al., 1977; Brouet et al., 1976), and between T-CLL (chronic lymphocytic leukemia), B-CLL, and its subtypes (Uchiyama et al., 1977; Hamblin and Hough, 1977). Continued effort in this area of immunological markers has recently led to the identification of two forms of non-T/non-B ALL (Chessells et al., 1977) of pre-B ALL (Vogler et al., 1978) and of several subtypes of T-ALL (Barrett et al., 1977; Kadin and Billing, 1977; Moretta et al., 1977; Richie et al., 1978). This work has also contributed significantly to our understanding of lymphocyte differentiation.

Although cell surface phenotyping of hematopoietic cells and other procedures discussed above have been very useful in the recognition of neoplasma involving specific cell types, insufficient attention has been given to the study of biochemical markers which could aid in more accurate diagnosis and perhaps better treatment of leukemias and lymphomas. These markers could also be useful in defining patterns of differentiation and in dissecting the biochemical and pharmacological basis of immune function. Recently, considerable interest has been generated in biochemical markers for the differential diagnosis of leukemias, and this review is an attempt to summarize the developments in this rapidly developing field.

II. PURINE AND PYRIMIDINE METABOLIZING ENZYMES

Adenosine deaminase is the most interesting enzyme in this group. This enzyme is found in most human tissues, and its deficiency in severe combined immunodeficiency disease of childhood was the first recognized association of a specific enzyme defect with an inherited disorder of T- and B-cell function (Giblett et al., 1972; Parkman et al., 1975; Van der Weyden and Kelley, 1976). Elevated adenosine deaminase activity has been found in thymocytes and T-cell disorders such as T-ALL, T-CLL, Sezary syndrome, and infectious mononucleosis, whereas low activity of this enzyme is found in B-CLL cells (Smyth and Harrap, 1975; Tung et al., 1976; Meier et al., 1976; Sullivan et al., 1977; Adams and Harkness, 1976; Ramot et al., 1977; Dietz and Czebotar, 1977). Low adenosine deaminase activity in normal B lymphocytes as compared to normal T lymphocytes has been reported (Tung et al., 1976; Huang et al., 1976).

Initial hope that this enzyme may help to distinguish ALL in which mean adenosine deaminase values are high from AML in which they are low (Smyth and Harrap, 1975) has not been fulfilled by findings of cases of AML with high activity (Meier et al., 1976). There was considerable overlap in adenosine deaminase activity between ALL and AML in both of the above studies. Similarly, adenosine deaminase cannot distinguish T-ALL from non-T/non-B ALL due to a similar range of activity in these cases (Coleman and Hutton, 1975).

However, Sullivan et al. (1977) recently reported low adenosine deaminase activity in one case of non-T/non-B ALL which expressed C_3 receptors, compared with three other cases which did not. Work done in this laboratory (Srivastava and DiCiccio, 1978) and others (Tritsch and Minowada, 1978) has shown low adenosine deaminase activity in non-T/non-B ALL cell lines, in Ph^1 chromosome positive lymphoblastic and myeloblastic cell lines, and in B-cell lines of normal origin, ALL, and other malignant origins, compared with high activity in all T-cell lines of leukemia/lymphoma origin (Minowada et al., 1978). These results indicate that adenosine deaminase may be of some value in classifying ALL and other lymphoid malignancies as either of B- or T-lymphocyte origin.

In contrast to adenosine deaminase, adenosine kinase (Dietz and Czebotar, 1977), pyrimidine nucleoside monophosphate kinase (Hande and Chabner, 1978), cytidine deaminase (Ho, 1973; Chabner et al., 1974; Coleman et al., 1975) and deoxycytidine kinase (Coleman et al., 1975; Scholar and Calabresi, 1973) do not appear to be of significance for the diagnosis of leukemias. However, the last two enzymes appear to be related to granulocyte maturation: Deoxycytidine kinase decreases and cytidine deaminase increases during maturation of normal and leukemic granulocytes (Chabner et al., 1974; Coleman et al., 1975).

Another enzyme, purine nucleoside phosphorylase, whose deficiency is associated with severely defective T-cell immunity (Giblett et al., 1975; Stoop et al., 1977), has been examined and found to have comparable activity in T-, B-, and non-T/non-B leukemic cell lines (Sullivan et al., 1977; Tritsch and Minowada, 1978). On the other hand, thymidine phosphorylase activity examined in this laboratory (Srivastava and DiCioccio, 1978) was found to be high (300-600 units/mg protein, where 1 unit equals 1 nmol of thymidine formed in 1 hr) in 12 B-cell lines of normal origin, ALL origin, and other malignant origins examined, whereas non-T/non-B cell lines and nine T-ALL cell lines had very low activity (20-60 units/mg protein). Although no clear-cut differences in this enzyme in normal and some leukemic leukocytes have been found (Marsh and Perry, 1964; Gallo and Perry, 1969), these studies were carried out using whole leukocyte populations, without cell separations or immunological characterizations.

III. DNA POLYMERASES

Three distinct DNA polymerases (α, β, and γ) (Weissbach et al., 1975) have been found in normal and leukemic human cells. And all these activities, particularly DNA polymerase α, have been found to be high in immature or proliferating cells (Srivastava, 1974a; Srivastava, 1975b; Coleman et al., 1974a; Bertazzoni et al., 1976; Mayer et al., 1975; Srivastava et al., 1978a). The leukocytes from CLL (T- or B-cell type), CML and HCL (hairy cell leukemia) patients, in line with their mature status, had DNA polymerase α and β activity comparable to that of unstimulated nonproliferating normal leukocytes or leukocytes from ALL patients in remission, whereas blast cells from acute leukemias or blastic phase CML patients had 80- to 30-fold higher DNA polymerase activity (Srivastava, 1975b; Coleman et al., 1974a; Srivastava et al., 1977; Srivastava, 1978b; Srivastava et al., 1978a), which was comparable to that of PHA-stimulated normal lymphocytes (Srivastava, 1975b; Mayer et al., 1975; Srivastava et al., 1980a).

Thus DNA polymerase estimates, although only of limited value in diagnosing leukemias, could be useful in determining perturbances in a proliferating leukemic cell population during therapy, or alterations in the stage of the disease. For example, a three- to fourfold increase in DNA polymerase α or β activity, but not in terminal deoxynucleotidyl transferase (TdT) activity 12-24 hours after leukapheresis in two ALL patients (Srivastava, 1978a) may be related to changes in the proportion of proliferating cells. And considerable variations in DNA polymerase activity among 15 CLL and 18 chronic phase CML patients examined in this laboratory (Srivastava et al., 1980a) could be related to the heterogeneity or the stage of the disease.

In addition to biochemical estimation of DNA polymerase activity, an autoradiographic technique which detects DNA polymerase and primer template DNA by measuring the in vitro incorporation of ^{3}H-TTP into nuclei has been developed recently to detect the proliferating blast cells (Nelson and Schiffer, 1973; Wantzin, 1977). A drawback of this procedure is its dependence primarily on DNA polymerase β activity, whereas it is DNA polymerase α which is particularly high in proliferating cells. An alternative approach would be the use of DNA polymerase antisera for the development of an immunofluorescence procedure to detect proliferating blast cells. Antisera that could be used for this purpose are to calf thymus DNA polymerase α obtained by Chang and Bollum (1972), and antisera against calf thymus and ALL cell DNA polymerase α prepared in this laboratory. These antisera inhibit both DNA polymerase α and β, but not DNA polymerase γ and TdT, or the antisera against DNA polymerase α from HeLa (Spadari et al., 1974) and RPMI-1788 (Smith et al., 1975) human cell lines, which inhibit DNA polymerase α but not β.

IV. CELL MEMBRANE ENZYMES

5'-Nucleotidase is the most extensively studied enzyme in this group. This enzyme, which catalyzes the phosphorylitic cleavage of 5'-nucleotides, is localized in the external surface of the plasma membrane of many cells, including lymphocytes. The precise function of this enzyme in vivo is not known. Low 5'-nucleotidase activity has been consistently found in lymphocytes from CLL (Lopes et al., 1973; Quagliata et al., 1974; La Mantia et al., 1975) and other B-immunoproliferative disorders (Kramers et al., 1976), in certain immunodeficiency states (Johnson et al.,

1977), as well as in cord blood lymphocytes (Kramers et al., 1977). Immunofluorescence test using antisera to purified 5'-nucleotidase indicates the absence of detectable 5'-nucleotidase immunoreactive protein from most patients with CLL (LaMantia et al., 1977), which correlates well with the biochemical assay of the enzyme.

Since no differences in the 5'-nucleotidase level have been found between normal B or T lymphocyte populations (Quagliata et al., 1974; Kramers et al., 1976; Kramers et al., 1977) it has been suggested that 5'-nucleotidase may be related to the degree of lymphocyte maturation irrespective of its type (Kramers et al., 1976). Thus low levels of 5'-nucleotidase in cord blood lymphocytes (Kramers et al., 1977), CLL, other B-cell disorders, and in three cases of T-cell ALL and one case of Sezary syndrome (Kramers et al., 1976) may reflect their immature state (Preud'Homme et al., 1974; Hamburg et al., 1976). It seems that 5'-nucleotidase levels could be helpful in subtyping lymphocytic malignancies similar to the finding of 67% of CLL patients with low activity and 10% with supranormal activity (Quagliata et al., 1974; LaMantia, et al., 1977). Whether such a subgrouping has any clinical relevance remains to be determined.

Several cell membrane enzymes which serve functions in amino acid and sugar transport were examined, and γ-glutamyl transpeptidase, leucine aminopeptidase, maltase, and trehalase were found to be low in T- or non-T/non-B ALL, AML, and B-CLL compared to normal peripheral blood lymphocytes, whereas they were normal or above normal in Sezary syndrome and T-CLL (Krammers et al., 1978). Sialyl transferase, which has been recently reported to be four- to eightfold higher in two B-cell lines as compared with two T-cell lines (Maca and Hakes, 1978), should be examined further on additional cell lines as well as clinical samples to determine its value as a marker enzyme.

V. ACID HYDROLASES

Several acid hydrolases have been considered useful in the differential diagnosis of leukemias. Cytochemical demonstration of tartarate-resistant acid phosphatase is considered to be a unique characteristic of HCL cells (Li et al., 1970; Burke et al., 1974; Catovsky et al., 1974b; Yam et al., 1971, 1972). Although weak to moderate acid phosphatase activity is present in lymphocytes, neutrophils, eosinophils, platelets, and erythroblasts, and strongly positive activity is present in monocytes, histiocytes, megakaryocytes, and HCL cells, this activity from only last cells is tartarate-resistant (Janckila et al., 1978), which is also identifiable by polyacrylamide gel electrophoresis as isoenzyme 5 (Yam et al., 1971).

V. Heyden et al. (1977) recently reported the detection of tartarate-resistant isoenzyme 5 with components a and b in monocytes, lymphocytes,

CLL cells, and B-cell lines from a healthy donor, and especially in hairy cells. Yet hairy cells were found to be positive only in the cytochemical procedure. Janckila et al. (1978) have recommended the use of the highly sensitive naphthol-ASBI phosphoric acid-fast garnet GBC method for the cytochemical demonstration of tartarate-resistant acid phosphatase in smears and imprints, whereas they consider the naphthol-ASBI phosphoric acid-pararosaniline method as highly specific and best for histochemical demonstration in tissue sections.

In addition to the above useful markers for HCL, increased activity of acid phosphatase and β-glucuronidase in certain T-lymphocytic malignancies (Catovsky, 1975; Catovsky et al., 1974a; Flandrin and Daniel, 1974; Stein et al., 1976) and a reduced acid phosphatase and/or β-glucuronidase activity in CLL (Konig et al., 1970) and cells from lymph node imprints of non-Hodgkin's lymphoma (Pangalis et al., 1977) have been reported using cytochemical procedures. Although a strong focal acid phosphatase activity in a patchy paranuclear (Golgi) distribution has been consistently found to be associated with T-ALL and T-cell lymphoblastic lymphoma (Ritter et al., 1975; Stein et al., 1976; Brouet et al., 1976; Chessells et al., 1977; Gralnick et al., 1977; Stein and Muller-Hermelink, 1977; Kumar et al., 1978; Stass et al., 1978b; Srivastava et al., 1980a) and absent in B-cell ALL (Gralnick et al., 1977), it is not entirely specific since it is absent in some cases of T-ALL (Kumar et al., 1978) and present in about 13% of the cases of non-T/non-B ALL (Chessells et al., 1977; Kumar et al., 1978). Positive acid phosphatase reaction has also been observed in some AML patients (Gralnick et al., 1977; Srivastava et al., 1980a); this reaction may be diffuse over the whole cytoplasm and not localized to the Golgi zone.

In spite of these exceptions, the strong focal acid phosphatase reaction appears to be useful for the characterization of T-ALL, particularly its variants in which blasts may be E-/anti-T serum positive, E+/EAC+, or predominantly EAC+ (Barrett et al., 1977; Kadin and Billing, 1977; Stein and Müller-Hermelink, 1977), Strong focal acid phosphatase activity has been demonstrated in lymphoblastic lymphoma cells which were simultaneously E+ and EAC+ and in E-/EAC+ or E+/EAC+ normal fetal thromocytes with decline during thymocyte maturation (Stein and Müller-Hermelink, 1977). In contrast to acid phosphatase reaction, the β-glucuronidase reaction does not appear to differ significantly between T- and non-T/non-B ALL (Brouet et al., 1976). Acid hydrolases have also been examined by biochemical procedures and no significant differences in the level of β-glucuronidase (Mouseis et al., 1976; Pangalis et al., 1978) or acid phosphatase (Pangalis et al., 1978) between normal T and B cells were found, whereas α-mannosidase activity was significantly higher and pH-4 β-glucosidase lower in B cells compared to T cells (Pangalis et al., 1978).

Different isoenzyme patterns of hexosaminidase have been observed in lymphocyte, granulocytes, and thymocytes (Ellis et al., 1978) and in a study with leukemias a hexosaminidase component I, separated on a DEAE-cellulose column, was greatly increased in 23 of 27 cases of non-T/non-B ALL which were positive for c-ALL antigen (Ellis et al., 1978). This component I was very low in a wide range of other leukemias, including AML, T-ALL, and CLL as well as in normal leukocytes and thymocytes. Low to variable amounts of I were found in AMMoL (acute myelommonocytic leukemia), whereas it was elevated in some blastic phase CML. Further work on this marker is desirable to determine its usefulness for the study of leukemias and leukocyte differentiation.

VI. ESTERASES

Cytochemical reactions designated to detect activities of specific esterases (naphthol-AS-D-chloroacetate as substrate) and nonspecific esterases (α-naphthyl acetate, α-naphthyl butyrate, or naphthol AS-D acetate as substrate) around neutral pH have been of particular value in differentiating lymphoblastic from nonlymphoblastic leukemias, and especially in distinguishing myeloblastic from monocytic leukemias (Hayhoe and Cawley, 1972; Bennett et al., 1976; Gralnick et al., 1977). The nonspecific esterase reaction is strongly positive for monocytes and their precursors, but reactions are negative to weak for granulocyte precursors, lymphoblasts, and erythroblasts. This reaction for a monocyte series of cells is strongly inhibited by NaF. The specific esterase reaction is strongly positive for granulocytes and their precursors, whereas monocytic, lymphoid, and erythroid lines are negative or only weakly positive.

These cytochemical observations have been recently confirmed by electrophoretic analysis in which specific or nonspecific esterase activity was not detected in ALL cells, and the NaF-sensitive band 5 of nonspecific esterase present in AMMoL and AMoL was weak or missing in AML (Kass and Peters, 1978). Several laboratories have recently reported an α-naphthyl butyrate reaction at pH 8.0 (Higgy et al., 1977) and an α-naphthyl acetate reaction at pH 5.8-6.1 (Kulenkampff et al., 1977; Horwitz et al., 1977; Ranki et al., 1976; Sher et al., 1977) to distinguish T cells from B cells. In both of these procedures, the normal peripheral blood T cells gave heavy well-defined dotlike esterase activity, whereas B cells, granulocytes, and mitogen-stimulated T and B cells were negative and monocytes gave an intense diffuse cytoplasmic reaction. Faint localized acid esterase reaction, as given by a subpopulation of thymocytes, was observed in 4 of 5 cases of T-ALL and in 1 of 11 of non-T/non-B ALL, whereas blasts from AML and CML showed an intense diffuse reaction product (Kulenkampff et al., 1977). The acid esterase reaction thus appears to be a marker for mature human T lymphocytes and may also be helpful in distinguishing T-ALL from non-T/non-B ALL or AML.

VII. DRUG AND RADIATION SENSITIVITY OF LEUKEMIC CELLS

Since variations in the response to therapy have been observed in leukemic patients depending on the target cell involved (Greaves, 1975; Bloomfield et al., 1976), the determination of drug and radiation sensitivity of leukemic cells could help in their characterization. Of 30 drugs examined in this laboratory for their differential cytotoxicity against several T-ALL cell lines and B-cell lines of ALL and other origin (Srivastava, 1978a; Srivastava, 1978b), 12 (ara-A, ara-C, cyclocytidine, 5-azacytidine, 5-aza-2'-deoxycytidine, 3'-decanyl cyclocytidine, 3'-decanyl ara-C, ara-T, 8-azahypoxanthine, maytansine, pretazettine, and asparaginase) were 10 to 1000 times more cytotoxic to T- than to B-cell lines. Seven (adriamycin, N^6-benzyladenosine, bleomycin, daunomycin, dexamethasone, vincristine, and N^6-isopentenyladenosine) were moderately selective against T-cell lines. One drug (5-fluorouracil) was 20-fold more cytotoxic against B- than T-cell lines, and 10 (actinomycin D, BCNU, hydroxyurea, tazettine, 6-mercaptopurine, methyl-mercaptopurine riboside, prednisone, rifamycin, streptovaricin C, streptoval C) were not selective.

In agreement with these results, Ohnuma et al. (1978) independently reported greater cytotoxicity of asparaginase and ara-C against T-cell lines, and of 5-fluorouracil against B-cell lines. Some of the drugs showing differential cytotoxicity against T-ALL cell lines in our study were examined further and found to be equally cytotoxic against non-T/non-B ALL cell lines (Srivastava, 1978a), and against cells from one T- and one non-T/non-B ALL patient. But they were far less cytotoxic against unstimulated or PHA-stimulated normal lymphocytes. In addition to the differential effect of drugs discussed above Nakazawa et al. (1978) have found profound radiosensitivity in T-ALL, leukemic T-cell lines, and thymocytes as compared to non-T/non-B ALL, B-cell lines, and unstimulated or PHA-stimulated lymphocytes. These results indicate that drugs and radiation sensitivity of leukemic cells could help in characterizing the target cell involved in neoplastic transformation, and such studies may also be useful in making rational therapy decisions.

VIII. MISCELLANEOUS MARKERS

Several probable markers which have not yet been critically evaluated have been proposed for the characterization of normal or some malignant hematopoietic cells. A human granulocyte ribonuclease which has a unique specificity toward secondary phosphate esters of uridine 3'-phosphate has been proposed as a diagnostic tool for certain granulocyte disorders (Reddi, 1976). ALL cells have been found to contain a protein kinase designated S-L which was separated on a DEAE-cellulose column and which phosphorylates casein and is not activated by C-AMP (Horenstein et al., 1976). Little or no activity of this enzyme was detected in unstimulated or PHA-stimulated normal lymphocytes.

Lactate dehydrogenase isoenzyme patterns have been found to differ between B and T cells, with the latter cells having a significantly higher amount of lactate dehydrogenase-1 isoenzyme than B cells (Plum and Ringoir, 1975). A histone H 1° which has a high content of lysine was not detected in normal B cells but was present to the extent of 3% of total histone in all six CLL patients examined (Maraldi et al., 1978). This histone has been found to be present in many normal tissues (e.g., thyroid, adrenal) and cancerous human tissues, but absent from thymus. And its amount appears to be inversely related to tissue replicating rate (Marks et al., 1975). Anderson et al. (1977) and Nilsson et al. (1977) recently found that surface glycoprotein pattern of T-cell lines was easily distinguished from those of other hematopoietic cell lines. These authors have suggested that surface glycoprotein analysis might be useful for the identification of cell lines and for the differential diagnosis of hematopoietic malignancies. About a seven-fold higher rate of RNA synthesis or turnover has been reported in normal or CLL lymphocytes containing C3 complement receptor as compared to normal lymphocytes which lacked this receptor (Huang et al., 1976).

Andreeff et al. (1978) found that non-T and T lymphocyte subpopulations from normal human peripheral blood have different mean quantities of RNA per cell. High protein content in B-cell lines of various origin as compared to T-cell lines of leukemia/lymphoma origin was found in this laboratory (Srivastava, 1978a). Gianni et al. (1978), using hybridization of human globulin complementary DNA versus total nuclear RNA from leukemic cells found globin RNA sequences in three of five cases of AML, some cases of CML and CLL, but not in four cases of ALL. Huang et al. (1975) reported that CLL lymphocytes contain a DNA-helix destabilizing protein (Alberts and Sternglanz, 1977) which was not detected in normal lymphocytes.

A new isoenzyme of alkaline phosphatase designated N-alkaline phosphatase was found in the sera of patients with lymphoproliferative diseases (Neuman et al., 1974). Unlike usual alkaline phosphatase and its isoenzymes, which hydrolyze monoesters of orthophosphoric acid and S-substituted monoesters of thiophosphoric acid with equal efficiency, the N-alkaline phosphatase could hydrolyze only the former and was inhibited by cysteamine-S-phosphate. N-alkaline phosphatase was also found to constitute 50-90% of the total alkaline phosphatase activity in B-cell lines and CLL cells, was present in a T-cell line, but was not detected in myeloblast cell line K-562, AML cells and untreated or PHA-stimulated normal lymphocytes (Neuman et al., 1976). Unfortunately, Dulis and Wilson (1978) have recently reported that 5,5'-dithiobis (2-nitrobenzoic acid) coupling assay method used by Neuman et al. was not suitable for serum, and they were unable to confirm the existence of N-alkaline phosphatase in serum by the above method, by polyacrylamide gel electrophoresis, or by inhibition by cysteamine-S-phosphate.

No significant differences in the usual type of alkaline phosphatase in sera of patients with various lymphoproliferative or myeloid disorders were noted by Dulis and Wilson (1978). Hayhoe et al. (1977) have, however, found high leukocyte alkaline phosphatase scores (range 160-280, normal 15-100) in 15 of 19 cases of HCL, whereas the somewhat atypical cases had lower scores (range 50-130). These authors suggested that the presence of high leukocyte alkaline phosphatase scores, although not helpful in distinguishing HCL from either CLL or ALL which often have above normal scores, could help in separating atypical cases of HCL from the typical ones.

Measurements of lysozyme activity in serum have proved to be of some value in the characterization of AML, which has low activity from AMMoL and AMoL, which usually have elevated activity (Catovsky et al., 1975). A cytochemical and cytobacterial method for the simultaneous demonstration of peroxidase and lysozyme activities in individual cells has been reported (Kageoka et al., 1977) to differentiate AMMoL and AMoL with higher lysozyme activity from AML with higher peroxidase activity.

Receptor assays for hormones and drugs are also being developed for the characterization of leukemic cells (Rosen et al., 1976). For example, higher values for glucocorticoid receptors (Konior et al., 1976; Yarbro et al., 1977) and insulin binding (Esber et al., 1976) were found for non-T/non-B ALL blasts as compared to T-ALL blasts, and total serum vitamin B_{12} binding capacity was correlated with the degree of promyelocytic differentiation in AML (Catovsky et al., 1975).

IX. TERMINAL DEOXYNUCLEOTIDYL TRANSFERASE (TdT)

This enzyme, which has received the greatest recent interest as a marker for certain leukemias, was first identified in calf thymus (Bollum, 1962; Krakow et al., 1962) and subsequently reported to be present in rat, rabbit, and procine thymus, but absent from spleen, bone marrow, and other tissues (Chang, 1971). However, later its presence in nonthymic tissues was appreciated (Srivastava, 1974a; Srivastava et al., 1976). Until recently, this enzyme was of interest primarily for its practical use in the synthesis and modification of polydeoxynucleotides (Bollum, 1974; Bollum et al., 1974). However, TdT gained major importance after it was found in ALL cells (McCaffrey et al., 1973) and in T-ALL cell line MOLT-4 (Srivastava and Minowada, 1973). Due to profound current interest in the function and distribution of TdT, the properties of this enzyme and the procedures for its assay will be discussed first, followed by some animal studies and its status as a marker for certain leukemias.

A. Properties of TdT

TdT has no template requirement and it catalyzes the addition of deoxynucleotide residues from deoxyribonucleoside triphosphates to the 3'-

hydroxyl groups of preformed polydeoxynucleotide chains, which can be as short as trinucleotide or as long as linear ØX-174 DNA (Bollum 1974; Bollum, et al., 1974). The 3'-OH groups found at nicks, gaps, and ends of linear DNA duplexes are accessible to TdT when secondary structure involving 3'-OH is destroyed. Ribo-initiators are, however, not used by TdT (Bollum, 1974).

Homogeneous TdT purified from ALL cells (Siddiqui and Srivastava, 1978) resembles the enzyme from calf thymus (Chang and Bollum, 1971; Kung et al., 1976) in having a sedimentation coefficient of 3-4S, a molecular weight of about 33,000, and a composition consisting of two subunits of molecular weights 2800 and 8500. Although there was some doubt (Kato et al., 1967) about the similarity of TdT from soluble (Chang and Bollum, 1971) and chromatin fraction (Krakow et al., 1962; Wang, 1968) with that from calf thymus, these enzymes from calf thymus or human cells are now considered to be the same (Srivastava, 1974a; Coleman, 1977). The reports of a high molecular weight TdT from calf thymus (Johnson and Morgan, 1976) and of two TdT activity peaks from phosphocellulose column which resemble in all other respects (Kung et al., 1975) are most likely artifacts. Similarly, the reports of retroviral TdT (Ashley et al., 1977) and a B-cell-specific TdT (Baltimore et al., 1976) were shown to be artifacts (Marcus and Sarkar, 1978; Modak et al., 1978). The reported presence of TdT-like activity in plant tissues (Srivastava, 1972; Brodniewicz-Proba and Buchowicz, 1976) needs more rigorous characterization. Antisera to homogeneous TdT from calf thymus (Bollum, 1975; Kung et al., 1976; Chan and Srivastava, 1976) and ALL cells (Siddiqui and Srivastava, 1978) have been prepared, and these antisera inhibit TdT activity from both calf thymus and human cells, but not DNA polymerase α, β, or γ from these cells or viral reverse transcriptase.

TdT catalyzes only a limited polymerization of ribonucleotides at the 3' end of oligodeoxynucleotides and at the terminal end of duplex DNA, particularly in the presence of cobalt ions (Bollum, 1974; Roychoudhury et al., 1976; Modak, 1978). Thus this enzyme is also finding use in 3'-end labeling and DNA sequence analysis (Roychoudhury et al., 1976). In contrast to ribonucleotides, the deoxyribonucleoside triphosphates (e.g., dATP, dGTP and dTTP), are extensively used by TdT, although some such as dXTP and dUTP (Hansbury et al., 1970; Srivastava, 1978a) are used very poorly, and dBrUTP and dBrCTP are not used at all (Hansbury et al., 1970). Although ara-CTP was reported not to be used by calf thymus TdT (Hansbury et al., 1970) both ara-CTP and ara-ATP were found to be used as efficiently as dCTP and dATP by TdT from calf thymus or human leukemic cells for polynucleotide synthesis (Srivastava, 1978b). Compared with natural substrates, these ara derivatives were, however, used only to an extent of 8-18% by DNA polymerase α and β and 1-2% by RNA polymerase II in the presence of the other three complementary nucleotides and DNA as the template. Moreover, human T-cell lines and null cell lines which have high TdT activity (Srivastava, 1976; Srivastava and Minowada, 1977; Minowada et al., 1978) were 30- to 40-fold more sensitive to the cytotoxic action of ara-C and ara-A (Srivastava, 1978c) compared with myeloblast cell line K-562 or B-cell lines of normal or malignant origin which have low TdT activity. No correlation between the sensitivity of the above cells to ara-A and ara-C and their deaminase, kinase (Srivastava, 1978c; Srivastava and DiCioccio, 1978), and DNA polymerase levels (Srivastava, 1976; Srivastava and Minowada, 1977) or ara-C uptake and nucleotide pools were observed.

These results indicate that TdT activity of cells could be an important factor in determining sensitivity of cells to ara-A and ara-C. In fact, some cases have been observed both in this laboratory (Srivastava et al., 1980a) as well as in one other (Hutton and Coleman, 1976), where TdT decreased as leukemia became refractory to therapy. Although the exact function of TdT in vivo is not known, it has been proposed that this enzyme acts as a somatic mutagen in lymphocytes (Baltimore, 1974). The original finding of TdT activity in B cells (Srivastava, 1974a), and recently in pre-B cells such as NALM-6-13 which have cIgM and SmIgM (Minowada, personal communications), and the reports of TdT activity in B-cell and pre-B-cell-ALL (Srivastava and Minowada, 1977; Shaw et al., 1978; Vogler et al., 1978) may support the above supposition. At the same time, the observation from this laboratory (Srivastava, 1974b) that contamination of DNA polymerase α and β by TdT made fidelity in copying the polydeoxynucleotides less accurate has been confirmed (Saffhill and Chudhuri, 1976). This confirmation indicates that TdT has the potential of introducing errors during DNA replication.

B. TdT Assay and Characterization

Essentially three procedures involving direct assay on crude extracts (Hutton and Coleman, 1976; Greenwood et al., 1977; Donlon et al., 1977; Hoffbrand et al., 1977)—phosphocellulose chromatography of whole extract (McCaffrey et al., 1975; Marks et al., 1978a) and assay of soluble and chromatin fraction enzyme separately following glycerol gradient centrifugation (Srivastava et al., 1976; Srivastava et al., 1977)—have been used for TdT assay. The disadvantages of TdT assay on crude extracts, even for high-activity tissue such as thymus, were obvious that an analytical fractination on sucrose gradients followed by TdT assay with oligodeoxynucleotide initiator and dGTP substrate was preferred (Chang, 1971). It was considered that fractionation on gradients may remove inhibitory activities as well as endogenous acceptors, and may add molecular weight constraint to fractionation and concentrate the enzyme peak in four to five 0.2 ml fractions (Chang, 1971). In view of these considerations it is obvious that

TdT assay on crude extracts is least reliable particularly, since high-salt extracts which would invariably contain DNA and high DNA polymerase activity are used. In addition, because of inhibitors, dilute extracts required for assay lower the sensitivity and the reliability due to a large multiplication factor for few counts.

The shortcomings of the procedure involving phosphocellulose chromatography of high-salt-detergent cell extracts containing DNA are apparent by detection of two (McCaffrey et al., 1975) or even three (Penit and Chapeville, 1977) TdT activity peaks in this procedure due to artifacts. Moreover, this procedure also results in considerable dilution of the enzyme instead of concentration, as well as loss of enzyme activity during chromatography, especially in small clinical samples.

The glycerol gradient procedure for TdT assay as used in our laboratory is tedious but it has distinct advantages, some of which were discussed above. The procedure involves (1) extraction of cells with Buffer A [25 mM Tris-sulfate, pH 8.3; 1 mM $MgSO_4$; 6 mM NaCl; 4 mM dithiothreitol; 0.1 mM EDTA; and 1 mM phenylmethyl sufonyl fluoride], (2) separation of soluble (30,000 × g supernatant containing all DNA polymerase α and <25% of TdT and DNA polymerase β and γ) and chromatin fractions, and (3) solubilization of TdT from chromatin (extraction of chromatin with 2 M NaCl followed by dialysis of the extract to 0.15 M NaCl to precipitate DNA-histone and leave TdT and DNA polymerase β and γ in the supernatant). This is followed by the precipitation of TdT and DNA polymerases from soluble and solubilized chromatin fractions by 70% saturation with ammonium sulfate (or concentration by some other suitable means) and their fractionation on 10-30% glycerol gradient, prepared in Buffer A containing only 0.1 M NaCl (which separates 3-4S TdT from 7S DNA polymerase α and γ and 3-4S DNA polymerase β which aggregate to dimeric forms).

Thus TdT could be assayed with oligodeoxynucleotide (e.g., dA_{12-18}), or even with activated DNA as the initiator. The assay with activated DNA could be advantageous when significant DNase contamination is present, although any contribution by residual DNA polymerase β should be rigorously ruled out, for example (1) by demonstrating around 90% reduction in incorporation of labeled deoxynucleotide on adding other three unlabeled deoxyribonucleoside triphosphates, (2) by assay in the presence of 0.3 M NaCl which completely inhibits β but is not inhibitory to TdT, and (3) perhaps by chromatography of 3-4S TdT peak fractions on a small DEAE-Sephadex A-25 column which binds DNA polymerase β but not TdT as given for PHA-stimulated normal human lymphocytes in Figure 1. Additional characterization of TdT could also be carried out by the use of TdT antisera or other characteristics of TdT given in Table 1.

Selective inhibition of TdT by ATP has also been proposed for its detection (Bhalla et al., 1977). N-ethylmaleimide alone or with 10% ethanol

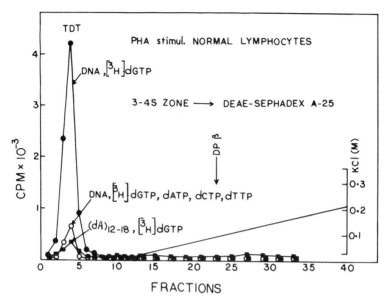

FIGURE 1 DEAE-Sephadex A-25 chromatography of glycerol gradient fractions containing TdT from chromatin of PHA-stimulated normal human lymphocytes. TdT-containing fractions from the 3-4S region of several glycerol gradients were pooled, dialyzed against Buffer B (50 mM Tris · HCl, pH 7.5, 1 mM dithiothreitol, 0.1 mM EDTA, 20% glycerol, 0.02 M KCl) and loaded on a column (1.1 × 8 cm) of DEAE-Sephadex (A-25) equilibrated with Buffer B. After a brief wash with Buffer B the column was eluted with a linear gradient formed with 40 ml of Buffer B and 40 ml of Buffer B containing 0.4 M KCl. Aliquots (30 μl) from each fraction were assayed for TdT with $(dA)_{12-18}$ or activated DNA as the initiator. Where indicated, 0.16 μmol each of cold dATP, dCTP, and dTTP were added to the assay mixture. The position of DNA polymerase β (DP β) is marked by an arrow.

as used by some workers (Coleman et al., 1974b) is, however, not satisfactory for the characterization of TdT because it inhibits not only TdT but also DNA polymerase β significantly (Srivastava, 1976; Dube et al., 1977; Mosbaugh et al., 1976). Separating the soluble and chromatin fraction in the procedure described here provides the possibility to cross-check the results, since about 20% of the TdT activity is found in the soluble fraction and about 80% in the chromatin fraction. Moreover, the latter fraction, which has a bulk of TdT and very little protein and inhibitory substances, could be greatly concentrated before gradient fractionation, and six samples could be analyzed at the same time.

Although cacodylate buffer and magnesium have been reported (Chang and Bollum, 1971) to give much higher activity of TdT compared with the Tris-manganese system, many laboratories (Kung et al., 1976; Sarin and

TABLE 1 Effect of Various Treatments on TdT and DNA Polymerase α, β, and γ Activity from Human Cells

Treatment	Enzyme activity (percent of control)				Reference
	TdT	DNA polymerase α	DNA polymerase β	DMA polymerase γ	
0.3 M NaCl	90	0	0	0	Srivastava (1976)
Six minutes at 50°C	0	30–60	50–60	30–60	Srivastava (1976)
Streptolydigin (200 µg/ml)	23	75	90	98	DiCioccio and Srivastava (1976a)
Calf thymus histone (200 µg/ml)	162 (198)	7	3	4	Srivastava (1978a)
Anti-calf-thymus or ALL TdT IgG (100 µg/protein)	20	100	100	100	Chan and Srivastava (1976) Siddiqui and Srivastava (1978)

Bleomycin (100 μg/ml)	82 (79)	15	12	75	DiCioccio and Srivastava (1976b)
Pyran (10 μg/ml)	17	82	91	77	DiCioccio and Srivastava (1978)
Poly A (100 μg/ml)	40 (50)	100	100	100	Srivastava (1975a)
Poly U (100 μg/μl)	100 (100)	10	17	0	Srivastava (1975a)
Poly 2'-O-methyl-adenylic acid (100 μg/ml)	30	98	96	92	DiCioccio and Srivastava (1977)
Poly 1,N^6-ethenadenylic acid (100 μg/ml)	24	100	100	82	DiCioccio and Srivastava (1977)

Values in parentheses are for TdT assays carried out with activated DNA instead of $(dA)_{12-18}$ as the initiator.

Gallo, 1974) including our own have been unable to reproduce these results. In our hands both cacodylate buffer and Tris buffer have given comparable activity, which is higher with manganese than with magnesium. Similar results have also been recently obtained by Coleman (1977). Whereas the metal requirements for TdT may be complicated (Chirpich, 1977; Chirpich, 1978; Coleman, 1977), cacodylate buffer (Coleman, 1977) or the Tris buffer (Srivastava et al., 1976) together with 0.5 mM $MnCl_2$, 100 µM ^{3}H-dGTP of about 1 Ci/mmol specific activity and dA_{12-18} initiator, should provide a satisfactory and sensitive assay for TdT.

In addition to the above approaches for the assay and characterization of of TdT, an oligo $(dT)_{12-18}$ cellulose column has recently been described (Okamura et al., 1978) for the purification of TdT. An anti-TdT IgG Sepharose column for the purification of TdT has been developed in this laboratory. Figures 2 and 3 describe the behavior of TdT and DNA polymerase from ALL cells on this column and the demonstration of TdT in B-ALL cell line BALM-2. A radioimmunoassay for the detection of TdT has been reported by Kung et al. (1976), and immunofluorescent procedures for the detection of TdT in cells are being utilized by this (Fig. 4) and other laboratories.

C. Animal Studies with TdT

TdT activity has been detected in leukemic cells of AKR mice (Harrison et al., 1976) and in leukemic spleen from BDFl mice (Saffhill and Chaudhuri, 1976). In AKR leukemic cells separated by 1 × g sucrose gradient sedimentation, TdT activity did not change with the cell cycle (Harrison et al., 1976). However, researchers have observed changes in TdT activity of thymus during early thymic development in calf (Chang, 1971) and chicken (Penit and Chapeville, 1977), and a decline in TdT activity in thymus with age in NIH and AKR mouse (Pazmino and Ihle, 1976). Although Hutton and Bollum (1977) detected a normal level of TdT activity in bone marrow from congenitally athymic nude mice, this activity was not detected in bone marrow of nude mice or of thymectomized C57BL/6 mice by Pazmino et al. (1978). But these workers did induce activity in vivo and in vitro by thymosin in bone marrow cells from athymic mice.

Cell fractionation on BSA or Ficoll gradient as well as hydrocortisone treatment have been used to delineate thymus cells with high TdT activity. Kung et al. (1975) found that Tdt phosphocellulose peak I was highest in cortisone-resistant low-density mouse thymocytes (medullary?) which were unaffected by cortisone, whereas TdT peak II was high in cortisone-sensitive thymocytes which declined after cortisone treatment. Contrary to these results, Pazmino et al. (1977) found no difference in the ratio of peak I to peak II in mice thymocytes separated as above on BSA gradients.

Barton et al. (1976) separated rat thymus cells on Ficoll-Hypaque density gradients and suggested that TdT was not present in medullary thymocytes but was present in a subset of cells that comprise 65% of cortical thymocytes, since TdT activity was not correlated with the percentage of

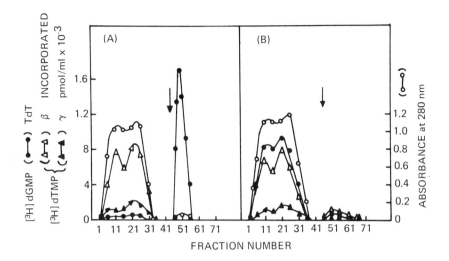

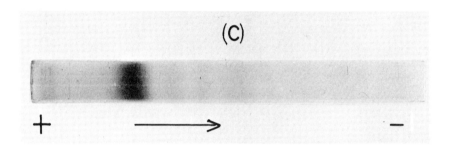

FIGURE 2 Chromatography of a crude extract of ALL cell chromatin on anti-TdT and control IgG Sepharose columns (1.5 × 12 cm). Coupling of Sephadex G-200 purified anti-TdT or of control IgG to Sepharose 4B was carried out according to Livingston et al. (1972) and the instructions provided by Pharmacia Fine Chemicals; the final product contained 25 mg protein/g Sepharose. Then 55 mg of ALL non-histone chromatin protein fraction containing TdT and DNA polymerase β and γ (Srivastava et al., 1978a) were applied to anti-TdT IgG column (A) and control IgG column (B) equilibrated with TDG buffer (0.05 M Tris-HCl, pH 8.0, 1 mM dithiothreitol and 15% glycerol). Next 2,4-ml fractions were collected and aliquots were assayed as described by Srivastava et al. (1978a). Fractions 1–21 represent the loading of the crude extract in TDG buffer; 22–45 represent the TDG buffer + 0.3 M KCl wash; and 45–65 represent TDG buffer and 0.6 M KCl + 0.2 M NH_4 OH eluate. Nondenaturing polyacrylamide gel electrophoresis at pH 4.5 (Reisfeld et al., 1962) of ALL cell TdT was obtained after anti-TdT IgG column (C). After electrophoresis the gel was stained with amido black. When chromatographed on anti-TdT or control IgG columns, DNA polymerase α from the soluble cytoplasmic fraction was not retained (data not shown). (From Srivastava et al., 1980b.)

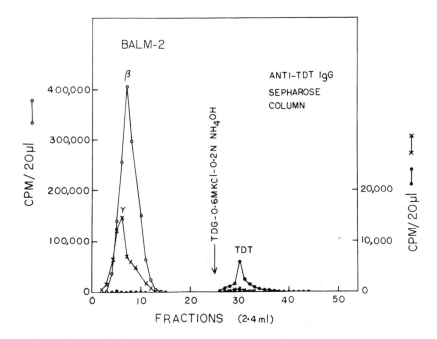

FIGURE 3 Chromatography of non-histone chromatin protein fraction containing TdT and DNA polymerase β and DNA polymerase γ from a B-ALL cell line BALM-2 (Srivastava and Minowada, 1977) on anti-TdT IgG Sepharose column. Fractions before arrow represent loading of the crude extract in TDG buffer and wash with TDG buffer + 0.3 M KCl. Rest as in Figure 2.

cortical thymocytes in the gradient fractions. Barton et al. (1976) and Kung et al. (1975) also proposed that greater declines in TdT activity compared with DNA polymerase activity on cortisone treatment indicate the location of TdT in cortisone-sensitive cortical thymocytes. However, it should be noted that simultaneous sharp declines in TdT and thymidine kinase (92% loss) have been found in thymus on cortisone treatment (Sobhy and Chirpich, 1975), and the latter enzyme is not specific for any particular type of cells. We have examined DNA polymerase α, β, and γ and TdT activity in cortisol-sensitive and cortisol-resistant P1798 tumors and, as Table 2 shows, we found no significant differences in these activities in sensitive and resistant tumors.

Several of the above studies are difficult to interpret because some workers (Penit and Chapeville, 1977) obtained two or even three phosphocellulose peaks of TdT, because assays were done on crude extracts (Batron et al., 1976), because unusually low values of TdT were obtained even for thymus cells (Pazmino et al., 1977), and because of limitations of the cell separation procedures. Perhaps the immunofluorescence studies

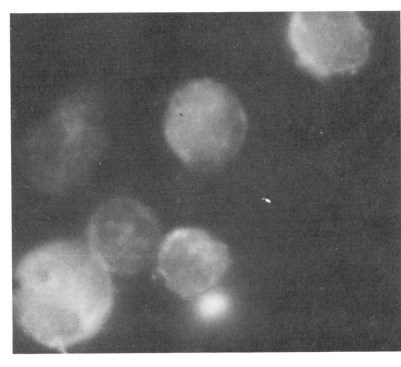
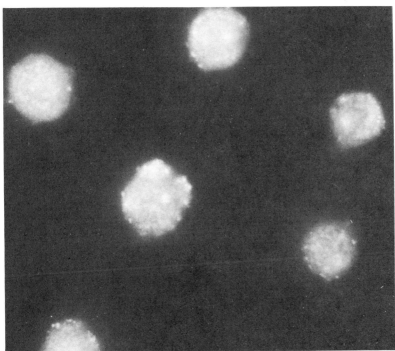

FIGURE 4 Immunofluorescence of human leukemic cell lines RPMI 8402 (left) and K-562 (right) which have high and low TdT activity respectively. Air dried smears were fixed with cold acetone followed by treatment with rabbit anti-ALL cell TdT which had been extensively absorbed against peripheral blood lymphocytes and B- and myeloid human cell lines. The smears were subsequently treated with FITC-goat anti-rabbit IgG.

TABLE 2 TdT and DNA Polymerase α, β, and γ Activity[a] of Cortisol-Sensitive and Cortisol-Resistant P1798 Tumors Examined 14-16 Days After Subcutaneous Implantation into DBA/2 × BABL/CF$_1$ Mice

Cells	TdT activity	DNA polymerase activity		
		α	β	γ
Sensitive tumors	33	58	8.2	5.6
Resistant tumors	29	65	7.3	5.2

[a] Expressed as units/mg of cellular DNA where 1 unit equals 1 nmole of nucleotide (from ^{3}H-dGTP for TdT, and from ^{3}H-dTTP for DNA polymerase α, β, and γ) polymerized in 1 hr. All enzyme activities were calculated by summing the activity of appropriate fractions from the glycerol gradients for both soluble and chromatin fractions except DNA polymerase α, which was present only in the soluble fraction. (dA)$_{12-18}$, poly (A) · (dT)$_{10}$, and activated DNA were used for the assay of TdT, DNA polymerase γ, and DNA polymerase α or β, respectively (Srivastava and Minowada, 1977).

using antisera, which are being carried on in this and other laboratories (Goldschneider et al., 1977; Gregoire et al., 1977) will be more helpful.

D. Clinical Studies with TdT

Although what constitutes a detectable or high TdT activity varies between laboratories due to differences in the sensitivity and efficacy of the assay procedures used, a certain consensus seems to have emerged regarding the distribution of TdT in normal and malignant hematopoietic cells. Disagreements regarding the distribution of TdT or its correlation with some parameters will be pointed out later.

TdT appears to be present in low concentrations in PHA-stimulated normal human lymphocytes (Srivastava, 1974a; Srivastava and Minowada, 1980; Srivastava et al., 1978a) in normal bone marrow (Hutton and Coleman, 1976; Barr et al., 1976; Greenwood et al., 1977; McCaffrey et al., 1975; Hoffbrand et al., 1977; Srivastava et al., 1980a), in B-cell lines of normal and malignant origin (Srivastava, 1974a; Srivastava, 1976; Srivastava and Minowada, 1977; Minowada et al., 1978), and in leukocytes from patients with B-ALL (Srivastava and Minowada, 1977), T-CLL (Srivastava and

Minowada, 1980; Penit et al., 1977; Marks et al., 1978c), and in B-CLL, CML, AML, Sezary syndrome, mycosis fungoides, multiple myeloma, Burkitt's lymphoma, and infectious mononucleosis (Srivastava, 1975b, Srivastava and Minowada, 1980; Srivastava et al., 1978a; Donlon et al., 1977).

In contrast, this enzyme has been found in high concentration in normal thymus or thymoma (Coleman et al., 1974b; McCaffrey et al., 1975; Sarin et al., 1976; Srivastava and Minowada, 1977), in T- and non-T/non-B ALL cells (McCaffrey et al., 1973; Srivastava, 1974a; McCaffrey et al., 1975; Srivastava and Minowada, 1980; Sarin et al., 1976; Hutton and Coleman, 1976; Coleman et al., 1976; Greenwood et al., 1977; Hoffbrand et al., 1977; Donlon et al., 1977; Srivastava et al., 1978a; Srivastava et al., 1980a; Kung et al., 1978; Janossy et al., 1977), AUL (Srivastava and Minowada, 1980; Marcus et al., 1976; Kung et al., 1978), in lymphoblastic Lymphoma (Donlon et al., 1977; Kung et al., 1978; Srivastava et al., 1980a; Mertelsmann et al., 1978b), in T-ALL/lymphoma cell lines (Srivastava and Minowada, 1973; Srivastava et al., 1975; Srivastava, 1976; Srivastava and Minowada, 1980; Srivastava and Minowada, 1977; Minowada et al., 1978; Sarin et al., 1976; Coleman and Hutton, 1975), and in non-T/non-B cell lines of ALL origin (Srivastava and Minowada, 1977; Minowada et al., 1978).

High TdT activity has also been found in some cases of the blastic phase of CML (Sarin and Gallo, 1974; McCaffrey et al., 1975; Sarin et al., 1976; Srivastava and Minowada, 1977; Hutton and Coleman, 1976; Rausen et al., 1977; Srivastava et al., 1977; Donlon et al., 1977; Hoffbrand et al., 1977) and the demonstration that TdT activity was not correlated with the lymphoid or myeloid morphology of the blasts or with the presence of C-ALL antigen in CML (Srivastava et al., 1977) has been established (Marks et al., 1978a; Marks et al., 1978b; Srivastava et al., 1980a). Furthermore, the reports of high TdT activity in rare cases of adult AML (Srivastava et al., 1976) and two AML pediatric patients, one containing Auer rods (Srivastava et al., 1978a), have been confirmed (Grodeon et al., 1977; Kung et al., 1978; Stass et al., 1978a) and about one-fifth of adult AML patients (some containing Auer rods) examined over a period of 3 years in our laboratory have been found to have high TdT activity (Srivastava et al., 1980a).

Recently, high TdT activity has also been reported in lymphoblastic-like leukemic transformation of polycythemia vera (Hoffman et al., 1978), in hand mirror variant of ALL (Stass et al., 1978b), and in B-cell ALL (Shaw et al., 1978), although in the last report the production of immunoglobulin by blasts, a characteristic of true B-lymphoid cells (Eden and Innes, 1978), was not ascertained. Nevertheless, high TdT activity has been found in one pre-B-ALL patient in whom blasts contained cIgM but no SmIg or Fc receptors (Vogler et al., 1978), and in one non-T/non-B ALL patient in whom about 30% of the blasts expressed surface IgM (Srivastava et al., 1978a). Moreover, NALM-6 to 13 cell lines established from this last patient, and Ph^1 chromosome-positive NALM-1 cell lines which contain high TdT activity and were considered to be of the non-T/non-B type (Minowada et al., 1978) have recently been found to contain cIgM and NALM-6 to 13, and to produce some surface IgM (Minowada, personal communication). Thus it seems that B-ALL as reported by us (Srivastava and Minowada, 1977) and by Penit et al. (1977) and B-cell lines (Minowada et al., 1978) may have low TdT activity, whereas pre-B-ALL have high TdT activity. Although high TdT activity has also been found in leukocytes from one Sezary syndrome (Penit et al., 1977) and one chronic phase CML patient (Saffhill et al., 1976), these observations have not been confirmed by us or by others. Lastly, we have rigorously characterized the presence of low TdT activity in spleen from some Hodgkins disease patients, and in spleen and peripheral blood leukocytes from some HCL patients (Srivastava et al., 1978b).

Thus there is general agreement that high TdT activity in peripheral blood and bone marrow leukocytes is present at initial diagnosis and relapse in almost all cases of ALL, whereas low-to-undetectable activity is found in normal leukocytes, ALL patients in remission, and most patients with AML as well as other leukemias except some with blastic-phase CML. Rare cases of ALL have been found by us (Srivastava et al., 1978a) as well as by others (McCaffrey et al., 1975; Greenwood et al., 1977) to have low TdT activity, and it is not clear whether they represent a distinct entity which may have a different prognosis compared with the ALL patients with high TdT activity. As reported initially from this laboratory (Srivastava, 1974a), B cells have been shown to contain some TdT activity and some non-T/non-B ALL, and cell lines could be considered to represent a pre-B-ALL phenotype (Vogler et al., 1978) if they contain cIgM.

Since some AML patients (around 20%) contain high TdT activity, the estimates of TdT activity—while valuable in the initial diagnosis and monitoring of ALL—may have limitations in separating ALL from AML. Nevertheless, the utility of TdT as a marker for ALL is illustrated by high TdT activity in two patients who presented as ALL patients but had low activity when blasts at relapse showed morphological and cytochemical characteristics of myeloblasts (Srivastava et al., 1980a). It remains to be seen whether the recently developed immunofluorescent assay for TdT (Kung et al., 1978; Gregoire et al., 1977; Srivastava, 1978a) will be superior to classical examination of bone marrow for detecting residual disease or predicting relapse.

Of patients with the several histologic types of malignant lymphoma, only those with T- or non-T/non-B lymphoblastic lymphoma were found to have TdT activity comparable to that of ALL cells (Donlon et al., 1977; Kung et al., 1978; Koziner et al., 1977; Mertelsmann et al., 1978b; Srivastava et al., 1980a). Thus TdT estimates could help in distinguishing lymphoblastic lymphoma from poorly differentiated lymphocytic lymphoma or Burkitt tumor. The reportedly high TdT activities in a patient with giant follicular lymphoma and in one with diffuse histiocytic lymphoma (Mertelsmann et al., 1978b) remain to be confirmed.

As outlined above, high TdT activity of T-ALL/lymphoma cell lines could be useful in their differentiation from B-cell lines which have low activity, although non-T/non-B cell lines of ALL origin, whether they contain cIgM or not, also have high TdT activity. In addition, the Ph^1-chromosome-positive lymphoid cell line NALM-1 has high TdT activity in contrast with the Ph^1-chromosome-positive cell line K562 of blastic phase CML origin, which has low activity (Minowada et al., 1978). Significant amounts of TdT activity have also been detected in a T-cell line recently established from pleural effusion of a Hodgkin's disease patient (Srivastava and Minowada, 1978).

In spite of earlier claims to the contrary (McCaffrey et al., 1975; Sarin et al., 1976), it is now established that TdT activity in the blastic phase of CML is not correlated with the lymphoid or myeloid morphology of blasts (Srivastava et al., 1977; Marks et al., 1978a; Marks et al., 1978b), although it has been suggested that cases of high TdT activity, irrespective of blast cell morphology, may respond to vincristine and prednisone therapy (Marks et al., 1978a). The response to this treatment was, however, very poor in their studies (Marks et al., 1978a; Marks et al., 1978b) and in other studies (Srivastava et al., 1977; Janossy et al., 1978b) compared with adult ALL.

In contrast to the blastic phase of CML with the preceding chronic phase, we have observed (Srivastava et al., 1980a) in Ph^1-chromosome-positive adult leukemia patients a greater frequency of patients with lymphoid type blasts, high TdT activity, and response to vincristine and prednisone. In fact, a second remission induced in one patient with vincristine, prednisone, and adriamycin resulted in bone marrow devoid of Ph^1-chromosome-positive cells. Disappearance of Ph^1 chromosomes from cells during remission, although observed in several cases of Ph^1-chromosome-positive acute leukemia with blasts of lymphoid morphology (Janossy et al., 1978b; Gibbs et al., 1977; Bloomfield et al., 1977) and in one case with meyloblast morphology (Gustavsson et al., 1977) has not been observed for the blastic phase of CML. These results indicate that Ph^1-chromosome-positive acute leukemia may be distinct, although related in its origin to both non-T/non-B ALL and CML blastic crisis.

Many of the above results, as well as the occurrence of mixed population of lymphoid and myeloid blasts in some patients (Peterson et al., 1976; Srivastava et al., 1977; Janossy et al., 1976a; Mertelsmann et al., 1978a), the conversion of CML to ALL (Janossy et al., 1978b; Srivastava and Minowada, 1977; Marks et al., 1978b) of ALL to AML (Srivastava et al., 1980a), and of ALL to CML (Gibbs et al., 1977; Mauri et al., 1977; Janossy et al., 1978b; Paolino et al., 1978) could perhaps be better understood on the basis of clonal evolution of leukemic cells from a stem cell pleuripotent for both lympoid and myeloid cells (Janossy et al., 1976b). Recent evidence based on the presence of Ph^1 chromosome in both myeloid cells and PHA-stimulated lymphocytes in CML patients (Secker, Walker, and Hardy, 1976; Barr and Watt, 1978) and the presence of only one glucose-6-phosphate dehydrogenase isoenzyme type A in erythrocytes, granulocytes, platelets, macrophages, and B and T lymphocytes—in contrast to both isoenzyme A and B in skin fibroblast and salivary epithelial cells in a patient with acquired idiopathic sideroblastic anemia (Prchal et al., 1978)—demonstrates the existence of a common progenitor for human myeloid and lymphoid cells.

X. PROTEASE AND DNase ACTIVITY

While working on TdT, we found (Srivastava, 1978a; Srivastava, 1978b) that normal leukocytes and leukocytes from AML and other leukemias except ALL have high activity of 3-4S DNase and serine type protease associated with the chromatin, as well as protease and DNase activities found in the soluble fraction. Intermediate to high activity of these enzymes are found in rare AML and blastic phase CML patients who have high TdT activity, in contrast with little or no protease and DNase activities in ALL cells. Thus tests based on these latter enzymes could be useful in discriminating ALL from AML. Significant protease activity in normal leukocytes as well as in AML, CML, and CLL cells was reported by Weisenthal and Ruddon (1973) who had, however, not examined the ALL cells.

XI. CONCLUSION

It is apparent from the foregoing discussion that a combination of cytological-cytogenetic analysis and cell surface phenotyping of leukemic cells, together with analysis of biochemical markers, all closely integrated within a clinical context, could give new dimensions to the diagnosis, prognosis and monitoring of leukemias and lymphomas. Such analyses could also be invaluable in the study of hematopoietic cell differentiation, as well as the study of their role in immune function. Lastly, the study of biochemical markers should help in better understanding the leukemic process, and this could lead to more rational therapeutic management of the patient as well as to the development of new therapies.

ACKNOWLEDGMENTS

The author's work in this area was supported by U.S. Public Health Service Grants CA-17140, CA-16045, and CA-13038 from the National Cancer Institute.

REFERENCES

Adams, A., and Harkness, R. A. (1976). Adenosine deminase activity in thymus and other human tissues, Clin. Exp. Immunol. 26:647-649.

Alberts, B., and Sternglanz, R. (1977). Recent excitement in the DNA replication problem, Nature 269:655-661.

Anderson, L. C., Gahmberg, C. G., Nilsson, K., and Wigzell, H. (1977). Surface glycoprotein patterns of normal and malignant human lymphoid cells, I. T cells, T blasts and leukemic T cell lines, Int. J. Cancer 20:702-707.

Andreeff, M., Beck, J. D., Darzynkiewicz, Z., Traganos, F., Gupta, S., Melamed, M. R., and Good, R. A. (1978). RNA content in human lymphocyte subpopulation, Proc. Natl. Acad. Sci. USA 75:1938-1942.

Ashley, R. L., Cardiff, R. D., and Manning, J. S. (1977). Characterization of terminal deoxynucleotidyl transferase activity in mouse mammary tumor virus, Virology 77:367-375.

Bakkeren, J. A. J. M., DeVaan, G. A. M., and Hillen, H. F. P. (1976). Proliferation of leukemic blood cells in short-term liquid culture. Relevance to the differential diagnosis of acute myeloblastic and lymphoblastic leukemia, Acta Haemat. 56:321-327.

Baltimore, D. (1974). Is terminal deoxynucleotidyl transferase a somatic mutagen in lymphocytes? Nature 248:409-411.

Baltimore, D., Silverstone, A. E., Kung, P. C., Harrison, T. A., and McCaffrey, R. P. (1976). Specialized DNA polymerases in lymphoid cells, in Cold Spring Harbor Symposia on Quantitative Biology (1977), vol. XLI, Coldspring Harbor Laboratory, Coldspring, New York, part 1, pp. 63-72.

Barr, R. D., and Watt, J. (1978). Preliminary evidence for the common origin of lympho-myeloid complex in man, Acta Haemat. 60:29-35.

Barr, R. D., Sarin, P. S., and Perry, S. M. (1976). Terminal transferase in human bone marrow lymphocytes, Lancet 1:508-509.

Barrett, S. G., Schwade, J. G., Ranken, R., and Kadin, M. E. (1977). Lymphoblasts with both T and B markers in childhood leukemia and lymphoma, Blood 50:71-79.

Barton, R., Goldschneider, I., and Bollum, F. J. (1976). The distribution of terminal deoxynucleotidyl transferase (TdT) among subsets of thymocytes in the rat, J. Immunol. 116:462-468.

Bennett, J. M., Catovsky, D., Daniel, M. T., Flandrin, G., Galton, D. A. G., Cralnick, H. R., and Sultan, C. (1976). Proposals for the classification of the acute leukemias, Br. J. Haematol. 33:451-458.

Berard, C. W., Jaffe, E. S., Braylan, R. C., Mann, R. B., and Nanba, K. (1978). Immunologic aspects and pathology of the malignant lymphomas, Cancer 42:911-921.

Bertazzoni, U., Stefanini, M., Noy, G. P., Giulotto, E., Nuzzo, F., Falaschi, A., and Spadari, S. (1976). Variations of DNA polymerase $-\alpha$ and $-\beta$ during prolonged stimulation of human lymphocytes, Proc. Natl. Acad. Sci. USA 73:785-789.

Bhalla, R. B., Schwartz, M. K., and Modak, M. J. (1977). Selective inhibition of terminal deoxynucleotidyl transferase (TdT) by adenosine ribonucleoside triphosphate (ATP) and its application in the detection of TdT in human leukemia, Biochem. Biophys. Res. Commun. 76:1056-1061.

Bloomfield, C. D., Kersey, J. H., Brunning, R. D., and Gajl-Peczalska, K. F. (1976). Prognostic significance of lymphocyte surface markers in adult non-Hodgkin's malignant lymphoma, Lancet 2:1330-1333.

Bloomfield, C. D., Peterson, L. C., Yunis, J. J., and Brunning, R. D. (1977). The Philadelphia chromosome (Ph^1) in adults presenting with acute leukemia: A comparison of Ph^1+ and Ph^1- patients, Br. J. Haematol. 36:347-358.

Bollum, F. J. (1962). Oligodeoxyribonucleotide-primed reactions catalyzed by calf thymus polymerase, J. Biol. Chem. 237:1945-1949.

Bollum, F. J. (1974). Terminal deoxynucleotidyl transferase, in Enzymes (P. D. Boyer, ed.), vol. 10, Academic Press, New York, pp. 145-171.

Bollum, F. J. (1975). Antibody to terminal deoxynucleotidyl transferase, Proc. Natl. Acad. Sci. USA 72:4119-4122.

Bollum, F. J., Chang, L. M., Tsiapalis, C. M., and Dorson, J. W. (1974). Nucleotide polymerizing enzymes from calf thymus gland, Methods Enzymol. 29:70-81.

Brodniewicz-Proba, T., and Buchowicz, J. (1976). Terminal deoxyribonucleotidyl transferase activity of germinating wheat embryo, FEBS Lett. 65:183-186.

Brouet, J. C., and Seligmann, M. (1978). The immunological classification of acute lymmphoblastic leukemias, Cancer 42:817-827.

Brouet, J. C., Valensi, F., Daniel, M. T., Flandrin, G., Preud'Homme, J. L., and Seligmann, M. (1976). Immunological classification of acute lymphoblastic leukemias: Evaluation of its clinical significance in a hundred patients, Br. J. Haematol. 33:319-328.

Burke, J. S., Byrne, G. E., and Rapport, H. (1974). Hairy cell leukemia (leukemic reticuloendotheliosis), Cancer 33:1399-1410.

Catovsky, D. (1975). T-cell origin of acid-phosphatase-positive lymphoblasts, Lancet 2:327-328.

Catovsky, D., Galetto, J., Okos, A., Miliani, E., and Falon, D. A. G. (1974a). Cytochemical profile of B and T leukaemic lymphocytes with special reference to acute lymphoblastic leukaemia, J. Clin. Pathol. 27:767-771.

Catovsky, D., Pettit, J. E., Galton, D. A. G., Spiers, A. S. D., and Harrison, C. V. (1974b). Leukemic reticuloendotheliosis ("hairy" cell leukemia): A distinct clinicopathological entity, Br. J. Haematol. 26:9-27.

Catovsky, D., Hoffbrand, A. V., Ikoku, N. B., Petrie, A., and Galton, D. A. G. (1975). Significance of cell differentiation in acute myeloid leukemia, Blood Cells 1:201-211.

Chabner, B. A., Johns, D. G., Coleman, C. N., Drake, J. C., and Evans, W. H. (1974). Purification and properties of cytidine deaminase from normal and leukemic granulocytes, J. Clin. Invest. 53:922-931.

Chan, J. Y. H., Srivastava, B. I. S. (1976). Antigenic relationships in calf thymus and human leukemic cell terminal deoxynucleotidyl transferase, Biochem. Biophys. Acta 447:353-360.

Chang, L. M. S. (1971). Development of terminal deoxynucleotidyl transferase activity in embryonic calf thymus gland, Biochem. Biophys. Res. Commun. 44:124-131.

Chang, L. M. S., and Bollum, F. J. (1971). Deoxynucleotide-polymerizing enzymes of calf thymus gland. V. Homogenous terminal deoxynucleotidyl transferase, J. Biol. Chem. 246:909-916.

Chang, L. M. S., and Bollum, F. J. (1972). Antigenic relationships in mammalian DNA polymerase, Science 175:116-117.

Chessells, J. M., Hardisty, R. M., Rapson, N. T., and Greaves, M. F. (1977). Acute lymphoblastic leukemia in children: Classification and prognosis, Lancet 2:1307-1309.

Chirpich, T. P. (1977). Factors affecting terminal deoxynucleotidyl transferase activity in cacodylate buffer, Biochem. Biophys. Res. Commun. 78:1219-1226.

Chirpich, T. P. (1978). The effect of different buffers on terminal deoxynucleotidyl transferase activity, Biochem. Biophys. Acta 518:535-538.

Coccia, P. F. (1977). Characterization of the blast cell in childhood non-Hodgkin's lymphoproliferative malignancies, Semin. Oncol. 4:287-298.

Coleman, M. S. (1977). Terminal deoxynucleotidyl transferase: Characterization of extraction and assay conditions from human and calf tissue, Arch. Biochem. Biophys. 182:525-532.

Coleman, M. S., and Hutton, J. J. (1975). Terminal deoxynucleotidyl transferase and adenosine deaminase in human lymphoblastoid cell lines, Exp. Cell Res. 94:440-442.

Coleman, M. S., Hutton, J. J., and Bollum, F. J. (1974a). DNA polymerases in normal and leukemic human hematopoietic cells, Blood 44:19-32.

Coleman, M. S., Hutton, J. J., Simone, P. D., and Bollum, F. J. (1974b). Terminal deoxynucleotidyl transferase in human leukemia, Proc. Natl. Acad. Sci. USA 71:4404-4408.

Coleman, C. N., Stoller, R. G., Drake, J. C., and Chabner, B. A. (1975). Deoxycytidine kinase: Properties of the enzyme from human leukemic granulocytes, Blood 46:791-803.

Coleman, M. S., Greenwood, M. F., Hutton, J. J., Bollum, F. J., Lampkin, B., and Holland, P. (1976). Serial observations on terminal deoxynucleotidyl transferase activity and lymphoblast surface markers in acute lymphoblastic leukemia, Cancer Res. 36:120-127.

DiCioccio, R., and Srivastava, B. I. S. (1976a). Selective inhibition of terminal deoxynucleotidyl transferase from leukemic cells by streptolydigin, Biochem. Biophys. Res. Commun. 72:1343-1349.

DiCioccio, R., and Srivastava, B. I. S. (1976b). Effect of bleomycin on deoxynucleotide-polymerizing enzymes from human cells, Cancer Res. 36:1664-1668.

DiCioccio, R. A., and Srivastava, B. I. S. (1977). Structure activity relationships and kinetic analyses of polyribonucleotide inhibition of human cellular deoxyribonucleotide-polymerizing enzymes, Biochem. Biophys. Acta 478:274-285.

DiCioccio, R. A., and Srivastava, B. I. S. (1978). Mechanism of inhibition of deoxyribonucleic acid polymerases from human cells and from simian sarcoma virus by pyrans, Biochem. J. 175:519-524.

Dietz, A. A., and Czebotar, V. (1977). Purine metabolic cycle in normal and leukemic leukocytes, Cancer Res. 37:419-426.

Donlon, J. A., Jaffe, E. S., and Braylan, R. C. (1977). Terminal deoxynucleotidyl transferase activity in malignant lymphomas, N. Engl. J. Med. 297:461-466.

Dube, D. K., Seal, G., and Loeb, L. A. (1977). Differential heat sensitivity of mammalian DNA polymerases, Biochem. Biophys. Res. Commun. 76:483-487.

Dulis, B., and Wilson, I. B. (1978). Serum alkaline phosphatase isoenzymes in lymphoproliferative diseases, Cancer Res. 38:2519-2522.

Eden, O. B., and Innes, E. M. (1978). Cell-surface markers in lymphoblastic leukaemia, Lancet 2:378.

Elias, L., and Greenberg, P. (1977). Divergent patterns of marrow cell suspension culture growth in the myeloid leukemias: Correlation of in vitro findings with clinical features, Blood 50:263-274.

Ellis, R. B., Rapson, N. T., Patrick, A. D., and Greaves, M. F. (1978). Expression of hexosaminadase isoenzymes in childhood leukemia, N. Engl. J. Med. 298:476-480.

Esber, E. C., Buell, D. N., and Leikin, S. L. (1976). Insulin binding of acute lymphocytic leukemia cells, Blood 48:33-39.

Filippa, D. A., Lieberman, P. H., Erlandson, R. A., Koziner, B., Siegal, F. P., Turnbull, A., Zimring, A., and Good, R. A. (1978). A study of malignant lymphomas using light and ultramicroscopic, cytochemical and immunologic technics. Correlation with clinical features, Am. J. Med. 64:259-268.

Flandrin, G., and Daniel, M. T. (1974). β-Glucuronidase activity in Sezary cells, Scand. J. Haematol. 12:23-31.

Gallo, R. C., and Perry, S. (1969). The enzymatic mechanisms for deoxythymidine synthesis in human leukocytes. IV. Comparisons between normal and leukemic leukocytes, J. Clin. Invest. 48:105-116.

Gianni, A. M., Favera, R. D., Polli, E., Merisio, I., Giglioni, G., Comi, P., and Ottolenghi, S. (1978). Globin RNA sequences in human leukaemic peripheral blood, Nature 274:610-612.

Gibbs, T. J., Wheeler, M. V., Bellingham, A. J., and Walker, S. (1977). The significance of the Philadelphia chromosome in acute lymphoblastic leukemia: A report of two cases, Br. J. Haematol. 37:447-453.

Giblett, E. R., Anderson, J. E., Cohen, F., Pollara, B., and Meuwissen, H. J. (1972). Adenosine deaminase deficiency in two patients with severely impaired cellular immunity, Lancet 2:1067-1069.

Giblett, E. R., Amman, A. J., Wara, D. W., Sandman, R., and Diamond, L. K. (1975). Nucleoside phosphorylase deficiency in a child with severely defective T-cell immunity and normal B-cell immunity, Lancet 1:1010-1013.

Goldschneider, I., Gregoire, K. E., Barton, R. W., and Bollum, F. J. (1977). Demonstration of terminal deoxynucleotidyl transferase in thymocytes by immunofluorescence, Proc. Natl. Acad. Sci. USA 74:734-738.

Gordon, D., Hutton, J., Meyer, L. M., and Metzger, R. (1977). Terminal transferase, membrane markers and leukemic antigens in adult acute leukemia, Proc. Am. Assoc. Cancer Res. 18:59.

Gralnick, H. R., Galton, D. A. G., Catovsky, D., Sultan, C., and Bennett, J. M. (1977). Classification of acute leukemia, Ann. Intern. Med. 87:740-753.

Greaves, M. F. (1975). Clinical applications of cell surface markers, Prog. Hematol. 9:255-303.

Greaves, M. F. (1977). Recent progress in the immunological characterization of leukemic cells, Blut 34:349-356.

Greenwood, M. F., Coleman, M. S., Hutton, J. J., Lampkin, B., Krill, C., Bollum, F. J., and Holland, P. (1977). Terminal deoxynucleotidyl transferase distribution in neoplastic and hematopoietic cells, J. Clin. Invest. 59:889-899.

Gregoire, K. E., Goldschneider, I., Barton, R. W., and Bollum, F. J. (1977). Intracellular distribution of terminal deoxynucleotidyl transferase in rat bone marrow and thymus, Proc. Natl. Acad. Sci. USA 74:3993-3996.

Gustavsson, A., Mitelman, F., and Olsson, I. (1977). Acute myeloid leukemia with the Philadelphia chromosome, Scand. J. Haematal. 19:449-452.

Hamblin, T., and Hough, D. (1977). Chronic lymphatic leukemia: Correlation of immunofluorescent characteristics and clinical features, Br. J. Haematol. 36:359-365.

Hamburg, A., Brynes, R. K., Reese, C., and Golomb, H. M. (1976). Human cord blood lymphocytes—Ultrastructural and immunologic surface marker characteristics: A comparison with B and T-cell lymphomas, Lab. Invest. 34:207-215.

Hande, K. R., and Chabner, B. A. (1978). Pyrimidine nucleoside monophosphate kinase from human leukemic blast cells, Cancer Res. 38:579-585.

Hansbury, E., Kerr, V. N., Mitchell, V. E., Ratliff, R. L., Smith, D. A., Williams, D. I., and Hayes, F. N. (1970). Synthesis of polydeoxynucleotides using chemically modified subunits, Biochim. Biophys. Acta 199:322-329.

Harrison, T. A., Barr, R. D., McCaffrey, R. P., Sarna, G., Silverstone, A. F., Perry, S., and Baltimore, D. (1976). Terminal deoxynucleotidyl transferase in AKR leukemic cells and lack of relation of enzyme activity to cell cycle phase, Biochem. Biophys. Res. Commun. 69:63-67.

Hayhoe, F. G. J., and Cawley, J. C. (1972). Acute leukemia: Cellular morphology, cytochemistry and fine structure, in Clinics in Haematology (S. Roth, ed.), vol. 1, W. B. Saunders, Philadelphia, pp. 49-94.

Hayhoe, F. G. J., Flemans, R. J., Burns, G. F., and Cawley, J. C. (1977). Leucocyte alkaline phosphatase scores in hairy cell leukemia, Br. J. Haematol. 37:158-159.

V. Heyden, H. W., Weber, R., Stuckstedte, H., Saal, J. G., and Fresen, K. O. (1977). Isoenzyme pattern of acid phosphatase in Epstein-Barr-virus-DNA positive permanent growing lymphoid cell lines, Blut. 35:395-404.

Higgy, K. E., Burns, G. F., and Hayhoe, F. G. J. (1977). Discrimination of B, T and null lymphocytes by esterase cytochemistry, Scand. J. Haematal. 18:437-448.

Ho, D. H. W. (1973). Distribution of kinase and deaminase 1-β-D-arabinofuranosylcytosine in tissues of man and mouse, Cancer Res. 33:2816-2820.

Hoffbrand, A. V., Ganeshaguru, K., Janossy, G., Greaves, M. F., Catovsky, D., and Woodruff, R. K. (1977). Terminal deoxynucleotidyl transferase levels and membrane phenotypes in diagnosis of acute leukemia, Lancet 2:520-523.

Hoffman, R., Estren, S., Kopel, S., Marks, S. M., and McCaffrey, R. P. (1978). Lymphoblastic-like leukemic transformation of polycythemia vera, Ann. Intern. Med. 88:89-71.

Horenstein, A., Piras, M. M., Mordoh, J., and Piras, R. (1976). Protein phosphokinase activities of resting and proliferating human lymphocytes, Exp. Cell Res. 101:260-266.

Horwitz, D. A., Allison, A. C., Ward, P., and Knight, N. (1977). Identification of human mononuclear leucocyte populations by esterase staining, Clin. Exp. Immunol. 30:289-298.

Huang, A. T., Logue, G. L., and Engelbrecht, H. L. (1976). Two biochemical markers in lymphocyte subpopulations, Br. J. Haematol. 34:631-638.

Huang, A. T., Riddle, M. M., and Koons, L. S. (1975). Some properties of a DNA-unwinding protein unique to lymphocytes from chronic lymphocytic leukemia, Cancer Res. 35:981-986.

Hutton, J. J., and Bollum, F. J. (1977). Terminal deoxynucleotidyl transferase in athymic nude mice, Nucleic Acid Res. 4:457-460.

Hutton, J. J., and Coleman, M. S. (1976). Terminal deoxynucleotidyl transferase measurements in the differential diagnosis of adult leukemias, Br. J. Haematol. 34:447-456.

Janckila, A. J., Li, C. Y., Lam, K. W., and Yam, L. T. (1978). The cytochemistry of tartarate resistant acid phosphatase, Am. J. Clin. Pathol. 70:45-55.

Janossy, G., Greaves, M. F., Revesz, T., Lister, T. A., Roberts, M., Durrant, J., Kirk, B., Catovsky, D., and Beard, M. F. J. (1976a). Blast crisis of chronic myeloid leukemia (CML). II. Cell surface marker analysis of "lymphoid" and myeloid cases, Br. J. Haematol. 34:179-192.

Janossy, G., Roberts, M., and Greaves, M. F. (1976b). Target cell in chronic myeloid leukemia and its relationship to acute lymphoid leukemia, Lancet 13:1508-1061.

Janossy, G., Greaves, M. F., Sutherland, R., Durrant, J., and Lewis, C. (1977). Comparative analysis of membrane phenotypes in acute lymphoid leukemia and in lymphoid blast crisis of chronic myeloid leukemia, Leukemia Res. 1:289-300.

Janossy, G., Greaves, M. F., Capellaro, D., Minowada, J., and Rosenfeld, C. (1978a). Membrane antigens of leukaemic cells and lymphoid cell lines, in Protides of the Biological Fluids (H. Peeters, ed.), Pergamon Press, Oxford, pp. 591-600.

Janossy, G., Woodruff, R. K., Paxton, A., Greaves, M. F., Capellaro, D., Kirk, B., Inunes, E. M., Eden, O. B., Lewis, C., Catovsky, D., and Hoffbrand, A. V. (1978b). Membrane marker and cell separation studies in Ph[1]-positive leukemia, Blood 51:861-877.

Johnson, D., and Morgan, A. R. (1976). The isolation of a high molecular weight terminal deoxynucleotidyl transferase from calf thymus, Biochem. Biophys. Res. Commun. 72:840-849.

Johnson, S. M., North, M. E., Ashershon, G. L., Allisop, J., Watts, R. W. E., and Webster, A. D. B. (1977). Lymphocyte purine 5'-nucleotidase deficiency in primary hypogammaglobulinaemia, Lancet 1:168-170.

Kadin, M. E., and Billing, R. J. (1977). Immunofluorescent method for positive identification of null cell type acute lymphocytic leukemias: Use of heterologous antisera, Blood 50:771-782.

Kageoka, T., Nakashima, K., and Miwa, S. (1977). Simultaneous demonstration of peroxidase and lysozyme activities in leukemic cells, Am. J. Pathol. 67:481-484.

Kass, L., and Peters, C. L. (1978). Esterases in acute leukemias. A cytochemical and electrophoretic study, Am. J. Clin. Pathol. 69:57-61.

Kato, K. I., Goncalves, J. M., Houts, G. E., and Bollum, F. J. (1967). Deoxynucleotide-polymerizing enzymes of calf thymus gland. II. Properties of the terminal deoxynucleotidyl transferase, J. Biol. Chem. 242:2780-2789.

Konig, E., Cohnen, G., Aberle, H. G., and Brittinger, G. (1970). Lysosomale Enzyme in lymphozyten. II. Chronische lymphatische Leukamie (CLL): Veranderungen des Enzymgehaltes (Saure phosphatase, β-glucuronidase) von Blutlymphozyten wahrend einer mehrtagigen, Stimulierung mit Phytohamagglutinin in vitro, Acta Haemat. 44:265-278.

Konior, G., Lippman, M., and Johnson, G. (1976). Glucocorticoid receptors in childhood acute lymphocytic leukemia, Clin. Res. 24:377.

Koziner, B., Filippa, D. A., Mertelsmann, R., Gupta, S., Clarkson, B., Good, R. A., and Siegal, F. P. (1977). Characterization of malignant lymphomas in leukemic phase by multiple differentiation markers of mononuclear cells, Am. J. Med. 63:556-567.

Krakow, J. S., Coutsogeorgopoulos, C., and Canellakis, E. S. (1962). Studies on the incorporation of deoxyribonucleotides and ribonucleotides into deoxyribonucleic acid, Biochim. Biophys. Acta 55:639-650.

Kramers, M. T. C., and Catovsky, D. (1978). Cell membrane enzymes: L-γ-Glutamyl transpeptidase, leucine aminopeptidase, maltase and trehalase in normal and leukaemic lymphocytes, Br. J. Haematol. 38:453-461.

Kramers, M. T. C., Catovsky, D., Foa, R., Cherchi, M., and Galton, D. A. G. (1976). 5'Nucleotidase activity in leukaemic lymphocytes, Biomedicine 25:363-365.

Kramers, M. T. C., Catovsky, D., Cherchi, M., and Galton, D. A. G. (1977). 5'Nucleotidase in cord blood lymphocytes, Leukemia Res. 1:279-281.

Kulenkampff, J., Janossy, G., and Greaves, M. F. (1977). Acid esterase in human lymphoid cells and leukaemic blasts: A marker for T lymphocytes, Br. J. Haematol. 36:231-240.

Kumar, S., Carr, T. F., Evans, D. I. K., Jones, P. M., and Hann, I. M. (1978). Cell surface-markers in childhood acute lymphoblastic leukaemia, Lancet 1:164-165.

Kung, P. C., Silverstone, A. E., McCaffrey, R. P., and Baltimore, D. (1975). Murine terminal deoxynucleotidyl transferase: Cellular distribution and response to cortisone, J. Exp. Med. 141:855-865.

Kung, P. C., Gottlieb, P. D., and Blatimore, D. (1976). Terminal deoxynucleotidyl transferase. Serological studies and radioimmunoassay, J. Biol. Chem. 251:2399-2404.

Kung, P. C., Long, J. C., McCaffrey, R. P., Ratliff, R. L., Harrison, T. A., and Baltimore, D. (1978). Terminal deoxynucleotidyl transferase in the diagnosis of leukemia and malignant lymphoma, Am. J. Med. 64:788-794.

LaMantia, K., Conklyn, M., Quagliata, F., and Silber, R. (1975). Immunologic studies of 5'nucleotidase in normal and chronic lymphocytic leukaemia patients, Blood 46:1042.

LaMantia, K., Conklyn, M., Quagliata, F., and Silber, R. (1977). Lymphocyte 5'-nucleotidase: Absence of detectable protein in chronic lymphocytic leukemia, Blood 50:683-689.

Li, C. Y., Yam, L. T., and Lam, K. W. (1970). Studies of acid phosphatase isoenzymes in human leukocytes: Demonstration of isoenzyme-cell specificity, J. Hitochem. Cytochem. 18:901-910.

Livingston, D. M., Scolnick, E. M., Parks, W. P., and Todaro, G. J. (1972). Affinity chromatography of RNA-dependent DNA polymerase from RNA tumor viruses on a solid phase immunoadsorbent, Proc. Natl. Acad. Sci. USA 69:393-397.

Lopes, J., Zucker-Franklin, D., and Silber, R. (1973). Heterogeneity of 5'nucleotidase activity in lymphocytes in chronic lymphocytic leukaemia, J. Clin. Invest. 52:1279-1300.

Maca, R. D., and Hakes, A. D. (1978). Differences in sialyl transferase activity and in concanvalin—A agglutination between T and B lymphoblastoid cell lines, Biochem. Biophys. Res. Commun. 81:1124-1130.

Mangalik, A., Robinson, W. A., and Holton, C. P. (1977). Granulopoietic studies in acute lymphocytic leukemia of children, Blut 34:77-88.

Maraldi, N. M., Cocco, L., Papa, S., Capitani, S., Mazzotti, G., and Manzol, F. A. (1978). Presence of H1° histone in human CLL lymphocytes, Immunol. Allerg. Pathol. 6:78.

Marcus, S. L., and Sarkar, N. H. (1978). Retroviral "terminal deoxynucleotidyl transferase" activity in reverse transcription, Virology 84:247-259.

Marcus, S. L., Smith, S. W., Jarowski, C. I., and Modak, M. J. (1976). Terminal deoxyribonucleotidyl transferase activity in acute undifferentiated leukemia, Biochem. Biophys. Res. Commun. 70:37-44.

Marks, D. B., Kanefsky, T., Keller, B. J., and Marks, A. D. (1975). The presence of histone H1° in human tissues, Cancer Res. 35:886-889.

Marks, S. M., Baltimore, D., and McCaffrey, R. (1978a). Terminal transferase as a predictor of initial responsiveness to vincristine and prednisone in blastic chronic myelogenous leukemia, N. Engl. J. Med. 298:812-814.

Marks, S. M., McCaffrey, R., Rosenthal, D. S., and Maloney, W. C. (1978b). Blastic transformation in chronic myelogenous leukemia: Experience with 50 patients, Med. Pediatric Oncol. 4:159-167.

Marks, S. M., Yanovich, S., Rosenthal, D. S., Moloney, W. C., and Schlossman, S. F. (1978c). Multimarker analysis of T-cell chronic lymphocytic leukemia, Blood 51:435-438.

Marsh, J. C., and Perry, S. (1974). Thymidine catabolism by normal and leukemic human leukocytes, J. Clin. Invest. 43:267-278.

Mathé, G. (1977). Integration of modern data in W.H.O. categorization of lymphosarcomas. Its value for prognosis prediction and therapeutic adaptation to prognosis, Biomedicine 26:377-384.

Mauri, C., Torelli, U., DiPrisco, U., Silingardi, V., Artusi, T., and Emilia, G. (1977). Lymphoid blastic crisis at the onset of chronic granulocytic leukemia, Cancer 40:865-870.

Mayer, R. J., Smith, R. G., and Gallo, R. C. (1975). DNA metabolizing enzymes in normal human lymphoid cells. VI. Induction of DNA polymerases α, β, and γ following stimulation with phytohemagglutinin, Blood 46:509-518.

McCaffrey, R., Smoller, D. F., and Baltimore, D. (1973). Terminal deoxynucleotidyl transferase in a case of childhood acute lymphoblastic leukemia, Proc. Natl. Acad. Sci. USA 70:521-525.

McCaffrey, R., Harrison, T. A., Parkman, R., and Baltimore, D. (1975). Terminal deoxynucleotidyl transferase activity in human leukemic cells in normal human thymocytes, N. Engl. J. Med. 292:775-804.

Meier, J., Coleman, M. S., and Hutton, J. J. (1976). Adenosine deaminase activity in peripheral blood cells of patients with haematological malignancies, Br. J. Cancer 33:312-319.

Mertelsmann, R., Koziner, B., Ralph, P., Filippa, D., McKenzie, S., Arlin, Z. A., Gee, T. S., Moore, M. A. S., and Clarkson, B. D. (1978a). Evidence for distinct lymphocytic and monocytic populations in a patient with terminal transferase-positive acute leukemia, Blood 51:1051-1056.

Mertelsmann, R., Mertelsmann, I., Koziner, B., Moore, M. A. S., and Clarkson, B. D. (1978b). Improved biochemical assay for terminal deoxynucleotidyl transferase in human blood cells: Results in 89 adult patients with lymphoid leukemias and malignant lymphomas in leukemic phase, Leukemia Res. 2:57-69.

Meusers, P., König, E., Fink, U., and Brittinger, G. (1976). Lysosomal acid phosphatase: Differences between normal and chronic lymphocytic leukemia T and B lymphocytes, Blut 33:313-318.

Minowada, J., Tsubota, T., Nakazawa, S., Srivastava, B. I. S., Huang, C. C., Oshimura, M., Sonta, S., Han, T., Sinks, L. F., and

Sandberg, A. A. (1977). Establishment and characterization of leukemic T-cell lines, B-cell lines, and null-cell line: A progress report on surface antigen study of fresh lymphatic leukemias in man, in Hematology and Blood Transfusion. Immunological Diagnosis of Leukemias and Lymphomas (S. Thierfelder, H. Rodt, and E. Thiel, eds.), vol. 20, Springer-Verlag, Berlin, pp. 241-250.

Minowada, J., Janossy, G., Greaves, M. F., Tsubota, T., Srivastava, B. I. S., Morikawa, S., and Tatsumi, E. (1978). Expression of an antigen associated with acute lymphoblastic leukemia lymphoma cell lines, J. Natl. Cancer Inst. 60:1269-1277.

Modak, M. J. (1978). Biochemistry of terminal deoxynucleotidyl transferase: Mechanism of inhibition by adenosine 5'-triphosphate, Biochemistry 17:3116-3120.

Modak, M. J., Bhat, H., Seidner, S., Hahn, E. C., Gupta, S., and Good, R. A. (1978). DNA polymerases of human tonsil and chicken bursa: Absence of a distinct B cell specific terminal deoxynucleotidyl transferase, Biochem. Biophys. Res. Commun. 83:266-273.

Moore, M. A. S. (1975). Marrow culture—A new approach to classification of leukemias, Blood Cells 1:149-158.

Moore, M. A. S. (1976). Prediction of relapse and remission in AML by marrow culture criteria, Blood Cells 2:109-124.

Moore, M. A. S. (1977). In vitro culture studies in chronic granulocytic leukemia, Clinics Haematol. 35:411-418.

Moretta, L., Mingari, M. C., Moretta, A., and Lydyard, P. M. (1977). Receptors for IgM are expressed on acute lymphoblastic leukemic cells having T-cell characteristics, Clin. Immunol. Immunopathol. 7:405-409.

Morris, T. C. M., Butler, M., Muldrew, J. G., McNeill, T. A., and Bridges, J. M. (1977). Changes in granulopoiesis detection by in vitro colony formation in acute lymphocytic leukemia, Br. J. Cancer 35:868-874.

Mosbaugh, D. W., Kunkel, T. A., Stalker, D. M., Tcheng, J. E., and Meyer, R. R. (1976). Novikoff hepatoma deoxyribonucleic acid polymerase. Sensitivity of the β-polymerase to sulfhydryl blocking agents, Nucleic Acid Res. 3:2341-2352.

Nakazawa, S., Minowada, J., Tsubota, T., and Sinks, L. F. (1978). Profound radiosensitivity to "leukemic" T-cell lines and T-cell type acute lymphoblastic leukemia demonstrated by sodium [^{51}Cr] chromate labelling, Cancer Res. 38:1661-1666.

Nelson, D. A. (1976). Cytomorphological diagnosis of the acute leukemias, Semin. Oncol. 3:201-208.

Nelson, J. S. R., and Schiffer, L. M. (1973). Autoradiographic detection of DNA polymerase containing nuclei in sarcoma 180 ascites cells, Cell Tissue Kinet. 6:45-54.

Neuman, H., Moran, E. M., Russell, R. M., and Rosenberg, I. H. (1974). Distinct alkaline phosphatase in serum of patients with lymphatic leukemia and infectious mononucleosis, Science 186:151-153.

Neumann, H., Klein, E., Hauck-Granoth, R., Yachnin, S., and Ben-Bassat, H. (1976). Comparative study of alkaline phosphatase activity in lymphocytes, mitogen-induced blasts, lymphoblastoid cell lines, acute myeloid leukemia and chronic lymphatic leukemia cells, Proc. Natl. Acad. Sci. USA 73:1432-1436.

Nilsson, K., Anderson, L. C., Gahmberg, C. G., and Wigzell, H. (1977). Surface glycoprotein patterns of normal and malignant human lymphoid cells. II. B cells, B blasts and Epstein-Barr virus (EBV)-positive and negative B lymphoid cell lines, Int. J. Cancer 20:708-716.

Ohnuma, T., Arkin, H., Minowada, J., and Holland, J. F. (1978). Differential chemotherapeutic susceptibility of human T-lymphocytes and B-lymphocytes in culture, J. Natl. Cancer Inst. 60:749-752.

Okamura, S., Crane, F., Messner, H. A., and Mak, T. W. (1978). Purification of terminal deoxynucleotidyl transferase by oligonucleotide affinity chromatography, J. Biol. Chem. 253:3765-3767.

Pangalis, G. A., Yataganas, X., and Fessas, P. (1977). β-Glucuronidase activity on lymph node imprints of malignant lymphomas and chronic lymphocytic leukemia, J. Clin. Pathol. 30:812-816.

Pangalis, G. A., Kuhl, W., Waldman, S. R., and Beutler, E. (1978). Acid hydrolases in normal B and T blood lymphocytes, Acta Haematol. 59:285-292.

Paolino, W., Levis, A., Caramellino, L., and Paolino, F. (1978). Chronic myeloid leukemia initiating as acute lymphoid leukemia, Acta Haematol. 60:56-58.

Parkman, R., Gelfand, E. W., Rosen, R. S., Sanderson, A., and Hirschhorn, R. (1975). Severe combined immunodeficiency and adenosine deaminase deficiency, N. Engl. J. Med. 292:714-719.

Pazmino, N. H., and Ihle, J. N. (1976). Strain, age and tumor-dependent distribution of terminal deoxynucleotidyl transferase in thymocytes of mice, J. Immunol. 117:620-625.

Pazmino, N. H., McEwan, R. N., and Ihle, J. N. (1977). Distribution of terminal deoxynucleotidyl transferase in bovine serum albumin gradient fractionated thymocytes and bone marrow cells of normal and leukemic mice, J. Immunol. 119:494-499.

Pazmino, N. H., Ihle, J. N., and Goldstein, A. L. (1978). Induction in vivo and in vitro of terminal deoxynucleotidyl transferase by thymosin in bone marrow cells from athymic mice, J. Exp. Med. 147:708-718.

Penit, C., and Chapeville, F. (1977). Developmental changes in terminal deoxynucleotidyl transferase of the chicken thymus, Biochem. Biophys. Res. Commun. 74:1096-1101.

Penit, C., Brouet, J. C., and Rouget, P. (1977). Terminal deoxynucleotransferase in acute lymphoblastic leukemias and chronic T-cell proliferations, Leukemia Res. 1:354-350.

Peterson, L. C., Bloomfield, C. D., and Brunning, R. D. (1976). Blast crisis as an initial or terminal manifestation of chronic myeloid leukemias. A study of 28 patients, Am. J. Med. 60:209-220.

Plum, J., and Ringoir, S. (1975). A characterization of human B and T lymphocytes by their lactate dehydrogenase isoenzyme pattern. Eur. J. Immunol. 5:871-874.

Polliack, A., Froimovici, M., Pozzoli, E., and Deliliers, G. L. (1976). Acute lymphoblastic leukemia: A study of 25 cases by scanning electron microscopy, Blut 33:359-366.

Prchal, J. T., Throckmorton, D. W., Carroll, A. J., Fuson, E. W., Gams, R. A., and Prchal, J. F. (1978). A common progenitor for human myeloid and lymphoid cells, Nature 274:590-591.

Preud'Homme, J. L., Brouet, J. C., Clauvel, J. P., and Seligmann, M. (1974). Surface IgD in immuno-proliferative disorders, Scand. J. Immunol. 3:853-859.

Quagliata, F., Faig, D., Conklyn, M., and Silber, R. (1974). Studies on lymphocyte 5'-nucleotidase in chronic lymphocytic leukemia, infectious mononucleosis, normal subpopulations and phytohaemagglutinin stimulated cells, Cancer Res. 34:3917-3202.

Ramot, B., Brok-Simoni, F., Barnea, N., Bank, I., and Holtzmann, F. (1977). Adenosine deaminase (ADA) activity in lymphocytes of leukemia, Br. J. Haematol. 36:67-70.

Ranki, A., Tötterman, T. H., and Häyry, P. (1976). Identification of resting human T and B lymphocytes by acid α-naphthyl acetate esterase staining combined with rosette formation with Staphtococcus accreus strain Cowan I, Sand. J. Immunol. 5:1129-1134.

Rausen, A. R., Kim, H. J., Burstein, Y., Rand, S., McCaffrey, R. M., and Kung, P. C. (1977). Philadelphia chromosome in acute lymphoblastic leukemia of childhood, Lancet 1:432.

Reddi, K. K. (1976). Human granulocyte ribonuclease, Biochem. Biophys. Res. Commun. 68:1119-1125.

Reisfeld, R. A., Lewis, V. J., and Williams, D. E. (1962). Disc electrophoresis of basic proteins and peptides on polyacrylamide gels, Nature 195:281-283.

Richie, E., Culbert, S., Sullivan, M. P., and VanEys, J. (1978). Acute lymphoblastic leukemia and complement receptors, Blood 52:467.

Ritter, J., Gaedicke, G., Winkler, K., Beckmann, H., and Landbeck, G. (1975). Possible T-cell origin of lymphoblasts in acid phosphatase-positive acute lymphatic leukemia, Lancet 2:75.

Roath, S., Newell, D., Polliack, A., Alexander, E., and Lin, P. S. (1978). Scanning electron microscopy and the surface morphology of human lymphocytes, Nature 273:15-18.

Roberts, M., Greaves, M. F., Janossy, G., Sutherland, R., and Pain, C. (1978). Acute lymphoblastic leukemia (ALL) associated antigen. I. Expression in different haematopoietic malignancies, Leukemia Res. 2:105-114.

Robinson, W. A., and Stonington, J. (1976). Clinical uses of semisolid bone marrow culture, Blut 32:1-12.

Rosen, F., Kaiser, N., Mayer, M., and Milholland, R. J. (1976). Glucocorticoids: Receptors and mechanism of action in lymphoid tissues and muscle, Methods Cancer Res. 13:67-100.

Rowley, J. D. (1978). The cytogenetics of acute leukemia, in Clinics in Haematology (J. V. Simone, ed.), vol. 7, W. B. Saunders, London, pp. 386-406.

Rowley, J. D., and DeLa Chapelle, A. (1978). General report on the first international workshop on chromosomes in leukemia, Int. J. Cancer 21:307-308.

Roychoudhury, R., Jay, E., and Wu, R. (1976). Terminal labeling and addition of homopolymer tracts to duplex DNA fragments by terminal deoxynucleotidyl transferase, Nucleic Acid Res. 3:863-877.

Saffhill, R., and Chaudhuri, L. (1976). The presence of terminal deoxynucleotidyl transferase in the N-methyl-N-nitrosourea induced leukemia in BDF1 mice and its effect on the accuracy of the DNA polymerases, Nucleic Acids Res. 3:277-284.

Saffhill, R., Dexter, T. M., Muldal, S., Testa, N. G., Morris Jones, P., and Joseph, A. (1976). Terminal deoxynucleotidyl transferase in a case of Ph1 positive infant chronic myelogenous leukemia, Br. J. Cancer 33:664-667.

Sarin, P. S., and Gallo, R. C. (1974). Terminal deoxynucleotidyl transferase in chronic myelogenous leukemia, J. Biol. Chem. 249:8051-8053.

Sarin, P. S., Anderson, P. N., and Gallo, R. C. (1976). Terminal deoxynucleotidyl transferase activities in human blood leukocytes and lymphoblast cell lines: High levels in lymphoblast cell lines and in blast cells of some patients with chronic myelogenous leukemia in acute phase, Blood 47:11-20.

Scholar, E., and Calabresi, P. (1973). Identification of enzymatic pathways of nucleotide metabolism in human lymphocytes and leukemic cells, Cancer Res. 33:94-103.

Secker Walker, L. M., and Hardy, J. D. (1976). Philadelphia chromosome in acute leukemia, Cancer 38:1619-1624.

Shaw, M. T. (1976). The cytochemistry of acute leukemia: A diagnostic and prognostic evaluation, Semin. Oncol. 3:219-228.

Shaw, M. T., Dwyer, J. M., Allaudeen, H. S., and Weitzman, H. A. (1978). Terminal deoxyribonucleotidyl transferase activity in B cell acute lymphoblastic leukemia, Blood 51:181-187.

Sher, R., Fripp, P. J., and Wadee, A. A. (1977). Esterase activity as a marker for human T lymphocytes, Br. J. Haematol. 37:301-302.

Siddiqui, F. A., and Srivastava, B. I. S. (1978). Terminal deoxynucleotidyl transferase from acute lymphoblastic leukemia cells and production of antisera, Biochem. Biophys. Acta 517:150-157.

Smith, R. G., Abrell, J. W., Lewis, B. J., and Gallo, R. C. (1975). Serological analysis of human deoxyribonucleic acid polymerases. Preparation and properties of antiserum to deoxyribonucleic and polymerase I from human lymphoid cells, J. Biol. Chem. 250:1702-1709.

Smyth, J. F., and Harrap, K. R. (1975). Adenosine deaminase in leukaemia, Br. J. Cancer 31:544-549.

Sobhy, C., and Chirpich, T. P. (1975). Decreased thymus terminal deoxynucleotidyl transferase activity following hydrocortisone treatment, Biochem. Biophys. Res. Commun. 64:1270-1273.

Spadari, S., Muller, R., and Weissbach, A. (1974). The dissimilitude of the low and high molecular weight deoxyribonucleic acid dependent deoxyribonucleic acid polymerases of HeLa cells, J. Biol. Chem. 249:2991-2992.

Spitzer, G., Dicke, K. A., McCredie, K. B., and Barlogie, B. (1977). The early detection of remission in acute myelogenous leukemia by in vitro culture, Br. J. Haematol. 35:411-418.

Srivastava, B. I. S. (1972). Association of terminal deoxynucleotidyl transferase activity with chromatin from plant tissue, Biochem. Biophys. Res. Commun. 48:270-273.

Srivastava, B. I. S. (1974a). Deoxynucleotide polymerizing enzymes in normal and malignant human cells, Cancer Res. 34:1015-1026.

Srivastava, B. I. S. (1974b). Fidelity of DNA polymerases from normal and leukemic human cells in polydeoxynucleotide replication, Biochem. J. 141:585-587.

Srivastava, B. I. S. (1975a). Modified nucleotide polymers as inhibitors of DNA polymerases, Biochem. Biophys. Acta 414:126-132.

Srivastava, B. I. S. (1975b). Terminal deoxynucleotidyl transferase and human leukemia, Res. Commun. Chem. Pathol. Pharmacol. 10:715-724.

Srivastava, B. I. S. (1976). Deoxynucleotide polymerizing enzyme activities in T- and B-cells of acute lymphoblastic leukemia origin, Cancer Res. 36:1825-1830.

Srivastava, B. I. S. (1978a). Unpublished data.

Srivastava, B. I. S. (1978b). Biochemical markers for the differential diagnosis of leukemias, in Advances in Comparative Leukemia Research (P. Bentvelzen, J. Hilgers, and D. S. Yohn, eds.), Elsevier, North-Holland Biomedical Press, Amsterdam, pp. 385-386.

Srivastava, B. I. S. (1978c). Utilization of araATP and araCTP by terminal deoxynucleotidyl transferase (TdT) and its relevance to chemotherapy, Fed. Proc. 37:1305.

Srivastava, B. I. S., and DiCioccio, R. A. (1978). Unpublished data.

Srivastava, B. I. S., and Minowada, J. (1973). Terminal deoxynucleotidyl transferase activity in a cell line (MOLT-4) derived from the peripheral blood of a patient with acute lymphoblastic leukemia, Biochem. Biophys. Res. Commun. 51:529-535.

Srivastava, B. I. S., and Minowada, J. (1980). Terminal deoxynucleotidyl transferase and leukemia, in Prevention and Detection of Cancer Part II, Detection, vol. 2, Cancer Detection in Specific Sites (H. E. Nieburgs, ed.), Marcel Dekker, New York, pp. 2359-2371.

Srivastava, B. I. S., and Minowada, J. (1977). Terminal deoxynucleotidyl transferase activity and cell surface antigens of two unique cell lines (NALM-1 and BALM-2) of human leukemic origin, Int. J. Cancer 20:199-205.

Srivastava, B. I. S., and Minowada, J. (1978). Unpublished data.

Srivastava, B. I. S., Minowada, J., and Moore, G. E. (1975). High terminal deoxynucleotidyl transferase activity in a new T-cell line (RPMI 8402) of acute lymphoblastic leukemia origin, J. Natl. Cancer Inst. 55:11-14.

Srivastava, B. I. S., Khan, S. A., and Henderson, E. S. (1976). High terminal deoxynucleotidyl transferase activity in acute myelogenous leukemia, Cancer Res. 36:3847-3850.

Srivastava, B. I. S., Khan, S. A., Minowada, J., Gomez, G., and Rakowski, I. (1977). Terminal deoxynucleotidyl transferase activity in blastic phase of chronic myelogenous leukemia, Cancer Res. 37:3612-3618.

Srivastava, B. I. S., Khan, S. A., Minowada, J., and Freeman, A.

(1978a). High terminal deoxynucleotidyl transferase activity in pediatric patients with acute lymphocytic and acute myelocytic leukemias, Int. J. Cancer 22:4-9.

Srivastava, B. I. S., Khan, S. A., Minowada, J., Henderson, E. S., and Rakowski, I. (1980a). Terminal deoxynucleotidyl transferase activity and blast cell characteristics in adult acute leukemias. Leukemia Research 4:209-215.

Srivastava, B. I. S., Kahn, S. A., and Song, S. Y. (1978b). Terminal deoxynucleotidyl transferase in hairy cell leukemia and Hodgkin's disease, Br. J. Cancer 38:643-644.

Srivastava, B. I. S., Chan, J. Y. H., Siddiqui, F. A. (1980b). Affinity chromatography of terminal deoxynucleotidyl transferase from calf thymus and human leukemic cells on a solid phase immunoadsorbent, J. Biochem. Biophys. Methods 2:1-9.

Stass, S. A., Veach, S., Pasquale, S. M., Schumacher, H. R., Keneklis, T. P., and Bollum, F. J. (1978a). Terminal deoxynucleotidyl transferase positive acute lymphoblastic leukemia with Auer rods, Lancet 1:1042-1043.

Stass, S. A., Perlin, E., Jaffe, E. S., Simon, D. R., Creegan, W. J., Robinson, J. J., Holloway, M. L., and Schumacher, H. R. (1978b). Acute lymphoblastic leukemia-hand mirror cell variant: A detailed cytological and ultrastructural study with an analysis of the immunologic surface markers, Am. J. Hematol. 4:67-77.

Stein, H., and Müller-Hermelink, H. K. (1977). Simultaneous presence of receptors for complement and sheep red blood cells in human fetal thymocytes, Br. J. Haematol. 36:225-230.

Stein, H., Petersen, N., Gaedicke, G., Lennert, K., and Landbeck, G. (1976). Lymphoblastic lymphoma of convuluted or acid phosphatase type-A tumor of T precursor cells, Int. J. Cancer 17:292-295.

Stoop, J. W., Zegers, B. J. M., Hendrickx, G. F. M., Siegenbeer van Heukelom, L. H., Staal, G. E. J., DeBree, P. K., Wadman, S. K., and Ballieux, R. E. (1977). Purine nucleoside phosphorylase deficiency associated with selective cellular immunodeficiency, N. Engl. J. Med. 296:651-655.

Sullivan, J. L., Osborne, W. R. A., and Wedgwood, R. J. (1977). Adenosine deaminase activity in lymphocytes, Br. J. Haematol. 37:157-158.

Tan, H. K., and Lamberg, J. D. (1977). Diagnosis of acute leukemia. Variability of morphologic criteria, Am. J. Clin. Pathol. 68:440-448.

Tritsch, G. L., and Minowada, J. (1978). Differences in purine metabolizing enzyme activities in human leukemic T-cell, B-cell and null cell lines, J. Natl. Cancer Inst. 60:1301-1304.

Tung, R., Silber, R., Quagliata, F., Conklyn, M., Gottesman, J., and Hirschhorn, R. (1976). Adenosine deaminase activity in chronic lymphocytic leukemia, J. Clin. Invest. 57:756-761.

Uchiyama, T., Yodoi, J., Sagawa, K., Takatsuki, K., and Uchino, H. (1977). Adult T-cell leukemia: Clinical and hematologic features of 16 cases, Blood 50:481-491.

Van der Weyden, M. B., and Kelley, W. N. (1976). Adenosine deaminase and immune function, Br. J. Haematol. 34:159-165.

Vincent, P. C., Sutherland, R., Bradley, M., Lind, D., and Gunz, F. W. (1977). Marrow culture studies in adult acute leukemia at presentation and during remission, Blood 49:903-912.

Vogler, L. B., Crist, W. M., Bockman, D. E., Pearl, E. R., Lawton, A. R., and Cooper, M. D. (1978). Pre-B-cell leukemia. A new phenotype of childhood lymphoblastic leukemia, N. Engl. J. Med. 298:872-878.

Wang, T. Y. (1968). Isolation of a terminal DNA-nucleotidyl transferase from calf thymus non-histone chromatin proteins, Arch. Biochem. Biophys. 127:235-240.

Wantzin, G. L. (1977). Nuclear labeling of leukaemic blast cells with tritiated thymidine triphosphate in 35 patients with acute leukemia, Br. J. Haematol. 37:475-482.

Weisenthal, L. M., and Ruddon, R. W. (1973). Catabolism of nuclear proteins in control and phytohemagglutinin-stimulated human lymphocytes, leukemic leukocytes and Burkitt lymphoma cells, Cancer Res. 33:2923-2935.

Weissbach, A., Baltimore, D., Bollum, F., Gallo, R. C., and Korn, D. (1975). Nomenclature of eukaryotic DNA polymerases, Science 190:401-402.

Winchester, R. J., Ross, G. D., Jarowski, C. I., Wang, C. Y., Halper, J., and Brownmeyer, H. E. (1977). Expression of Ia-like antigen molecules on human granulocytes during early phases of differentiation, Proc. Natl. Acad. Sci. USA 74:4012-4016.

Yam, L. T., Li, C. Y., and Lam, K. W. (1971). Tartrate-resistant acid phosphatase isoenzyme in the reticulum cells of leukemic reticuloendotheliosis, N. Engl. J. Med. 284:357-360.

Yam, L. T., Li, C. Y., and Finkel, H. E. (1972). Leukemic reticuloendotheliosis. The role of tartrate-resistant acid phosphatase in diagnosis and splenectomy in treatment, Arch. Intern. Med. 284:357-360.

Yarbro, G. S. K., Lippman, M. E., Johnson, G. E., and Leventhal, B. G. (1977). Glucocorticoid receptors in subpopulations of childhood acute lymphocytic leukemia, Cancer Res. 37:2688-2695.

8

STEROID-RECEPTOR INTERACTIONS IN NORMAL
AND NEOPLASTIC MAMMARY TISSUES

JAMES L. WITTLIFF, RAJENDRA G. MEHTA,[*] WALTER M. LEWKO,[†]
and DANIEL C. PARK / Department of Biochemistry, Health Sciences
Center, University of Louisville, Louisville, Kentucky

PATRICIA A. BOYD-LEINEN[‡] / Department of Molecular Medicine, Mayo
Clinic, Rochester, Minnesota

I. Introduction 184
   A. Hormonal Control of Mammary Gland Development 184
   B. Morphology of the Mammary Gland 185
   C. Early Studies Indicating the Presence of Steroid Receptors in
      Mammary Gland 187
II. Characteristics of Steroid Receptors from Normal Mammary
   Gland 188
   A. Titration of Specific Estrogen Binding Sites 188
   B. Kinetics of Association of ^{3}H-Ligands with Specific Estrogen
      Binding Sites 190
   C. Kinetics of Dissociation of Estrogen-Receptor Complexes as a
      Function of Temperature 191
   D. Determination of the Dissociation Constant from Kinetic
      Parameters 192
   E. Ligand Specificity of the Estrogen Binding Site 192
   F. Nature of the Inhibition of [^{3}H]Estradiol-17β Binding 196
   G. Temperature Inactivation of Occupied and Unoccupied Estrogen
      Binding Sites 197
   H. Stability of Estrogen Binding Proteins at 25°C as a Function of Site
      Occupancy 197
   I. Stability of Estrogen Receptors of Lactating Mammary Gland as a
      Function of Storage Temperature 197

Present affiliation:
[*]Illinois Institute of Technology Research Institute, Chicago, Illinois.
[†]Laboratory of Pathophysiology, National Cancer Institute, Bethesda, Maryland.
[‡]<u>Chemical Abstracts</u> Service, Columbus, Ohio.

J. Molecular Properties of Estrogen Binding Proteins in Lactating Mammary Gland   200
K. Characterization of the Activation Process of Estrogen Receptors   204
III. Characteristics of Steroid Receptors in Experimental Mammary Tumors   211
    A. NMU-Induced Mammary Tumor   211
    B. MCCLX Mammary Adenocarcinomas   213
    C. Mammary Tumors of the C3H Mouse   213
    D. R3230AC Mammary Adenocarcinoma   214
    E. DMBA-Induced Mammary Tumors   214
IV. Estrogen Receptors in Human Breast Carcinoma and Their Clinical Significance   217
V. Summary   223
    References   224

I. INTRODUCTION

The mammary gland is a unique organ in that it possesses the potential for repeated sequences of cellular reorganization from a resting state to a structurally differentiated state during pregnancy. At parturition it becomes a secretory tissue which finally de-differentiates, returning to the involuted or resting state at weaning (Fig. 1). Furthermore, it is a common target organ for several steroids and polypeptide hormones (e.g., estrogens, progestogens, glucocorticoids, insulin, prolactin and possibly growth hormone) which are required for growth and differentiation.

A. Hormonal Control of Mammary Gland Development

The emphases of most early studies on the development of the mammary gland in vivo were directed toward elucidating the minimal hormonal requirements necessary for proliferation of and secretory activity in the mammary gland. These studies measured the responses of animals to various hormonal treatments following endocrinectomy. Lyons et al. [1], as well as Nandi and Bern [2], working with ovariectomized, adrenalectomized, and hypophysectomized (virgin) rats and mice, demonstrated that prolactin, growth hormone, and the ovarian steroids acted synergistically to stimulate alveolar growth in mammary glands. Growth hormone, estrogen, and corticoids were discovered to be the principal mammogens responsible for growth in the prepuberal gland. Postpuberty ductal branching required estrogen, prolactin, and corticoid, while growth

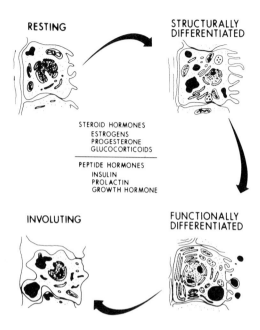

FIGURE 1 Stages in the differentiation of the mammary gland. The hormones known to influence mammary epithelial growth and differentiation are listed. (From Ref. 21.)

hormone enhanced the response to these three hormones. Prolonged treatment of the animals with this latter combination (plus progesterone) resulted in the development of significant lobuloalveolar structures, similar to those seen in pregnancy. Thus the minimal hormonal requirement for mammary gland development through the pregnancy stage appeared to be a combination of estradiol, progesterone, growth hormone, prolactin, and corticoids alone; thyroxine and growth hormone augmented milk production but were not absolutely essential.

B. Morphology of the Mammary Gland

In the mammary glands of the adult, complete morphological and functional development was not observed until pregnancy and lactation [cf. 3]. The resting (prepregnancy) gland of the rat consists of small epithelial lobules separated by connective tissue and infiltrated by adipose cells (Fig. 2A). The entire gland is surrounded by subcutaneous adipose cells. During pregnancy, the mammary gland attains maximal development under the

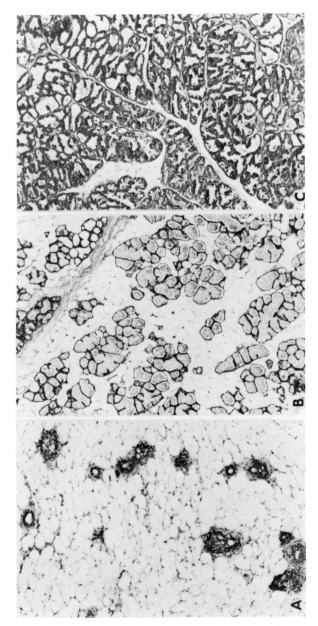

FIGURE 2 Alterations in the cellular composition of the mammary gland of the rat. Thick sections of mammary gland from a virgin (A), pregnant (B), and lactating (C) rat. Magnification was similar (50X) in all of the hematoxylin and eosin stained sections. (From Ref. 21.)

regulation of ovarian, adrenal, placental, and hypophyseal hormones [4].

In early pregnancy, the epithelial cells increase in size and in number (Fig. 2B) and begin the structural differentiation process. The interlobule connective tissue decreases in size as pregnancy advances. During mid-pregnancy, the amount of proliferation decreases and cells demonstrate increased differentiation, including cytoplasmic enlargement and the accumulation of organelles (mitochondria, rough endoplasmic reticulum , and glycogen deposits) and secretory products [3].

At parturition, the prepared cell machinery is triggered into operation by a multitude of neuroendocrine stimuli [5]. The lobules, containing enlarged alveoli, are surrounded by dilated vascular channels (Fig. 2C). At the ultrastructural level, the granular endoplasmic reticulum (RER) undergoes extensive hyperplasia and the Golgi apparatus becomes conspicuous. During lactogenesis the number of microvilli increase on the apical surface of the epithelial cells which exchange material with the lumen. Oxytocin apparently plays a role in the excretion of milk by the lactogenic cells as a result of a contraction process [6].

Figure 1 is a diagrammatic representation of the stages in the development and differentiation of a typical mammary cell. Following lactation, the organelles shrink and the microvilli become less conspicuous. Phagocytic vacuoles increase in size and number as the cell involutes.

C. Early Studies Indicating the Presence of Steroid Receptors in Mammary Gland

Glascock and Hoekstra [7] first demonstrated the accumulation in vivo of the synthetic estrogen [^{3}H]hexestrol in mammary tissues of goats and sheep. The same phenomenon was observed later in certain mammary tumors of patients with breast cancer [8]. Using C3H/HeJ mice, Bresciani and Puca [9] first reported the uptake of [^{3}H]estradiol by normal mammary glands. Shortly thereafter, the preferential localization of unmetabolized [^{3}H]-estradiol-17β in the epithelial cells of both the normal and malignant mammary gland was demonstrated [e.g., 10-13]. Autoradiography indicated that the radioactivity was present in both the cytoplasmic and nuclear fractions [13]. Cells of the mammary gland fat pad, which predominate in the virgin state, were unable to retain [^{3}H]estradiol-17β in a specific fashion [14]

Collectively, these findings suggested that breast tissue, like the uterus, was a target organ for estrogen action. However, due to the high proportion of adipose cells to epithelial cells in the quiescent gland, these studies did not detect the presence of specific binding components for estradiol-17β. However, using sucrose gradient analysis, Jensen et al. [15] and independently our laboratory [16, 17] demonstrated specific estrogen binding components in the cytosols of human breast tumors. Normal tissue gave no indication of the presence of these proteins.

With the knowledge that estrogens, progestins, and glucocorticoids influence the development of mammary epithelium, the authors' laboratory investigated the presence of specific binding sites for these steroids using normal mammary gland of the rat. Early studies indicated that the number of estrogen binding sites varied with the stage of differentiation of the mammary gland [18, 19]. Furthermore, it was found that although the estrogen and glucocorticoid binding capacities were low or virtually absent in mammary gland from both virgin and early pregnant animals, the number of binding sites in cytosol increased during lactation [18-20]. Although somewhat more variable, the progestin receptor appeared to increase during differentiation of the mammary gland [21]. Consequently the lactating mammary gland of the rat was used as a source of the steroid binding proteins for characterization studies.

II. CHARACTERISTICS OF STEROID RECEPTORS FROM NORMAL MAMMARY GLAND

The rates of association and dissociation of steroid hormones with specific binding sites in cytosol from lactating mammary gland are dependent upon incubation time and temperature according to the reaction:

$$\text{Ligand + receptor} \underset{k_{-1}}{\overset{k_1}{\rightleftarrows}} [\text{ligand - receptor}]$$
$$\downarrow k_3 \qquad\qquad\qquad \downarrow k_2$$
$$\text{degradation} \qquad\qquad \text{degradation}$$

For example, the binding of [^{3}H]estradiol to its receptor was maximal in 4 hr at 0-3°C and remained unchanged for 20 additional hours of incubation [22]. At 25°C, apparent equilibrium was reached in 30 min and was maintained for 30 additional minutes before a gradual loss of binding activity was observed. Presumably, this loss is due to degradation of the receptor or irreversible dissociation of the steroid-receptor complexes. As a result of the temperature sensitivity of the estrogen-receptor complex at 25°C, the majority of binding reactions are performed at 0-3°C.

A. Titration of Specific Estrogen Binding Sites

To demonstrate the affinity and concentration of estrogen receptors in cytosol from lactating mammary gland, aliquants are incubated with increasing concentrations of various tritium-labeled estrogenic compounds

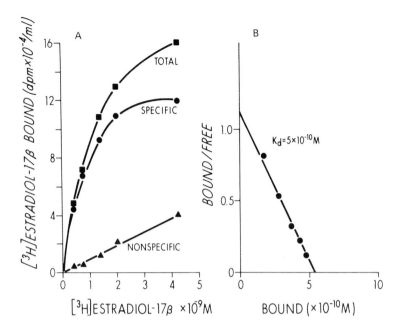

FIGURE 3 Titration of estrogen-binding sites in cytosol of lactating mammary gland using [^{3}H]estradiol-17β. A constant volume of cytosol (0.2 ml) from the mammary gland of a rat, 18 days postpartum, was incubated with increasing concentrations of [^{3}H]estradiol-17β either alone (■) or in the presence of 5 μM diethylstilbestrol (▲) as shown in panel A. The closed circles represent specific binding. Receptor activity was measured by the dextran-coated charcoal procedure. The number of binding sites was 54 fmol/mg cytosol protein and the K_d value was 5.0×10^{-10} M as determined from Scatchard analysis (B). (From Ref. 22.)

for 5-18 hr at 0-3°C (Fig. 3). Binding observed in the presence of an excess of unlabeled inhibitor is related to nonspecific (low affinity, high capacity) association of the ligand. Specific binding is estimated as the difference between total and nonspecific binding.

Sites binding [^{3}H]estradiol-17β specifically were saturated at approximately 2-4 nM. Scatchard analysis [23] of these titration data indicated a single class of binding sites with apparent dissociation constants of $4.6 \pm 0.8 \times 10^{-10}$ M. The range of the number of binding sites was 23-184 fmol/mg cytosol protein for glands from animals, 14-21 days postpartum.

Specific [^{3}H]estrone binding sites were saturated at 10-15 nM and gave apparent dissociation constants of 4-6 nM. The number of binding sites ranged from 50 to 70 fmol/mg cytosol protein as determined by Scatchard analysis of the titration data. Saturation of specific binding sites using [^{3}H]estriol occurred at 10-15 nM, similar to the saturation point of estrone. The number of binding sites was 80-100 fmol/mg cytosol protein, and the K_d value ranged from 3 to 5 nM.

When specific binding sites were titrated with [^{3}H]diethylstilbestrol, saturation was reached at 4-6 nM. Scatchard analyses yielded K_d values of 1-5 nM, and the number of binding sites ranged from 27 to 34 fmol/mg cytosol protein. These data clearly indicate the ligand preference for the estrogen receptor in mammary gland and suggest that they parallel the estrogenicity of the compounds.

B. Kinetics of Association of ^{3}H-Ligands with Specific Estrogen Binding Sites

The temperature dependence of the association of [^{3}H]estradiol with specific binding sites in cytosol from lactating mammary gland is known [18]. To determine the rate constant of association at 0-3°C, time course experiments were conducted using various [^{3}H]estradiol concentrations with a constant amount of receptor preparation. The initial velocity (V_i), estimated as the slope of the tangent to each association curve, was then plotted versus the concentration of [^{3}H]estradiol present. If a single steroid molecule complexed with one binding entity, this plot should be linear [24]. The corresponding experiment must be performed also in which the [^{3}H]estradiol concentration is kept constant and the protein concentration is varied. The plot of initial binding velocity versus binding site concentration is linear also. As a result of the linearity of these two relationships, the association of [^{3}H]estradiol with its receptor site exhibited second-order kinetics. Using the equation

$$k_1 = \frac{V_i}{[S][R]} \tag{1}$$

where [S] equals the concentration of ^{3}H-steroid and [R] equals the concentration of receptor binding sites, determined by Scatchard analysis, a second-order rate constant of association can be determined (Table 1). The association rate constants calculated for estradiol were 2-3 × 10^7 M^{-1} min^{-1}. The rate constant of association calculated for [^{3}H]estrone with its specific receptor was 8.8 × 10^6 M^{-1} min^{-1}, while the k_1 values calculated for [^{3}H]estriol were 3.0 × 10^7 M^{-1} min^{-1} to 9.6 × 10^6 M^{-1} min^{-1}. Using [^{3}H]diethylstilbestrol at varying concentrations while holding the receptor site concentration fixed, a k_1 value of 1.7 × 10^7 M^{-1} min^{-1} was obtained.

TABLE 1 Summary of Kinetic Parameters of [^{3}H]Estradiol-17β Binding Proteins in Cytosol of Lactating Mammary Gland of the Rat

Temperature (°C)	Kinetic measurements			Scatchard analysis
	k_1 ($M^{-1} min^{-1}$)	k_{-1} (min^{-1})	K_d (M)	K_d (M)
3	2.4×10^7	2.7×10^{-3}	1.1×10^{-10}	4.6×10^{-10}
25	3.7×10^7	6.5×10^{-2}	1.8×10^{-9}	6.6×10^{-10}

Source: Adapted from Ref. 22.

C. Kinetics of Dissociation of Estrogen-Receptor Complexes as a Function of Temperature

Two methods have been used in previous studies to determine the dissociation kinetics of steroid-receptor complexes. Adding either an excess of unlabeled ligand or of dextran-coated charcoal changes the concentration of unbound steroid in the reaction mixture. Each of these procedures should alter the equilibrium of the binding reaction and thus enable one to monitor the dissociation process.

The time courses of both degradation alone and degradation plus dissociation of ^{3}H-ligand-receptor complexes were followed at 0-3°C and 25°C. Since both of these processes followed first-order kinetics, the difference between their rate constants gave an estimate of the rate constant of dissociation of estradiol-17β from its specific binding sites. The rate constant of each first-order process was calculated from the equation

$$k = \frac{\ln 2}{t_{1/2}} \qquad (2)$$

where $t_{1/2}$ equals the half-life of the [^{3}H]estradiol-receptor complex. The calculated rate constant of complex degradation (k_2) was 0.6×10^{-3} min^{-1} at 0-3°C, while the rate constant of degradation plus dissociation ($k_2 + k_{-1}$) was 1.9×10^{-3} min^{-1}. Therefore the rate constant of dissociation (k_{-1}) at 0-3°C was calculated to be 1.3×10^{-3} min^{-1}. The mean k_{-1} value of three separate determinations was 2.7×10^{-3} min^{-1} (Table 1).

At 25°C, the values of k_2 and ($k_2 + k_{-1}$) were 0.8×10^{-2} min^{-1} and 5.8×10^{-2} min^{-1}, respectively. The rate constant of dissociation at 25°C was calculated to be 5.0×10^{-2} min^{-1}. The value obtained as the mean of three separate determinations at 25°C was 6.5×10^{-2} min^{-1}. These results demonstrate that increasing the incubation temperature from 0-3°C to 25°C caused a 20- to 40-fold increase in the rate of dissociation of [^{3}H]estradiol-17β from the receptor binding site (Table 1).

When dextran-coated charcoal was added to the incubation mixture to alter the equilibrium of the reaction, a dissociation constant could not be determined. At both 0-3°C and 25°C, the $t_{1/2}$ values of both the degradation process and the degradation plus dissociation process were equal. Since the k_2 values (degradation) were identical to those determined from the experiment described earlier, it was assumed that no ligand-receptor dissociation occurred in the presence of dextran-coated charcoal.

D. Determination of the Dissociation Constant from Kinetic Parameters

Using the association rate constant (k_1) and the dissociation rate constant (k_{-1}), an apparent dissociation constant (K_d) can be calculated using equation (3).

$$K_d = \frac{k_{-1}}{k_1} \tag{3}$$

Values of 1.1×10^{-10} M and 1.8×10^{-9} M were obtained for estradiol-17β binding at 0-3°C and 25°C, respectively. They are in good agreement with K_d determinations obtained by Scatchard analyses (Table 1).

E. Ligand Specificity of the Estrogen Binding Site

To gain some insight into the chemistry of the estrogen receptor binding site, a series of experiments was conducted in which varying concentrations of unlabeled compounds were permitted to compete for binding sites in the presence of a fixed concentration of [^{3}H]estradiol-17β. After the reactions reached equilibrium, the quantity of receptor-bound [^{3}H]estradiol-17β was measured by the dextran-coated charcoal assay, and plots of binding inhibition were generated (e.g., Fig. 4). The dissociation constant of unlabeled ligands can be estimated using the relationship

$$K_c \simeq \frac{K_s [I_{50}(c)]}{[I_{50}(s)]} \tag{4}$$

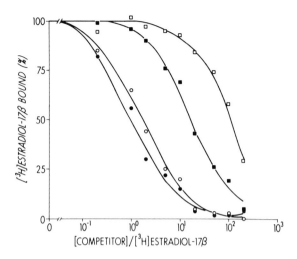

FIGURE 4 Inhibition of [^{3}H]estradiol-17β binding in cytosol of lactating mammary gland by <u>Fusarium</u> mycotoxins. A constant volume of cytosol (0.2 ml) from the mammary gland of a rat, 20 days postpartum, was incubated with 4.8 nM [^{3}H]estradiol-17β either alone or in the presence of various concentrations of estradiol-17β (●), diethylstilbestrol (○), dihydrozearalanol (■), or zearalenone (□) for 16 hr at 0-3°C. Receptor activity was measured by the dextran-coated charcoal procedure. Each point represents the mean of three determinations. The K_d value of estradiol-17β binding, calculated from separate Scatchard analysis of titration data, was 2.8×10^{-10} M. By means of the Rodbard relationship, the estimated K_d values of the competitors were 5.7×10^{-10} M for diethylstilbestrol, 5.4×10^{-9} for dihydrozearalanol, and 3.6×10^{-8} for zearalenone. (From Ref. 26.)

suggested by Rodbard [25], where K_c = the dissociation constant of competitor-receptor complexes; K_s = the dissociation constant of ^{3}H-steroid-receptor complexes, determined by separate Scatchard analysis; $I_{50}(c)$ = the [competitor]/[^{3}H-steroid] which inhibited the ^{3}H-steroid binding by 50%; $I_{50}(s)$ = the [unlabeled steroid]/[^{3}H-steroid] which inhibited the binding of ^{3}H-steroid by 50%.

Titration studies were conducted separately and Scatchard analyses yielded K_d values of 2-10 $\times 10^{-10}$ M for estradiol binding. The calculated K_d values were 4.3×10^{-10} M for diethylstilbestrol, 6.7 nM for estriol, 7.3 nM for estrone, and 1.6×10^{-8} for estradiol-17α. Triamcinolone acetonide did not significantly inhibit [^{3}H]estradiol-17β binding; therefore a K_d value could not be estimated. Progesterone and 5α-dihydrotestosterone were poor inhibitors of [^{3}H]estradiol-17β binding, also precluding the determination of their dissociation constants. At low concentrations of estrone, estriol, and estradiol-17α, an increase rather than a decrease in [^{3}H]estradiol-17β binding was observed. The significance of this observation is unknown.

The inhibition of [^{3}H]estradiol binding by the Fusarium mycotoxins, zearalenone and dihydrozearalenol, has been reported [26, 27]. These compounds have been classified as estrogens since they produce cornification of the vagina of adult mice, as in normal estrus [28]. The calculated K_d values were 5.4 nM for dihydrozearalenol and 3.6×10^{-8} M for zearalenone (Fig. 4).

A relative affinity (RA) value, comparing the binding affinity of the ^{3}H-steroid to that of each unlabeled competitor, was determined using either equation (5) or (6).

$$RA = \frac{I_{50}(s)}{I_{50}(c)} \times 100 \qquad (5)$$

$$RA = \frac{K_s}{K_c} \times 100 \qquad (6)$$

A summary of the I_{50}, K_d, and RA values of various competitive substances are presented in Table 2. Hexestrol, a synthetic estrogen, showed almost a threefold higher binding affinity than estradiol-17β, whereas diethylstilbestrol competed for sites equally as well as estradiol-17β. It has been suggested that these nonsteroidal compounds can assume a conformation similar to estradiol-17β in the receptor binding site [29].

We have demonstrated that any substitution of steric hindrance of the C_3-hydroxyl group drastically lowered or abolished the affinity of the ligand for the binding site. The $C_{17\beta}$-hydroxyl position appeared to be less critical to binding site recognition than the C_3 position. However, a change in the 17-hydroxyl group from the β to the α configuration caused a 25-fold reduction in the binding affinity. Compounds such as 17-desoxyestradiol, which has no functional group in the 17 position, were ineffective (RA = 2) in competing for [^{3}H]estradiol sites (Table 2).

It appears from our studies that the degree of saturation of the A ring is critical as illustrated using 5(10)-estran-3α, 17β-diol, which contains only one double bond ($\Delta^{5, 10}$). This substance had less than 1% the effectiveness of estradiol in complexing with the binding site. Compounds with additional double bonds in the B ring demonstrated varied binding affinities. One double bond in the $\Delta^{7, 8}$ position resulted in only a fourfold decrease in binding. However, when a double bond exists between

TABLE 2 Ligand Specificity of the Binding Site on the Estrogen Receptor from Cytosol of Lactating Mammary Gland[a]

Competitive substance	I_{50}[b]	K_d (M)[c]	RA[d]
Estradiol-17β	0.4–7.0 × 10⁰	2–10 × 10⁻¹⁰	100
Hexestrol	0.4 × 10⁰	1.8 × 10⁻¹⁰	265
Diethylstilbestrol	1.8 × 10⁰	4.3 × 10⁻¹⁰	107
7-Dehydroestradiol	8.4 × 10⁰	1.9 × 10⁻⁹	28
Estrone	2.1 × 10¹	4.2 × 10⁻⁹	23
Estriol	2.1 × 10¹	4.1 × 10⁻⁹	20
Dihydroequilenin	2.6 × 10¹	5.3 × 10⁻⁹	8
Dihydrozearalenol	1.8 × 10¹	5.4 × 10⁻⁹	5
Estradiol-17α	7.6 × 10¹	1.2 × 10⁻⁸	4
17-Desoxyestradiol	1.2 × 10²	2.5 × 10⁻⁸	2
Estradiol-3-methylether	8.7 × 10²	1.1 × 10⁻⁶	<0.1

[a] A constant volume of cytosol (0.2 ml) from the mammary glands of rats, 14–21 days postpartum, were incubated with a saturating concentration of [³H]estradiol-17β either alone or in concentrations of unlabeled competitors for 5–18 hr at 0–3°C. Triplicate determinations were made using the dextran-coated charcoal procedure.
[b] Defined as the value of [competitor]/[³H]estradiol-17β] at which the specific binding of [³H]estradiol-17β is inhibited 50%.
[c] The dissociation constant of each competitive substance was estimated using the Rodbard relationship. The K_d of estradiol-17β was determined by separate Scatchard analyses of titration data
[d] The relative affinity is defined as $[I_{50}(s)/I_{50}(c)] \times 100$.
Source: Adapted from Ref. 22.

carbons 6 and 7, the ability of the compound to compete for sites was only 2% that of estradiol-17β. Similar effects were observed when additional double bonds were present in the B ring, as in equilenin. Glucocorticoids, androgens, and progestins showed less than 1% of the binding affinity of estradiol.

F. Nature of the Inhibition of [^{3}H]Estradiol-17β Binding

To determine the nature of the inhibition by the compounds just discussed, cytosol preparations were titrated with various concentrations of [^{3}H]-estradiol-17β either alone or in the presence of a single concentration of unlabeled competitor. Double reciprocal plots of the titration data, using the quantity bound specifically versus the amount unbound, were generated. Double reciprocal plots demonstrated that the inhibition by estrogenic compounds was competitive (Fig. 5). The K_m value of estradiol-17β was 2.0×10^{-10} M, which agreed with K_d values determined by Scatchard analyses. The K_i values calculated from these types of experiments were 3.6×10^{-10} M for estrone, 3.2×10^{-10} M for estriol, 1.5×10^{-10} M for hexestrol, and 4.0×10^{-10} M for 7-dehydroestradiol as calculated from the equation:

FIGURE 5 Nature of the inhibition of estrogen binding sites in cytosol of lactating mammary gland by <u>Fusarium</u> mycotoxins. A constant volume of cytosol (0.2 ml) from the mammary gland of a rat, 20 days postpartum, was incubated with increasing concentrations of [^{3}H]estradiol-17β either alone or in the presence of 4 μM diethylstilbestrol (■), 0.4 μM zearalenone (□), or 0.4 μM dihydrozearalanol (●) for 19 hr at 0-3°C. Receptor activity was measured by the dextran-coated charcoal assay. Only specific binding was plotted. Double reciprocal plots of the data in panel A are shown in panel B. Each point represents the mean of three determinations. (From Ref. 26.)

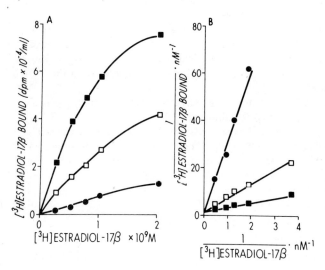

$$\text{Slope (I)} = \text{slope (S)} \left(1 + \frac{[I]}{K_i}\right) \tag{7}$$

where I = inhibitor and S = ^{3}H-steroid.

Likewise, the Fusarium mycotoxins which are nonsteroidal in structure, zearalenone and dihydrozearalenol, exhibited competitive inhibition of estradiol-17β binding [26]. Estradiol-17β had a K_m value of 5.8×10^{-10} M, and the K_i values were 1.2×10^{-7} M for zearalenone and 2.2×10^{-8} M for dihydrozearalenol.

G. Temperature Inactivation of Occupied and Unoccupied Estrogen Binding Sites

The thermostability of estrogen binding proteins with either occupied or unoccupied sites may be investigated using a method usually applied to enzyme studies. Briefly, cytosols are incubated at various temperatures for exactly 20 min, either before or after reacting with ^{3}H-estradiol for 16 hr at 0-3°C. From plots of the residual binding capacity versus the incubation temperature, one may calculate the temperature inactivation coefficient (T_i) of the receptor protein (Fig. 6). The T_i values of 29°C and 35°C were observed for unoccupied and occupied estrogen sites, respectively. Similar T_i values of 28°C and 35°C were obtained in the presence of monothioglycerol, indicating that monothioglycerol did not affect the temperature sensitivity of the binding component, whereas bond ligand protected the protein from temperature inactivation.

H. Stability of Estrogen Binding Proteins at 25°C as a Function of Site Occupancy

To investigate further the lability of occupied and unoccupied estrogen receptor sites, cytosols were incubated at 25°C for various times either before or after reacting with [^{3}H]estradiol-17β for 16 hr at 0-3°C. A semilogarithmic plot of the binding capacity versus incubation time at 25°C was generated. The half-life of the receptor protein with occupied sites was 210 min, whereas unoccupied sites degraded with a half-life of 95 min. Again, these results demonstrated the protection afforded to the receptor protein by estradiol-17β.

I. Stability of Estrogen Receptors of Lactating Mammary Gland as a Function of Storage Temperature

The stability of estrogen receptors in minces of lactating mammary gland stored at subzero temperatures was determined (Table 3). Lactating

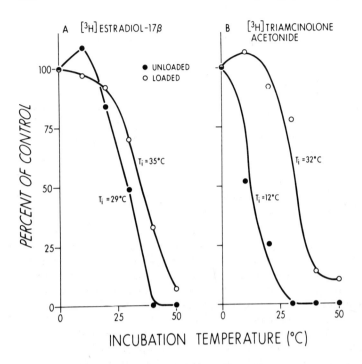

FIGURE 6 Thermostability of steroid-binding components in cytosol of lactating mammary gland of the rat. The specific binding capacities at 0°C were 32 fmol/mg cytosol protein for estradiol-17β and 180 fmol/mg protein for triamcinolone acetonide. The temperature inactivation coefficient (T_i) is defined as the temperature at which 50% of the binding capacity remains after heating 20 min. Note that the glucocorticoid-binding sites were considerably more sensitive to heating when unloaded than when loaded as compared with estrogen-binding sites. (From Ref. 21.)

mammary gland was mined and mixed homogeneously. One portion was used to prepare a fresh control cytosol, while the remaining portions were frozen in liquid nitrogen and stored at either -86°C or -20°C. When the tissue was stored at -86°C, the binding capacity steadily increased over the control level until a maximum value was reached at 3 weeks of storage time. After 4 weeks at -86°C, the binding capacity fell below that of the control. A similar elevation in the estrogen binding capacity was observed when the tissue was stored at -20°C. However, at this temperature the maximum increase occurred at 1 week of storage time, followed by a gradual return to the control level at 4 weeks of storage. At both the -86°C and -20°C storage temperatures, the molecular properties of the estrogen

TABLE 3 Stability of Estrogen Receptors in Minces of Lactating Mammary Gland as a Function of Storage Temperature

Time (wk)	Parameters of estrogen-binding capacity[b]			
	−86°C		−20°C	
	K_d (M)	n (fmol/mg cytosol protein)	K_d (M)	n (fmol/mg cytosol protein)
0	1.0×10^{-10}	38	7.8×10^{-10}	68
1	—	—	3.4×10^{-9}	184
2	4.5×10^{-10}	50	6.6×10^{-10}	77
3	6.5×10^{-10}	75	6.1×10^{-10}	66
4	1.1×10^{-9}	62	—	—
5	2.3×10^{-10}	25	3.4×10^{-10}	36

[a] Lactating mammary gland from 21-day postpartum rats was minced and mixed homogeneously. From one aliquot of tissue mince, a fresh cytosol was prepared, while the remaining portions were frozen in liquid nitrogen stored at −86°C or −20°C and from these at various times the cytosols were prepared. All cytosols were incubated with increasing concentrations of [³H]estradiol either alone or in the presence of excess unlabeled ligand for 16 hr at 0–3°C. Receptor activity was measured by the dextran-coated charcoal assay.
[b] Determined by Scatchard analysis of the titration data; K_d = apparent dissociation; n = number of specific estrogen binding sites.
Source: Adapted from Ref. 22.

binding proteins remained unchanged as determined by sucrose gradient analysis.

In an attempt to ascertain whether the observed elevation in binding capacity could be a result of the freezing step, tissue minces were stored at 0-3°C and 22°C without prior freezing (Table 4). At 0-3°C, an elevation in the binding capacity was observed at the 4-hr time point, followed by a gradual return to the control value by 24 hr. However, at 22°C the estrogen receptor apparently degraded in a first-order fashion with a $t_{1/2}$ value of ~13 hr.

To determine whether similar increases in the number of estrogen binding sites could be observed with frozen cytosol preparations, aliquots of the cytosol fraction were titrated immediately or frozen in liquid nitrogen and stored at -86°C for various times before titration experiments were performed. Table 5 demonstrates that frozen cytosol retained the same level of binding capacity as freshly prepared cytosol, for at least 3 weeks. The K_d values of estrogen receptors in cytosol remained unchanged during storage at -86°C. Furthermore, there was essentially no difference in K_d values of receptors in stored minces (Tables 3 and 4) when compared with those of cytosols (Table 5).

J. Molecular Properties of Estrogen Binding Proteins in Lactating Mammary Gland

Separation of specific steroid binding proteins on sucrose gradients has been reported for the normal mammary gland and several mammary tumors [cf. 19]. The [^{3}H]estradiol binding component in cytosol of rat lactating mammary gland sedimented in the 8S region of a linear sucrose gradient of low ionic strength [18]. To compare the sedimentation properties of other [^{3}H]estrogen receptor complexes, cytosols were incubated either with [^{3}H]estradiol-17β, [^{3}H]estrone, [^{3}H]estriol, or [^{3}H]diethylstilbestrol alone or in the presence of unlabeled competitor. Then the ^{3}H-ligand-receptor complexes were sedimented on linear 5-40% sucrose gradients.

As previously described, the [^{3}H]estradiol-17β binding components in these experiments had a sedimentation coefficient of approximately 8S (Fig. 7). Infrequently a peak of specific binding was also observed in the 4-5S regions of low ionic strength gradients. The [^{3}H]estrone-receptor complexes sedimented primarily in the 8S region of the gradient. However, some specific [^{3}H]estrone binding was observed in the ~4S region. The sedimentation profiles for [^{3}H]estriol binding components were similar to those observed for [^{3}H]estradiol complexes, containing a single 8S peak of specific binding. Gradients containing [^{3}H]diethylstilbestrol receptors demonstrated an 8S peak of specific binding and also a large peak (~4S) of binding which was not displaced by either unlabeled diethylstilbestrol or estradiol-17β. Therefore this latter binding component was assumed to be nonspecifically bound [^{3}H]diethylstilbestrol.

TABLE 4 Stability of Estrogen Receptors in Minces of Lactating Mammary Gland as a Function of Storage Temperature[a]

Time (hr)	Parameters of estrogen-binding capacity			
	0–3°C		22°C	
	K_d (M)	n (fmol/mg cytosol protein)	K_d (M)	n (fmol/mg cytosol protein)
0	1.9×10^{-10}	54	1.9×10^{-10}	54
4	1.1×10^{-10}	63	4.5×10^{-10}	50
6	1.0×10^{-10}	59	2.4×10^{-10}	36
24	2.0×10^{-10}	57	2.7×10^{-10}	17

[a] Lactating mammary gland from 21-day postpartum rats was minced and mixed homogeneously. Aliquots were maintained at 0–3°C or 22°C for various times before cytosols were prepared. Supernatants were incubated with increasing concentrations of [³H]estradiol either alone or in the presence of excess unlabeled competitor for 20 hr at 0–3°C. The dextran-coated charcoal assay was used to measure receptor activity.

Source: Adapted from Ref. 22.

TABLE 5 Stability of Estrogen and Glucocorticoid Receptors in Cytosol of Lactating Mammary Gland[a]

Time (wk)	Estrogen binding[b]		Glucocorticoid binding[b]	
	K_d (M)	n (fmol/mg cytosol protein)	K_d (M)	n (fmol/mg cytosol protein)
0	3.9×10^{-10}	36	7.6×10^{-9}	30
1	2.9×10^{-10}	34	1.4×10^{-8}	28
2	2.5×10^{-10}	36	1.6×10^{-8}	29
3	2.3×10^{-10}	37	—	—

[a] Lactating mammary gland from 14-day postpartum rats was minced and cytosol was prepared. Aliquots were frozen in liquid nitrogen and stored at -86°C. A fresh portion was left unfrozen to serve as a control. At various times the thawed cytosols were incubated with increasing concentrations of either [³H]estradiol or [³H]triamcinolone acetonide alone or in the presence of excess unlabeled competitor for 17 hr at 0-3°C.
[b] Receptor activity was measured by the dextran-coated charcoal assay using Scatchard analysis.
Source: Adapted from Ref. 22.

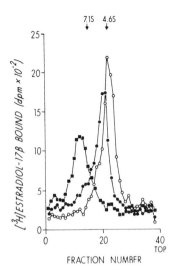

FIGURE 7 Effect of ionic strength on sedimentation properties of estrogen binding proteins. Constant volumes of cytosol from the mammary gland of a rat, 19 days postpartum, were incubated either with 5 nM [^{3}H]estradiol-17β alone or in the presence of 2.5 μM diethylstilbestrol for 4 hr at 0-3°C. After treatment with dextran-coated charcoal, 0.2 ml aliquots of the supernatant were layered onto 5-40% linear sucrose gradients containing either 0.0 M (■), 0.2 M (●), or 0.4 M KCl (○) and centrifuged at 308,000 × g for 16 hr. Only specific binding is illustrated. Human serum albumin (4.6S) and γ-globulin (7.1S) were used as marker proteins. (From Ref. 22.)

Earlier we reported the presence of a 4-5S binding species following dissociation of the 8S entity on sucrose gradients containing 0.4 M KCl [18]. A conservation of receptor-bound radioactivity was observed in the conversion of the 8S component into the 4-5S component. To determine the sedimentation properties of estrogen binding proteins in the presence of low, physiologic, and high concentrations of salt, [^{3}H]estradiol-receptor complexes were separated on sucrose gradients containing 0.0, 0.2, or 0.4 M KCl. Representative profiles were given in Figure 7. The hormone-binding entities which sedimented at 8S in the absence of salt shifted to components sedimenting at ~5S and ~4S in the presence of 0.2 M and 0.4 M KCl, respectively. A conservation of specifically bound radioactivity was observed.

We have proposed that the estrogen receptors of normal mammary gland and responsive mammary tumors are composed of two different subunits [30, 31]. This proposal was based on experiments using buffers of different ionic strengths to extract estrogen receptors from these tissues.

TABLE 6 Molecular Parameters of Estrogen Receptors Determined by Density Gradient Centrifugation and Gel Filtration Chromatography

Parameters	Charged (E • Rc)	Activated (E • R'c)	Nuclear (E • Rn)
Sedimentation coefficient	4.6 ± 0.1 S	4.6 ± 1 S	4.5 ± 0.1 S
Stokes radius	43.6 Å	37.6 Å	36.9 Å
Molecular weight	$87,000 \pm 2,100$	$76,000 \pm 1,800$	$72,000 \pm 1,700$
Fricational coefficient (f/fo)	1.37	1.24	1.23
Axial ratio			
Prolate	7.1	5.0	4.9
Oblate	8.0	5.4	5.2

When low salt was used, the estrogen receptor of lactating mammary gland migrated as a single component at 8S on linear gradients of sucrose. At ionic strengths of 0.15 M, the estrogen receptors sedimented as a 5.5S component, which appears to represent a dimer of two 4S components [32]. To determine if the 8S estrogen receptor in lactating mammary gland contained subunits with different ionic properties, ^{3}H-labeled-estradiol receptor complexes were separated on columns of DEAE-cellulose [33]. Following association with the ion exchange resin, two components binding ^{3}H-estradiol in a specific fashion were separated, suggesting the presence of subunits with different ionic properties. The molecular heterogeneity of the estrogen receptor has been confirmed through isoelectric focusing (Wheeler et al., in preparation). Table 6 summarizes the molecular parameters of estrogen receptors from lactating mammary gland.

K. Characterization of the Activation Process of Estrogen Receptors

Earlier we reported the association and translocation of estrogen receptors in vivo [18] and in slices of mammary gland [34]. To determine whether the temperature sensitivity of nuclear binding of estrogen receptors resided in cytosol, the following experiment was carried out. Cytoplasmic estrogen receptors charged with [^{3}H]estradiol at 0°C were either warmed at 28°C for 30 min or maintained at 0°C for an equivalent time. Time course studies of nuclear binding by the estradiol-receptor complexes in these cytosols were conducted at 0°C. It was found that prewarming of the cytosolic estrogen-receptor complexes caused a threefold increase in nuclear binding compared to that observed at 0°C (control). Also, nuclear association was more rapid when the estrogen receptors were prewarmed. In all similar experiments, incubation of [^{3}H]estradiol-receptor complexes at 28°C for 30 min consistently produced a 2.5 to sevenfold increase in receptor uptake by nuclei over the 0°C control level.

Hereafter, steps that facilitate receptors' interaction with nuclei will be described as the <u>activation</u> process. Cytosolic receptors that have been charged with saturating concentration of estradiol-17β at 0°C will be designated as <u>charged receptors</u>. On the other hand, cytosolic receptors that have been charged with hormones at 0°C and then warmed to facilitate their nuclear binding will be referred to as <u>temperature-activated receptors</u> or simply <u>activated receptors</u>.

Observations from earlier experiments [33, 34] indicated that the activation of estrogen receptors for translocation into nuclei was a temperature-sensitive step. To investigate further this temperature-dependent process, it was necessary to establish a rapid and reproducible method of assessing receptor activation. The use of isolated nuclei is complicated by (1) the requirement of considerable quantities of tissue, (2) lack of rapidity in the nuclear isolation procedure, and (3) poor reproducibility of nuclear uptake. Furthermore, only 50% of the activated estrogen receptors translocated into isolated nuclei were extracted with buffer containing KCl (Table 7). Thus we took advantage of the observation that the steroid-receptor complex binds to "naked" DNA [e.g., 35], as well as to DNA coupled to cellulose [e.g., 36]. Using DNA-cellulose binding, a rapid, sensitive and reproducible method was developed for assessing in vitro activation of estrogen receptors in mammary tissue [37].

Time course studies of estrogen-receptor association with DNA-cellulose were conducted to establish the conditions of the assay. Figure 8A shows that receptor incubated with [^{3}H]estradiol at 0°C (charged) associated poorly with DNA-cellulose. However, preincubation of the steroid-receptor complexes in cytosol at 28°C resulted in a markedly enhanced binding to DNA-cellulose at 0°C. This increase varied between three- and sevenfold above that of the control. Association of the activated hormone-receptor complexes reached a maximum within 10 to 20 min. Similar results were observed using estrogen receptors in cytosol from human breast carcinomas, except that the extent of increase upon activation was less (Fig. 8B).

Using the conditions of the DNA-cellulose binding assay established earlier, the temperature dependency of the activation process was examined more carefully. Charged estrogen receptors were incubated for 30 min at a series of temperatures varying between 0 and 35°C. The steroid-receptor complexes' ability to bind to DNA-cellulose was enhanced by elevated temperatures, reaching a maximum at 28°C. A decline was observed at temperatures greater than 30°C, presumably due to thermal denaturation or

TABLE 7 Extraction of Nuclear-Bound [^{3}H]Estradiol-Receptor Complexes with KCl[a]

Extraction medium	[^{3}H]estradiol-receptor complexes extracted (dpm/reaction)	Percentage of total extraction
Buffer alone	471 ± 69	6.8 ± 1.0
0.1 M KCl	741 ± 166	10.7 ± 2.4
0.3 M KCl	1795 ± 201	26.0 ± 2.9
0.5 M KCl	3325 ± 232	48.1 ± 3.4
1.0 M KCl	2329 ± 253	33.7 ± 3.7
Ethanol	6913 ± 97	100.0

[a] Cytosol prepared from rat mammary gland 18 days postpartum was charged with 10 nM [^{3}H]estradiol either alone or in the presence of 2 μM diethylstilbestrol for 2 hr at 0°C. The cytosol was then warmed at 28°C for 30 min, followed by cooling to 0°C. Aliquots of the charged and activated cytosol were incubated with purified nuclei (1.5 mg DNA/reaction) for 30 min at 0°C. After three washings, separate preparations of nuclear-bound [^{3}H]estradiol-receptor complexes were extracted for 30 min at 0°C with 0.4 ml of buffer containing KCl of the concentrations indicated. Values represent mean ± SEM of three separate determinations. The quantity of [^{3}H]estradiol-17β extracted with 99% ethanol was considered as the total.

dissociation of the steroid-receptor complexes. Similar results were observed using estrogen receptors from human breast cancer [33].

The time course of activation of estrogen receptors was determined by first charging the binding sites with [^{3}H]estradiol at 0°C and then incubating them at 28°C for increasing periods of time. The extent of activation increased with time, reaching a maximum at 20-30 min. Thus activation of estrogen receptor occurred as a function of both time and temperature. From these results, 30 min at 28°C was used in all further studies to bring about activation of charged estrogen receptors.

The effect of temperature-activation on the sedimentation profile of estrogen receptors from mammary gland was examined using estrogen receptors from uteri of rats as a control [37]. Figure 9 shows that warming of the uterine estrogen receptors for 30 min at 28°C in the presence of 1 M urea resulted in a shift of sedimentation behavior from ~4S to ~5S. In contrast with those of uterus, estrogen receptors from mammary glands, activated under identical conditions, did not exhibit an alteration in their sedimentation characteristics. Estrogen receptors of rat mammary gland,

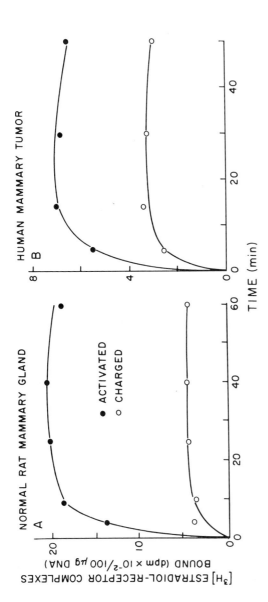

FIGURE 8 Time course of binding of [³H]estradiol-receptor complexes to DNA-cellulose. Cytosol was prepared from either rat mammary gland (A) or human mammary tumor (B) in 10 mM Tris · HCl buffer, pH 7.4, containing 1 mM EDTA. Cytosol was charged for 2 hr (4 hr for tumor cytosol) at 0°C with 10 nM [³H]estradiol-17β alone or in the presence of a 200-fold excess of unlabeled DES. One-half of the reactions containing charged cytosol were activated (●) by incubating at 28°C for 30 min (30°C for 45 min for tumor cytosol), while the others were kept at 0°C (○). Then all cytosols were incubated with DNA-cellulose for the times indicated at 0°C. (From J. L. Wittliff, W. M. Lewko, D. C. Park, T. E. Kute, D. T. Baker, Jr., and L. N. Kane, in Hormones, Receptors, and Breast Cancer (W. L. McGuire, ed.), vol. 10, p. 340, copyright 1978, by permission of Raven Press, New York.

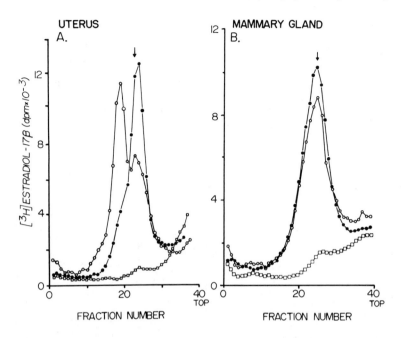

FIGURE 9 Influence of activation conditions on sedimentation profile of estrogen receptors in uterus (A) and mammary gland (B) of the rat. Uteri and mammary gland from 18-day lactating rats were homogenized separately in 2 and in 3 volumes (W/v), respectively, of 40 mM Tris • HCl buffer. Aliquots of the cytosols were charged with [^{3}H]estradiol for 2 hr at 0°C, and a portion was made 1 M with respect to urea and incubated at 28°C for 30 min, then cooled to 0°C (○) while others were incubated further at 0°C for 30 min in the absence of urea (●). Nonspecific binding (□) was measured in the presence of unlabeled diethylstilbestrol. The unbound [^{3}H]estradiol-17β in each reaction was adsorbed to dextran-coated charcoal pellets; then the cytosol preparations that did not receive urea earlier were made 1 M with respect to urea before separation by sucrose gradient configuration at 246,000 × g for 16 hr at 1-3°C. The arrows indicate the sedimentation of a 4.3S marker protein. (From Ref. 37.)

which either had been charged only or charged and activated in the presence of 1 M urea, sedimented as 4S species. Essentially the same sedimentation behavior was observed when estrogen receptors from mammary tissue were heat-activated in the absence of urea. However, when urea was omitted from the cytosol during the activation of uterine estrogen receptors, little or no alteration in sedimentation behavior was observed. Both activated

TABLE 8 Influence of Estrogen Receptors on RNA Polymerase Activities in Isolated Nuclei [a]

Receptor stage	DNA-dependent RNA polymerase activity (% of control)		
	$Mg^{2+} + Mn^{2+}$	Mn^{2+}	Mg^{2+}
Uncharged	100	100	100
Charged	136	123	135
Charged and activated	203	148	246

[a] Cytosol was prepared from mammary glands of rats 18-20 days postpartum and charged with 10 nM unlabeled estradiol at 0°C for 2 hr. Aliquots of the charged cytosol were either maintained at 0°C for 30 min or activated by incubation at 28°C for 30 min. Aliquots of intact nuclei from lactating mammary glands were incubated at 0°C for 30 min in the presence of either charged or activated estradiol-receptor complexes. After separation of nuclei from cytosol followed by washing, the RNA polymerase activity in each nuclear sample was assayed. Mg^{2+}- and Mn^{2+}-dependent RNA polymerase activities were measured by omitting either Mn^{2+} or Mg^{2+} from the assay mixture. Control values were obtained with cytosol which was not complexed with steroid (uncharged).
Source: Adapted from Ref. 33.

and charged estrogen receptors of uteri from adult, lactating rats also sedimented at ~4S when activation was carried out in the absence of urea.

To determine if the nuclear translocation of estradiol-receptor complexes observed in our studies had biological significance, the following experiment was carried out. Either charged or activated estradiol-receptor complexes prepared from mammary gland were incubated at 0°C for 30 min with isolated mammary gland nuclei. The effect of nuclear translocation of estrogen receptors on nuclear RNA polymerase activity was determined. Table 8 shows that preincubation of isolated nuclei from lactating mammary gland with charged estradiol-receptor complexes resulted in a small increase in either Mg^{2+}- or Mn^{2+}-dependent activities or total RNA polymerase activities. However, preincubation of the isolated nuclei with activated estradiol-receptor complexes gave a twofold increase in total RNA polymerase activities. The data in Table 8 also indicate that this increase was due mainly to an elevation of Mg^{2+}-dependent RNA polymerase activities, which presumably are responsible for the synthesis of stable RNA species.

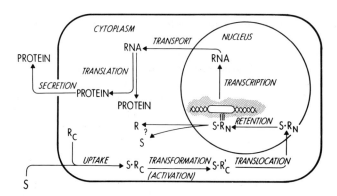

FIGURE 10 Proposed steps in the interaction of a steroid hormone with a breast cell. The cytoplasmic form of the receptor is designated as (R_c), the nuclear form as (R_N), and the steroid hormone as S. (From Ref. 19.)

Our current understanding of the sequence of events which follows the interaction of a steroid hormone with a target cell (Fig. 10) evolved from the original "two-step mechanism" suggested independently by Dr. Jack Gorski and co-workers [38], then at the University of Illinois, and by Dr. Elwood V. Jensen and colleagues [39] at the University of Chicago. These workers utilized uterine tissues from rodents for these early studies. Investigations from the authors' laboratory suggest that a similar cascade of events exists in normal and neoplastic mammary cells [19, 21, 30-34, 37, 40]. Steroid hormones are transported in the plasma compartment by a number of proteins including albumin, testosterone-estradiol-binding globulin, and corticosteroid-binding globulin, each with a characteristical affinity and capacity [cf. 41]. The unbound steroid enters the cell, apparently by passive diffusion, and combines with its specific receptor protein in a reaction termed uptake (Fig. 10). Before translocation into the nucleus, the steroid-receptor complex must undergo an activation step. After it enters the nucleus, the steroid hormone-receptor complex associates with the chromatin in an event called retention. This interaction stimulates RNA synthesis resulting in the formation of certain breast cell proteins. Thus the steroid receptor appears to be a prerequisite for responsiveness to hormonal perturbations; in its absence alterations in macromolecular synthesis do not occur at physiological hormone concentrations.

III. CHARACTERISTICS OF STEROID RECEPTORS IN EXPERIMENTAL MAMMARY TUMORS

In addition to steroid receptors of normal mammary gland, we have studied those in a variety of experimental mammary tumors of rodents that exhibit hormonal responsiveness. These investigations have been conducted with the assumption that these experimental tumors bear some resemblance to the human lesion. However, no single experimental mammary tumor exhibits all the properties of the carcinoma in man. This is not surprising since the disease is highly variable in a number of characteristics. These include the rates of proliferation and metastatic spread, cellular composition, and response to chemotherapy or endocrine manipulation among breast cancer patients of similar age and menopausal status. Thus we have selected the tumor model best fitted to our particular experimental question. The discussion of the properties of steroid receptors in tumors described in this section is not meant to be exhaustive, but rather to summarize briefly these properties in each model system.

A. NMU-Induced Mammary Tumor

N-nitrosomethylurea is an alkylating agent which is known to induce mammary adenocarcinomas in rats when administered by three intravenous injections of 50 mg/kg at monthly intervals [42]. Using Fischer 344 rats, we have found that these tumors appear with a mean latency period of approximately 160 days. Palpable tumors appeared in 70% of the animals treated in this manner. When animals with tumors are ovariectomized, 19% of the tumors continued to grow (ovarian independent), while 74% of the tumors regressed in size (ovarian dependent). The remainder of the tumors exhibited static growth.

Estradiol-receptor complexes from tumors of intact rats exhibited an apparent dissociation constant (K_d) of $5.2 \pm 0.5 \times 10^{-10}$M (mean ± SEM) and a binding capacity of 92 ± 13 fmol/mg cytosol protein. These receptors sedimented at 8S and 4S, but at 4S only in the presence of 0.4 M KCl using sucrose gradient centrifugation [33]. Binding sites exhibited high affinities for compounds with estrogenic activity. While ovarian-dependent tumors exhibited a binding capacity of 95 ± 14 fmol/mg protein, ovarian-independent

TABLE 9 Steroid Receptors in Cytosols of Mammary Tumors Induced by NMU

Growth response to ovariectomy	Specific steroid binding capacity [a] (fmol/mg cytosol protein)		
	Estradiol-17β	Triamcinolone acetonide	R5020
Dependent	95 ± 14	232 ± 40	100 ± 12
Independent	58 ± 9	321 ± 54	122 ± 17

[a] Determined from Scatchard analyses of titration data. Values represent mean ± SEM 7-14 determinations each.
Source: Adapted from Ref. 33.

tumors also contained 58 ± 9 fmol/mg cytosol protein (Table 9). Therefore, ovarian-dependent tumors contained a higher estrogen-binding capacity than ovarian-independent tumors.

When tumor slices were incubated with [^{3}H]estradiol in medium 199, estrogen receptors were differentially distributed between cytoplasm and nuclei of ovarian-dependent and ovarian-independent tumors. Furthermore, nuclear-bound receptors were extracted equally well with 0.4 M KCl. These results suggested that the estrogen receptor mechanism in ovarian-independent tumors was not defective with respect to the process of nuclear binding.

When triamcinolone acetonide was used as the labeled ligand, the glucocorticoid receptor exhibited a K_d of $1.3 \pm 0.2 \times 10^{-8}$ M and a binding capacity of 210 ± 23 fmol/mg cytosol protein in tumors from intact rats. Similar results were obtained with [^{3}H]dexamethasone. Glucocorticoid receptors sedimented at 8S, but shifted to 4S species in the presence of a 0.4 M KCl. Specificity studies showed that while synthetic and naturally occurring glucocorticoids exhibited high affinity, progestins were recognized by the receptor as well. When ovarian-dependent and independent tumors were compared, there were no statistical differences in glucocorticoid binding capacities or affinities (Table 9).

Progestin receptors, with R5020 as the labeled ligand, exhibited K_d values of $8.2 \pm 1.6 \times 10^{-9}$ M and 434 ± 45 fmol bond/mg cytosol protein in tumors from intact animals. A single component with a sedimentation coefficient of 4S was observed either in the presence of absence of KCl using sucrose gradient centrifugation. In studies of specificity, progestins associated with the receptor with high affinity; however, triamcinolone acetonide exhibited significant competition for this binding site as well.

There were no differences in progestin-binding capacities or affinities among ovarian-dependent and independent tumors (Table 9). Tumors from ovariectomized rats had a lower progestin-binding capacity than those of intact animals, suggesting an estrogen requirement for synthesis.

B. MCCLX Mammary Adenocarcinomas

MCCLX tumors are transplantable, prolactin-dependent mammary adenocarcinomas of rats [43]. These tumors exhibited a low estrogen-binding capacity of 31 fmol/mg protein (Table 10). In contrast, the binding capacity for R5020 was 382 ± 77 fmol/mg protein, and that for triamcinolone acetonide was 560 ± 85 fmol/mg protein. When dexamethasone was used as a ligand for the glucocorticoid receptor, a lower binding capacity was observed (Table 10). The affinities, specificities, and sedimentation properties of these receptors were similar to those of the NMU-induced tumors. These results suggest that this prolactin-dependent tumor, which exhibits low estrogen receptor levels and moderately high concentrations of progestin receptors, may contain a defective mechanism of estrogen-induced formation of progestin receptors.

C. Mammary Tumors of the C3H Mouse

Estrogenic compounds such as diethylstilbestrol administered orally at concentrations of 1000 ppb in feed enhanced mammary tumor frequency and decreased the latency period of mammary tumor appearance in C3H mice which were positive for murine mammary tumor virus [44]. Subsequent

TABLE 10 Steroid Binding Proteins in Cytosols of MCCLX Mammary Tumors

Steroid ligand	Dissociation constant (K_d) [a]	Specific binding capacity [a] (fmol/mg cytosol protein)
Estradiol-17β	8.9×10^{-10} M	31
Triamcinolone acetonide	9.4×10^{-9} M	560
Dexamethasone	1.2×10^{-8} M	278
R5020	4.1×10^{-9} M	382

[a] Determined by Scatchard analysis of titration data.
Source: Adapted from Ref. 62.

growth does not appear to be ovarian-dependent, although the induction of these tumors is estrogen-sensitive. Tumors from mice fed with 1000 ppb estrogen were analyzed for specific estrogen binding components 1 week after removal of the hormone-containing diet.

Mammary tumors of C3H mice which were fed a diet containing diethylstilbestrol until 1 week before sacrifice exhibited an estrogen-binding capacity of 16 ± 5 fmol/mg protein. Tumors from control mice bound 11 ± 5 fmol/mg protein. All of these mouse mammary tumors exhibited ovarian-independent growth in keeping with the low estrogen receptors. Mammary glands of virgin mice fed diethylstilbestrol bound 41 ± 8 fmol/mg cytosol protein, while those of control mice bound 24 ± 5 fmol/mg protein.

Mouse mammary tumors exhibited a glucocorticoid-binding capacity of 344 ± 48 fmol/mg protein in estrogen-fed mice, and 343 ± 14 in control mice. Glucocorticoid binding capacity of virgin mammary gland was 190 fmol/mg protein. Although a few mammary tumors exhibited 4S progestin receptors, the majority did not exhibit these proteins.

D. R3230AC Mammary Adenocarcinoma

The ovarian-independent R3230AC mammary tumor of the rat contains specific estrogen receptors in the cytoplasm [45]. The absence of estrogen-dependent growth has been attributed to a defect in the levels of cytoplasmic receptors necessary to produce significant nuclear binding [46]. However, we demonstrated the presence of nuclear bond [^{3}H]estradiol-17β following administration of the steroid in vivo [47], suggesting that the defect residues at a step after nuclear binding.

Although tumor growth and cytoplasmic estrogen-binding capacity [45] were not altered by ovariectomy, the R3230AC tumor may be classified as estrogen responsive since large doses of estrogen caused changes in its secretory state and in the activities of several enzymes [48]. The characteristics of specific estrogen binding components in the R3230AC were similar to those of the lactating mammary gland (Table 11).

In addition to these components, this tumor also contained specific glucocorticoid and progestin-binding proteins. The specific estrogen-binding components sedimented principally at 8S, while the glucocorticoid-binding proteins sedimented at approximately 7S. When tritiated progesterone was used as ligand, the progestin-binding component in the R3230AC mammary tumor sedimented as a single entity at approximately 4S. However, when [^{3}H]R5020 was used, two types of progestin binding components were observed using sucrose gradient centrifugation (Table 11).

E. DMBA-Induced Mammary Tumors

The Huggins mammary adenocarcinoma [49] induced in female rats by treatment with the cyclic hydrocarbon, dimethylbenzanthracene (DMBA),

TABLE 11 Specific Steroid Binding Proteins of the R3230AC Transplantable Mammary Adenocarcinoma of the Rat[a]

Receptor	Sedimentation coefficients (S)			Dissociation constant (M)	Number of binding sites (fmol/mg cytosol protein)
	Cytoplasm		Nucleus		
	Low salt	High salt			
Estradiol-17β	8-9	ND[b]	4-5	$1\text{-}4 \times 10^{-10}$	3-23
Triamcinolone acetonide	8 and 4.7	4.1	4-5	$1\text{-}6 \times 10^{-8}$	100-498
Progesterone	4.2	4.1	ND	$4\text{-}6 \times 10^{-8}$	154-663
R5020	4.3 and 3.5	ND	ND	$5\text{-}8 \times 10^{-9}$	106-387
Dihydrotestosterone	7-8	ND	ND	ND	17

[a] All experiments were conducted in the author's laboratory using cytosol preparation from tumors 18-24 days post-transplantation [45].
[b] Not determined.
Source: Data from Ref. 62.

TABLE 12 Specific Steroid Binding Proteins of the DMBA-Induced Mammary Tumors of the Rat[a]

Receptor	Sedimentation coefficients (S)			Dissociation constant (M)	Number of binding sites (fmol/mg cytosol protein)
	Cytoplasm		Nucleus		
	Low salt	High salt			
Estradiol-17β	8.4	4.3	4-5	$3-7 \times 10^{-10}$	5-76[b] 5-33[c]
Triamcinolone acetonide	7.1	4.6	4	$7-10 \times 10^{-9}$	56-508
Progesterone	4.1	4.0	ND[d]	$4-6 \times 10^{-8}$	70-426
R5020	4.3	4.3	ND	$3-7 \times 10^{-9}$	85-241

[a] All experiments were conducted in the author's laboratory using cytosol preparations from tumors induced as described in Ref. 47.
[b] Ovarian-dependent.
[c] Ovarian-independent; estrogen receptors were present in ~70% of tumors.
[d] Not determined.
Source: Data from Ref. 62.

has been used extensively as an animal model for the study of breast cancer. Most tumors exhibit ovarian-dependent growth; however, approximately 15% continue to proliferate after ovariectomy and may be classified ovarian- or estrogen-independent. It has been suggested that ovarian-independent tumors concentrated less labeled estrogen than ovarian-dependent tumors and possibly lack cytoplasmic and nuclear receptors specific for estrogen [50, 51]. However, we demonstrated the presence of intranuclear estrogen receptors sedimenting at 4-5S following the in vivo administration of labeled estradiol to tumor-bearing animals [52]. Furthermore, significant quantities of cytosolic estrogen-binding proteins were observed in DMBA-induced tumors exhibiting ovarian-independent growth (Table 12).

These data and our findings on the presence of estrogen-binding components in the ovarian-independent R3230AC mammary tumor [45, 47] cast doubt on the concept that there is an absolute correlation between estrogen receptors in a tumor and its growth response in ovariectomy. The presence of specific estrogen-binding proteins in cytosol does not necessarily indicate that tumor growth will be ovarian-dependent, while the absence of detectable estrogen binding in a tumor suggests that its growth will be ovarian-independent.

In summary, these data suggested that many experimental mammary tumors which exhibit ovarian-independent growth contained receptors for both estrogens and progestins. Although there was a direct relationship between the levels of estrogen and progestin receptors in ovarian-dependent NMU-induced tumors, this relationship was not observed in tumors exhibiting ovarian-independent growth. Furthermore, the presence of cytoplasmic estrogen receptors in ovarian-independent tumors, induced either by DMBA or by NMU which undergo nuclear translocation, suggests there may be a defect in the hormone response mechanism at the level of chromatin interaction. Thus the presence of progestin receptors in a tumor biopsy may <u>not</u> indicate that the estrogen response mechanism is intact.

IV. ESTROGEN RECEPTORS IN HUMAN BREAST CARCINOMA AND THEIR CLINICAL SIGNIFICANCE

The molecular basis of successful endocrine therapy appears to be related to the fact that certain tumors have retained the cellular mechanism to respond to hormonal stimuli. The presence of estrogen receptors in a tumor is known to provide a molecular basis for distinguishing hormonally responsive breast carcinomas from those that are not [e.g., 53-55].

Steroid receptors have been estimated in a variety of hormone target organs using methods ranging from in vivo administration of labeled steroid to in vitro procedures [e.g., 19, 56]. Two methods are used routinely for the determination of estrogen and progesterone receptors in human breast carcinomas. The titration procedure (Fig. 11), which utilizes dextran-coated charcoal to remove the unbound steroid from that associated with

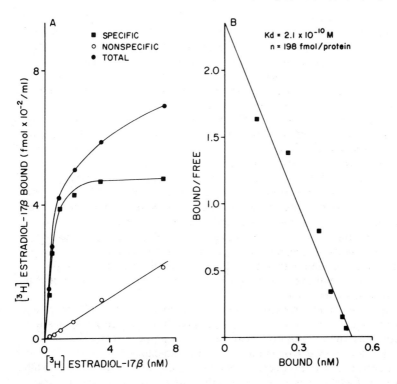

FIGURE 11 Titration analysis of estrogen receptors in human breast carcinoma. Aliquots (0.1 ml) of cytosol prepared from a frozen powder of human breast tumors were incubated in triplicate with 0.1 ml of [^{3}H]-estradiol-17β solutions in homogenization buffer containing increasing amounts of radioactive ligand in the absence (●) or presence (○) of a 200-fold excess of diethylstilbestrol (A). Specific binding (■) was estimated as the difference between total binding and binding in the presence of the competitor. Titration data from (A) were plotted according to the method of Scatchard (B). The dissociation constant (K_d) determined from the slope of the curve was 2.1×10^{-10} for this preparation. The binding capacity of the estrogen-receptor complexes, estimated from the intercept on the abscissa, was 198 fmol/mg of cytosol protein. (From J. L. Wittliff, P. H. Heidemann, and W. M. Lewko, in <u>The Clinical Chemistry of Cancer</u> (M. Fleisher, ed.), p. 196, 1979, by permission of the American Association of Clinical Chemistry, Washington, D.C.)

the intracellular receptor, provides a measure of the binding capacity and the affinity. Usually binding capacity is expressed as femtomoles of ^{3}H-labeled steroid bound per mg cytosol protein. Generally, it is accepted that less than 3 fmol/mg cytosol protein represents a quantity of estrogen-binding sites usually correlated with the lack of response by a breast cancer patient given endocrine therapy [cf. 55]. Although there appears to be a borderline range of values from 3 to 10 fmol/mg cytosol protein, estrogen-binding capacities >10 fmol of estrogen bound apparently represent a "receptor-positive" tumor. Estrogen-binding capacities ranging from 3 to 2500 fmol/mg cytosol protein have been observed in the authors' laboratory. We recommend that quantitative values of both estrogen and progestin receptors as well as affinity data be reported in order to develop the relationship between the specific binding capacity of a tumor biopsy and a patient's response to endocrine manipulation.

A typical titration curve of estrogen binding sites in a tumor biopsy is seen in Figure 11. To determine specific binding the analyses should be performed in the presence of a competitive inhibitor of the association of ^{3}H-ligand with the receptor. Routinely unlabeled diethylstilbestrol is used in a 200-fold excess since this competitor associates with the estrogen receptor with a $K_d \sim 10^{-9}$ M [40]. This compound has a low affinity for sex-steroid binding proteins of plasma, which may contaminate cytosol preparations for human breast tumors. Generally, the dissociation constant of estrogen-receptor complexes ranges from 10^{-10} to 10^{-11} M, indicating a high affinity. Our experience indicates that the titration assay is equally useful for the clinical determination of receptors for progestins, glucocorticoids, and androgens in human breast tumors, as long as consideration is given to the ^{3}H-ligand.

The sucrose gradient method, which separates the various forms of the steroid receptors, assesses certain molecular properties of those proteins in a tumor extract (Fig. 12). With this method it has been determined that the sedimentation profiles of both estrogen and progestin receptors in human breast carcinomas fall into four general categories [15, 54, 57, 58]. These are tumors which contain specific steroid-binding components migrating at either 8S, 4S, or both 8S and 4S (Fig. 12), and those in which receptors are undetectable. A more accurate estimation of the sedimentation coefficients of these molecular forms has been made using a number of marker proteins.

From these data the higher molecular weight species of the estrogen receptor sedimented at 7.6-8.1S, while the lower molecular weight form sedimented at 4.0-4.6S under conditions of low ionic strength [40]. Occasionally we have observed specific estrogen binding components sedimenting at <4S under these same conditions. The sucrose gradient method may be used to quantitate the steroid binding capacity of human breast tumors. When a saturating concentration of ^{3}H-ligand as well as an excess of a competitor are used in this assay, the estimation of specific binding capacity is in good agreement with that determined by the titration analyses [59].

A long-term study was initiated in our laboratory in 1969 to examine the original hypothesis set forth by Jensen and colleagues [11, 15] that the presence of specific estrogen-binding components in breast carcinomas was predictive of a patient's response to endocrine therapy. Using sucrose gradient separation of specific estrogen binding proteins in cytosol fractions of biopsies of primary and metastatic breast carcinomas, we demonstrated

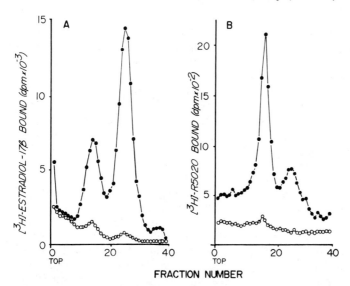

FIGURE 12 Sucrose gradient separation of estrogen and progestin (R5020) receptors in a human breast carcinoma. Tumor cytosol was reacted either with [^{3}H]estradiol (A) or with [^{3}H]R5020 (B) for 4 hr at 3°C in the presence (○) or absence (●) of a 200-fold excess of unlabeled competitor (A, diethylstilbestrol, DES; B, R5020). The double peaks indicate the presence of both 8S and 4S forms of these steroid receptors in the single breast carcinoma analyzed. (From J. L. Wittliff, P. H. Heidemann, and W. M. Lewko, in <u>The Clinical Chemistry of Cancer</u> (M. Fleisher, ed.), p. 198, 1979, by permission of the American Association of Clinical Chemistry, Washington, D.C.)

four types of receptor profiles, as discussed earlier. Most breast tumors containing estrogen receptors exhibited either the 8S species alone or both the 8S and 4S forms (Fig. 12). From 10 to 15% of breast carcinomas contained only the 4S type of estrogen receptor using sucrose gradients of low ionic strength [30, 31]. The ligand-binding specificities and affinities of the 8S and 4S species of estrogen receptors were similar in spite of the differences in sedimentation properties. The fourth category, comprising 50% of the infiltrating ductal carcinomas of the breast, included those that did not contain any type of estrogen-binding component.

Table 13 summarizes our results in relating a breast cancer patient's response to endocrine therapy to the presence of specific forms of the estrogen receptors in breast tumor biopsies. No objective remissions have been observed in breast cancer patients who had tumor biopsies which were estrogen-receptor negative, regardless of the type of hormone therapy administered [33]. This correlation is far better than that reported for the collective results presented at the International Workshop in 1974 [cf. 55]. Objective remissions were documented in approximately 75% of the patients who were given various types of hormone therapy in which the biopsies contained either the 8S species alone or both the 8S and 4S forms of estrogen receptors. As the footnote to Table 13 indicates, nine additional patients whose tumors were examined only by titration analysis also exhibited remissions to endocrine therapy; the molecular forms of estrogen receptors in these specimens were not distinguished. Of 23 patients whose tumors contained exclusively or predominantly the 4-5S forms of estrogen receptors and who were administered hormone therapy, only four responded objectively. These data suggest that these molecular forms of estrogen receptors in human breast cancer have clinical significance.

Our earlier results clearly showed that the 8S estrogen receptor of lactating mammary gland was composed of at least two components, each with different ionic properties as measured by DEAE-cellulose chromatography [33]. Unlike normal mammary cells, the 8S estrogen receptor human breast tumors separated into a variable number of components, each with different ionic properties [60]. Furthermore, the 4S estrogen receptor of human breast tumors also separated into at least two components, each with different ionic properties. These results indicate that the estrogen receptors of human breast cancers exhibit molecular heterogeneity. We have confirmed this molecular heterogeneity using isoelectric focusing (Wheeler et al., in preparation).

These data also suggest that the molecular properties of estrogen receptors in human breast biopsies may be related to the clinical

TABLE 13 Relationship between Patient Response to Endocrine Therapy and Presence of Estrogen Receptors in Tumor Biopsy

Therapy[a]	Objective remissions according to estrogen receptor species in tumor		
	8S, 8S, and 4S	4S	Undetectable
Hormone administration	16/23	3/15	0/33
Endocrine organ ablation	17/21	1/8	0/11
	33/44	4/23	0/44

[a] Nine additional patients exhibited remissions but had unclassified ER-positive tumors; five responded to estrogen, three to androgen therapy, and one to oophorectomy.
Source: Adapted from Ref. 33.

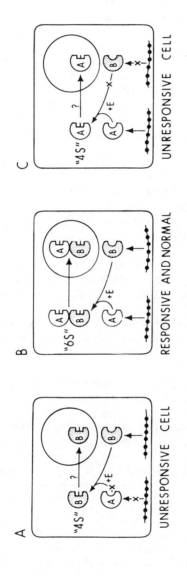

FIGURE 13 Proposed explanation of unresponsiveness to hormone therapy by breast cancer patients exhibiting estrogen-receptor-positive tumors. The subunits of the estrogen receptor in normal and responsive breast cells (B) are depicted as separate entities that sediment at 4S. In the presence of estradiol under physiological conditions, the subunits presumably combine to form a 6S dimer, which then translocates into the nucleus and stimulates a cellular response. This 6S form of the receptor has been observed in cytosols from human breast tumors [30]. In unresponsive cells (A and C), two types of defects, shown by the Xs, are postulated, either of which would result in detection of 4S components even under conditions of low ionic strength. (From Refs. 30 and 31.)

responsiveness of patients treated by hormonal manipulation. Earlier we proposed a cellular mechanism for this unresponsiveness [30, 31] to hormonal manipulation (Fig. 13). Allegra et al. [61] suggested that breast cancer patients with estrogen-receptor-<u>negative</u> tumors, in contrast to endocrine-responsive tumors, showed an increased response rate to cytotoxic chemotherapy. Thus both the quality (integrity) and quantity (number of binding sites) of estrogen receptors appear to be important predictive indices when the oncologist is faced with selecting a therapy likely to produce objective remission.

V. SUMMARY

Growth and differentiation of the mammary epithelium are regulated by a number of steroid hormones, particularly the estrogens, progestins, and glucocorticoids in concert with insulin, prolactin, and possibly growth hormone. The manner in which these hormones signal regulatory events appears to involve a class of proteins termed <u>receptors.</u> These receptors are found in small quantities (500-5000 sites per cell) in target tissues. They associate with their characteristic steroid hormones in reversible reactions exhibiting high ligand affinity and specificity, and they appear to be cellular prerequisites for a biological response to a hormonal perturbation.

Although steroid receptors are normal cell traits, they have been identified and characterized in a number of experimental mammary tumors of rodents. Often these tumors will exhibit quantities and properties of steroid receptors different from those of normal mammary gland. It appears their presence in tumor cells is related to the level of responsiveness or dependence on the steroid hormones.

Certain human breast carcinomas also contain receptors for these steroid hormones. Approximately 60% of primary breast tumors and 50% of metastatic breast tumors exhibit estrogen receptors, while approximately one-half of all breast carcinomas contain progestin receptors. There is a positive correlation between the presence of these receptors in a tumor and an objective clinical response by a patient treated by endocrine manipulation. More recent evidence suggests that both the quantity of the steroid binding sites and the integrity of the receptor protein itself are related to a breast cancer patient's response to hormone therapy.

ACKNOWLEDGMENTS

Recent investigations from the authors' laboratory have been supported in part by USPHS Grant CA-23848 and CA-12836 from the National Cancer Institute, a grant from the American Cancer Society, and funds from the National Surgical Adjuvant Breast Project (NCI-CB-23876) and the Southeastern Cancer Study Group (CA-19657).

The authors wish to acknowledge the valuable contributions of Drs. J. G. Leinen, T. E. Kute, P. H. Heidemann, and P. A. Wheeler-Feldhoff to our research efforts. The personal encouragement and support of our work by Mitzie Wittliff are deeply appreciated.

REFERENCES

1. W. R. Lyons, C. H. Li, and R. E. Johnson, Recent Prog. Horm. Res. 14:219-254 (1958).
2. S. Nandi and H. A. Bern, Gen. Comp. Endocrinol. 1:195-210 (1961).
3. H. Salazar and H. Tobon, in Lactogenic Hormones, Fetal Nutrition, and Lactation (J. B. Josimovich, M. Reynolds, and E. Cobo, eds.), vol. 2, John Wiley and Sons, New York, 1974, pp. 221-277.
4. R. D. Anderson, in Lactation: A Comprehensive Thesis. I. The Mammary Gland/Development and Maintenance (B. L. Larson and V. R. Smith, eds.), Academic Press, New York-London, 1974.
5. R. Caldeyro-Barcia, in Lactogenesis—Initiation of Milk Secretion at Partuition (M. Reynolds and S. J. Holley, eds.), University of Pennsylvania Press, Philadelphia, 1969, pp. 229-243.
6. B. A. Cross and G. W. Harris, J. Endocrinol. 8:148-161 (1952).
7. R. F. Glascock and W. G. Hoekstra, Biochem. J. 72:673-682 (1959).
8. P. J. Folca, R. F. Glascock, and W. T. Irvine, Lancet 2:796-798 (1961).
9. F. Bresciani and G. A. Puca, Atti. Soc. Ital. Pathol. 9:661-664 (1965).
10. R. J. B. King, J. Gordon, D. M. Cowan, and D. R. Inman, J. Endocrinol. 36:139-150 (1966).
11. E. V. Jensen, E. R. DeSombre, and P. R. Jungblut, in Endogenous Factors Influencing Host Tumors Balance (R. W. Wissler, T. L. Dao, and S. Wood, Jr., eds.), University of Chicago Press, Chicago, 1967.
12. B. G. Mobbs, J. Endocrinol. 41:339-344 (1968).
13. S. Sander, Acta Endocrinol. 58:49-56 (1968).
14. G. A. Puca and F. Bresciani, Endocrinology 85:1-10 (1969).
15. E. V. Jensen, G. E. Block, S. Smith, K. Kyser, and E. R. DeSombre, Natl. Cancer Inst. Monogr. 34:55-79 (1971).
16. R. Hilf, J. L. Wittliff, W. D. Rector, Jr., E. D. Savlov, T. C. Hall, and R. A. Orlando, Cancer Res. 33:2054-2062 (1973).
17. J. L. Wittliff, R. Hilf, W. F. Brooks, Jr., E. D. Savlov, T. C. Hall, and R. A. Orlando, Cancer Res. 32:1983-1992 (1972).
18. D. G. Gardner and J. L. Wittliff, Biochemistry 12:3090-3096 (1973).
19. J. L. Wittliff, in Methods in Cancer Research (H. Busch, ed.), vol. XI, Academic Press, New York, 1975, pp. 293-354.
20. J. E. Goral and J. L. Wittliff, Biochemistry 14:2944-2952 (1975).

21. J. L. Wittliff, R. G. Mehta, P. A. Boyd, and J. E. Goral, J. Toxicol. Environ. Health Suppl. 1:231-256 (1976).
22. P. A. Boyd, Kinetic and molecular properties of estrogen and glucocorticoid binding proteins in the lactating mammary gland of the rat, Ph.D. Dissertation, University of Rochester, New York, 1976.
23. G. Scatchard, Ann. N.Y. Acad. Sci. 51:660-672 (1949).
24. W. B. Pratt and D. N. Ishii, Biochemistry 11:1401-1410 (1972).
25. D. Rodbard, in Receptors for Reproductive Hormones (B. W. O'Malley and A. R. Means, eds.), vol. 36, Plenum Press, New York, 1973, pp. 289-364.
26. P. A. Boyd and J. L. Wittliff, J. Toxicol. Environ. Health 4:1-8 (1978).
27. D. L. Greenman, R. G. Mehta, and J. L. Wittliff, J. Toxicol. Environ. Health 5:593-598 (1979).
28. C. J. Mirocha, C. M. Christensen, and G. H. Nelson, in Microbial Toxins (S. Kadis, A. Ciegler, and S. J. Aji, eds.), vol. VII, Academic Press, New York, 1971, pp. 107-138.
29. M. Hospital, B. Busetta, R. Bucourt, H. Weintraub, and E. E. Baulieu, Mol. Pharmacol. 8:438-445 (1972).
30. J. L. Wittliff, B. W. Beatty, E. D. Savlov, W. B. Patterson, and R. A. Cooper, Jr., in Recent Results in Cancer Research (G. St. Arneault, P. Brand, and L. Israël, eds.), vol. 57, Springer-Verlag, Berlin, 1976, pp. 59-77.
31. J. L. Wittliff, B. W. Beatty, D. T. Baker, Jr., E. D. Savlov, and R. A. Cooper, Jr., in Research on Steroids (A. Vermeulen, A. Klopper, F. Sciarra, P. Jungblut, and L. Lerner, eds.), vol. VII, Elsevier/North-Holland Biomedical Press, Amsterdam, 1977, pp. 393-403.
32. A. C. Notides, D. E. Hamilton, and H. E. Auer, J. Biol. Chem. 250:3945 (1975).
33. J. L. Wittliff, W. M. Lewko, D. C. Park, T. E. Kute, D. T. Baker, Jr., and L. N. Kane, in Hormones, Receptors, and Breast Cancer (W. L. McGuire, ed.), vol. 10, Raven Press, New York, 1978, pp. 325-359.
34. J. L. Wittliff, R. G. Mehta, P. A. Boyd, and J. E. Goral, J. Toxicol. Environ. Health Suppl. 1:231-256 (1976).
35. T. A. Musliner and G. J. Chader, Biochim. Biophys. Acta 262:256-263 (1972).
36. L. E. Clemens and L. J. Kleinsmith, Nature (New Biol.) 237:204-206 (1972).
37. D. C. Park and J. L. Wittliff, Biochem. Biophys. Res. Commun. 78:251-258 (1977).
38. J. Gorski, D. Toft, G. Shyamala, D. Smith, and A. Notides, Recent Progr. Horm. Res. 24:45-80 (1968).

39. E. V. Jensen, T. Suzuki, T. Kawashima, W. E. Stumpf, P. W. Jungblut, and E. R. DeSombre, Proc. Natl. Acad. Sci. USA, 59:632-638 (1968).
40. J. L. Wittliff, P. H. Heidemann, and W. M. Lewko, in The Clinical Chemistry of Cancer (M. Fleisher, ed.), American Association of Clinical Chemistry, Washington, D.C., 1979, pp. 179-212.
41. U. Westphal, in Monographs on Endocrinology (F. Gross, A. Labhart, T. Mann, L. T. Samuels, and J. Sander, eds.), vol. 4, Springer-Verlag, Berlin-Heidelberg, 1971.
42. P. M. Gullino, H. M. Pettigrew, and F. H. Grantham, J. Natl. Cancer Inst. 54:401 (1975).
43. T. Kano-Sueoka and P. Hsieh, Proc. Natl. Acad. Sci. USA 70:1922 (1973).
44. G. H. Gass, D. Coats, and N. Graham, J. Natl. Cancer Inst. 33:971 (1964).
45. J. L. Wittliff, D. G. Gardner, W. L. Battema, and P. J. Gilbert, Biochem. Biophys. Res. Commun. 48:119-125 (1972).
46. W. L. McGuire, K. Huff, A. Jennings, and G. C. Chamness, Science 175:335 (1972).
47. E. S. Boylan and J. L. Wittliff, Cancer Res. 33:2903-2908 (1973).
48. R. Hilf, I. Michel, C. Bell, J. Freeman, and A. Borman, Cancer Res. 25:286 (1965).
49. C. Huggins, L. C. Grand, and F. P. Brillantes, Nature 189:204 (1961).
50. K. A. Kyser, The tissue subcellular and molecular binding of estradiol to dimethylbenzanthracene-induced rat mammary tumors, Ph.D. Dissertation, The University of Chicago, Illinois, 1970.
51. W. L. McGuire and J. Julian, Cancer Res. 31:1440 (1971).
52. E. S. Boylan and J. L. Wittliff, Cancer Res. 35:506-511 (1975).
53. E. V. Jensen, G. E. Block, S. Smith, K. Kyser, and E. R. DeSombre, Natl. Cancer Inst. Monogr. 34:55 (1971).
54. J. L. Wittliff, Semin. Oncol. 1:109-119 (1974).
55. W. L. McGuire, P. O. Carbone, and E. P. Vollmer (eds.), Estrogen Receptors in Human Breast Cancer, Raven Press, New York, 1975.
56. P. J. Folca, R. F. Glascock, and W. T. Irvine, Lancet 2:796-789 (1961).
57. K. B. Korwitz and W. L. McGuire, Steroids 25:497-505 (1975).
58. J. L. Wittliff, R. G. Mehta, and T. E. Kute, in Receptors in Normal and Neoplastic Tissues (W. L. McGuire, J. P. Raynaud, and E. E. Baulieu, eds.), Raven Press, New York, 1977, pp. 39-57.
59. J. L. Wittliff and E. D. Savlov, in Estrogen Receptors in Human Breast Cancer (W. L. McGuire, P. P. Carbone, and E. P. Vollmer, eds.), Raven Press, New York, 1975, pp. 73-91.

60. T. E. Kute, P. Heidemann, and J. L. Wittliff, Cancer Res. 38:4307-4313 (1978).
61. J. C. Allegra, M. E. Lippman, E. B. Thompson, and R. Simon, Cancer Res. 38:4299-4307 (1978).
62. J. L. Wittliff, Mol. Aspects. Med. 2:395-437 (1979).

9

ISOFERRITINS: ALTERATION IN HUMAN
MALIGNANCY AND CLINICAL SIGNIFICANCE

ELLIOT ALPERT* / GI Unit, Harvard Medical School, Boston
Massachusetts, and Medical Services, Massachusetts General
Hospital, Boston, Massachusetts

I. Introduction   229
II. Structure   229
III. Isoferritins   230
IV. Tumor Isoferritins   231
V. Serum Ferritin   233
VI. Serum Ferritin in Cancer   233
VII. Conclusion   235
References   235

I. INTRODUCTION

Ferritin is an iron-storage protein that is probably present in all eukaryotic cells. It is particularly abundant in liver, spleen, and bone marrow, where it plays a major role in iron metabolism and storage [1]. The molecule exists in multiple forms which we have called isoferritins, whose distribution changes with embryological development and in various pathological and malignant states [2-4]. The findings of a carcinofetal form of ferritin [2-4] and of elevations of serum ferritin in the sera of some cancer patients [5-7] have raised the possibility that ferritin may be useful as a tumor marker.

II. STRUCTURE

Ferritin consists of a spherical protein shell, 120-140 Å in diameter but only 30-40 Å thick, enclosing a core that can be as large as 65 Å. On electron microscopy, the molecule resembles a small icosahedral virus [1].

---

*Present affiliation: Department of Medicine, Baylor College of Medicine, Houston, Texas.

Inorganic iron is sequestered within the core of the molecule, which may contain up to approximately 4000 atoms of iron per molecule, present as a hydrous Fe (III) oxide phosphate complex. The iron can be removed from the core by reduction and inserted under oxidative conditions. Tiny channels visualized by x-ray diffraction have been thought to provide a structural basis for the ability of iron to traverse the protein shell without altering the protein [1]. The mechanism of iron deposition and removal from the intact shell is still largely undefined and under intensive investigation.

The molecular weight of the ferritin monomer ranges from 440,000 to over 650,000, depending on its source and iron content. The protein shell is highly symmetrical according to x-ray diffraction studies and consists of 24 subunits [1, 8, 9]. The amino acid composition is not unusual and shows slight differences between species [10], and its N-terminal amino acid appears to be an acetylated serine [11, 12]. These characteristics of ferritin appear to be common to all ferritins, regardless of species or tissue origin, and are probably the structural requirements necessary for its function.

III. ISOFERRITINS

The first suggestion of multiple molecular forms of ferritin was made in 1965 by Richter, who noted that a rat tumor ferritin had a more rapid electrophoretic mobility than normal tissue ferritin [13]. However, in most electrophoretic systems, the monomeric ferritin protein shells run as a broad single band. The marked resolution afforded by the new technique of isoelectric focusing demonstrated a great complexity in all monomeric ferritin preparations [14-20]. Most electrophoretically pure and homogeneous ferritins can be resolved into 6 to 10 distinct protein bands by analytic isoelectric focusing. The individual isoferritins, when separated and refocused, do not redistribute [2], and similar isoferritins can be resolved by other techniques such as ion exchange chromatography [16, 21]. The heterogeneity is not due to differences in the internal iron content or to polymeric forms. Each tissue has its characteristic isoferritin composition [20] which can change with iron loading [19, 22] or with malignancy [2, 14, 23].

Although once thought to be identical in all tissues, it is clear that ferritins from different tissues, which have different isoferritin profiles, are chemically different [20]. Ferritins from different tissues have a different amino acid composition and different tryptic peptide patterns [10, 19], and quantitative immunological differences have been shown between heart and liver ferritins [24]. The basis for the different tissue ferritins is uncertain. Some feel that each tissue has its own genetically distinct tissue-specific subunit, with the ferritin molecule comprised of a homopolymer of the unique subunit [1, 9, 10]. Others feel that the different tissue isoferritins are composed of different proportions of the same two genetically distinct subunits [25-27].

A variety of postsynthetic modifications, such as amidation, proteolytic cleavage, and conformational changes with interchain disulfide bonding, have all been proposed to explain the multiplicity of banding and the isoferrins described as an artifact [8, 28]. To complicate the problem further, small amounts of carbohydrate have been found recently in horse spleen ferritin [29]. However, increasing evidence has supported the concept of at least two distinct subunits. Two subunits can be demonstrated in all tissue ferritins by electrophoresis in acid urea or by gradient gel SDS electrophoresis [25-27]; these subunits have molecular weights of 19,000 and 21,000, respectively. Amino acid compositions recently performed on the two separated subunits shows small but distinct amino acid and tryptic peptide differences [30]. Therefore it appears that the isoferritins of different charge are composed of different proportions of two subunits. It is not yet clear whether the same two subunits account for all the isoferritin profiles in normal and pathological states in all tissues, or whether some tissues or tumors may contain additional unique subunits [26].

Despite the continued controversy concerning the structural basis for the isoferritin heterogeneity, it is clear that isoferritin profiles vary in biologically different situations. First, in both iron-overloaded rats and in humans with hemochromatosis [19] or prolonged hemosiderosis [22], the isoferritin profile shifts to the more basic end of the isoferritin spectrum. This shift occurs particularly in heart, kidney, and pancreas, which typically contain a high proportion of the more acidic isoferritins [20]. Changes are also reflected in the relative proportions of the two subunits. Indeed, studies in rats have shown preferential synthesis of one of the two subunits in response to increased iron concentrations [30]. Moreover, in hemochromatotic patients, whose excess tissue iron has been removed by venesection, the isoferritin patterns change, becoming similar to those in normal subjects [31].

Second, isoferritin profiles are altered during embryological development. Early fetal livers have more acidic isoferritin forms than found in normal adult liver, or even late gestational liver [2]. A similar change has also been reported in early placental tissue ferritin near term [32].

The third biological state where isoferritin forms are altered is malignancy.

IV. TUMOR ISOFERRITINS

Richer first showed in 1965 that ferritin from malignant HeLa cells in culture differed electrophoretically from normal adult rat liver ferritin [13]. A number of studies have confirmed the observation of electrophoretic variants of ferritin in tumor tissues [3, 33, 34]. Buffe and Rimbaut described an iron-containing glycoprotein purified from hepatoma tissue which they termed alpha$_2$ H globulin [35-37]. They found this α_2H globulin in sera of a large number of cancer patients as well as in patients with some benign diseases, but not in normal adult sera. They also found α_2H globulin in tumor and fetal tissue. We were able to identify this α_2H protein as ferritin, both immunologically and chemically [38].

Tissue ferritins from rat liver tumors were found to be separable into multiple isoferritins by electrofocusing, and ferritin from chemically induced rat hepatomas was found to contain more acidic isoferritins than found in normal liver [14, 23]. We were able to find a similar phenomenon in human hepatoma ferritin, isolated from human liver carcinoma, and found an identical profile in early fetal liver ferritin with more acidic isoferritins not seen in normal adult liver [2]. Subsequantly, similar acidic isoferritins were found in early placenta in HeLa cells [32] and in pancreatic and mammary carcinoma ferritin [39]. Since we found the same acidic isoferritins in fetal tissue, we called these altered ferritins carcinofetal isoferritins [2].

Tumor tissue ferritins are probably a mixture of ferritins from malignant cells as well as normal cells, mesenchymal cells, and inflammatory cells, all of which contain ferritin. It is of interest, therefore, that although the tumor tissue isoferritins overlap with the normal tissue isoferritins, ferritin derived from HeLa cells, and a human hepatoma cell line, both contained only the acidic isoferritins [32, 40, 41]. These data suggest that the acidic isoferritins are indeed synthesized by the malignant cell and are not due to inflammatory cells, or to proteolytic cleavage by lysozomal enzymes in the necrotic tumor tissue.

The structural basis for the tumor isoferritins remains unclear. It is still uncertain whether the carcinofetal acidic isoferritins are indeed unique to cancer and fetal tissues, or whether they are identical to acidic isoferritins, with a similar pI, found in certain normal tissues such as heart and kidney [20]. The carcinofetal isoferritins may arise from increased synthesis of one of the two normal subunits or from the insertion into the protein shell of a unique tumor-specific subunit.

In addition to the above biochemical evidence for an alteration in the structure of ferritin with malignancy, immunological differences between different isoferritins have become apparent. It was previously thought that isoferritins from different tissues of man were immunologically identical by immunodiffusion with fused lines of immunological identity [20]. However, quantitative and qualitative differences between tissue ferritins have been demonstrated by more sensitive competitive inhibition radioimmunoassays. Immunization with the acidic HeLa or heart ferritins induces antibodies which appear to recognize preferentially the acidic isoferritins in normal heart and in HeLa cells [24, 42, 43]. It is not clear whether the immunologically different antigenic determinants arise from an alteration in primary structure, from possible post-translational glycosylation, or from changes in conformational structure. It is also not yet clear whether all tumor isoferritins are similar antigenically.

V. SERUM FERRITIN

In 1956 Reissman and Dietrich first found ferritin in the serum of patients with a variety of inflammatory diseases, suggesting that ferritin in serum was due to tissue damage [44]. Recently, the development of sensitive immunoradiometric or competitive inhibition radioimmunoassays have shown, however, that ferritin is present in small amounts in all normal human sera [45-47]. Therefore one must now include ferritin as one of the normal beta globulins in sera.

The serum concentration in normal adults ranges between 10 and 150 ng/ml. The mean serum concentration in normal males is somewhat greater than in females, and the serum ferritin concentration appears to be an excellent measure of total body iron stores [45-47]. The serum level is clearly low in iron deficiency and markedly elevated in primary or secondary iron-overloaded states. Although intracellular ferritin synthesis is induced by iron [48], it is not yet clear whether the change in its serum concentration with iron storage is due solely to increased synthesis, or to changes in clearance. The source of normal serum ferritin is unclear, but it is thought to arise largely from reticuloendothelial cells and to be cleared by the liver. The clearance rate of exogenously labeled rat liver ferratin in rats is extremely rapid, in the order of 5-10 min, suggesting an enormously high turnover rate of serum ferritin [49]. Also, iron-rich isoferritins appear to be cleared more rapidly than iron-poor isoferritins [50].

It is of great interest and significance that the high ferritin levels found in the serum of patients with increased body Fe stores and in iron-overloaded states, contains little or no iron and appears to correspond to the more basic isoferritins normally found in liver and spleen [51, 52]. In addition, it has recently been shown that the serum isoferritin profile can change with alterations of body Fe stores [31], suggesting that the isoferritins may have a functional role related to iron transport or metabolism. Recent data have suggested that normal serum ferritin is glycosylated and able preferentially to bind to conconavalin A (Con A) in contrast to normal tissue ferritin [53]. The biochemical structure of serum ferritin in normal humans is still unclear, although it appears to be iron-poor [51, 52]. The site of synthesis and the physiological role of serum ferritin in iron storage remain unknown and are the subjects of active interest and investigation.

VI. SERUM FERRITIN IN CANCER

Elevated levels of serum ferritin occur in a variety of malignant and non-malignant diseases [55, 56]. Serum ferritin is elevated in most patients with Hodgkin's lymphoma and has been found to correlate with the stage of disease [55]. Extraordinarily high concentrations occur in acute myelocytic leukemia and several other myeloproliferative diseases, and fall to normal with successful chemotherapy [56]. Indeed, it was found that both leukemic cells and Hodgkin's lymphocytes synthesized ferritin as a faster rate than normal leukocytes [57, 58]. Serum ferritin is elevated in most patients with liver carcinoma [59], but also in the sera of patients with any type of acute or chronic liver inflammation. Serum ferritin is also elevated in most patients with pancreatic carcinoma, in some patients with acute pancreatitis, but not in those with chronic pancreatitis [60]. Ferritin is also elevated in patients with teratoblastoma [61] and breast cancer [62], and has proved helpful in elevating prognosis and response to therapy [61-63].

The mechanism responsible for the increase of serum ferritin in many of these diseases is not clear. Tissue ferritin is clearly released from damaged tissue in hepatitis [44, 38, 64]. Elevations of serum ferritin in patients with hepatitis correlate with the extent of liver damage and the SGOT levels [64]. But this cannot be the entire explanation since it is also elevated in patients with primary liver carcinoma with no correlation to inflammation or elevated SGOT [59]. Elevations have been found in patients with small localized tumors with little or no histological evidence of significant necrosis [6, 61]. The levels in breast cancer are similar to those found in benign rheumatoid arthritis [62], and it was thought that elevated serum ferritin levels in cancer and benign diseases reflected an increase in nonutilizable iron stores occurring with chronic disease. However, it is now clear that the malignant tumor synthesizes and secretes ferritin, since human tumors transplanted to nude mice released human ferritin into the circulation [60].

The structure of serum ferritin in cancer patients is also unclear. Different cell populations in tumor tissue may contribute to the serum isoferritin pool, and the serum ferritin level may be elevated by several of the mechanisms noted above. The isoelectric profiles of ferritin in cancer sera is clearly different from those of normal liver or spleen ferritin, but there are conflicting data concerning the precise isoferritin pattern [52, 65, 66]. Of great interest, however, was the finding that antisera relatively more specific for the acidic isoferritins of HeLa may be able to recognize selectively the acidic isoferritin in cancer sera [7]. The apparent level found in cancer sera may therefore depend, to a large extent, on the type and specificity of the antisera used in the assay. The serum level may very well be underestimated by using an antinormal liver ferritin or antinormal human spleen ferritin, and the level has been found to be increased significantly when using an anti-HeLa ferritin antibody [7]. The antigenic characteristics of the tumor ferritins need to be further clarified. It remains to be seen whether a more specific radioimmunoassay can be developed using reagents recognizing a tumor-specific isoferritin or subunit.

VII. CONCLUSION

Ferritin is a complex intracellular macromolecule, long known to be involved in iron storage and metabolism. It is now clear that ferritin is also a normal serum protein, present in trace amounts (15-200 ng/ml), and that the serum concentration directly reflects the total body iron stores in normal adults. Elevated serum ferritin levels occur in a variety of benign and malignant diseases. Therefore serum ferritin cannot be used as a screening or diagnostic test for cancer. Measurement of serial ferritin levels may be useful in monitoring certain myeloproliferative diseases, and helpful in assessing prognosis or response to therapy in a number of types of malignancy.

Several mechanisms probably contribute to the normal serum ferritin pool in cancer sera, including tissue ferritin released from necrotic cells, ferritin synthesized by inflammatory blood cells, or synthesized in response to increased nonutilizable iron stores, as well as to tumor cell synthesis and secretion. The ferritin synthesized by tumor cells is different from normal tissue ferritin, biochemically and immunologically, although the biochemical basis for the alteration of isoferritin profiles in embryonic and malignant tissue ferritin is still unclear. It remains to be seen whether the antigenic differences noted in tumor ferritins will lead to the development of a more specific radioimmunoassay to quantitate selectively the tumor isoferritins in cancer sera.

REFERENCES

1. P. M. Harrison, Ferritin: An iron-storage molecule, Semin. Hematol. 14:55-71 (1977).
2. E. Alpert, R. L. Coston, and J. W. Drysdale, Carcino-fetal human liver ferritins, Nature 242:194-196 (1973).
3. J. W. Halliday, L. V. McKerring, R. Tweedale, and L. W. Powell, Serum ferritin in hemochromatosis—Changes in the isoferritin composition during venesection therapy, Br. J. Hematol. 36:395-404 (1977).
4. J. W. Drysdale and E. Alpert, Carcinofetal isoferritins, Scand. J. Immunol. (Suppl 8) 8:65-71 (1978).
5. Y. Niitsu, Y. Kohgo, and M. Yokota, Radioimmunoassay of serum ferritin in patients with malignancy, Ann. N.Y. Acad. Sci. 259:450-452 (1975).
6. Y. Kohgo, Y. Niitsu, N. Watanable, S. Otsuka, J. Koseki, K. Shibata, and I. Urushizaki, Studies on radioimmunoassay of serum ferritin and its clinical implication in diseases of the digestive organs, Jap. J. Gastroenterol. 73:1553-1562 (1976).
7. J. T. Hazard and J. W. Drysdale, Ferritinaemia in cancer, Nature 265:755-756 (1977).

8. P. M. Harrison, R. J. Hoare, T. C. Hoy, and I. G. Macara, Ferritin and hemosiderin: Structure and function, in Iron in Biochemistry and Medicine (A. Jacobs and M. Worwood, eds.), Academic Press, London, 1974, pp. 73-109.
9. R. R. Crichton, Ferritin: Structure, synthesis and function, N. Engl. J. Med. 284:1413-1422 (1971).
10. R. R. Crichton, J. A. Millar, R. L. C. Cummings, and C. F. A. Bryce, The organ specificity of ferritin in human and horse liver and spleen, Biochem. J. 131:51-59 (1973).
11. W. I. P. Mainwaring and T. Hofmann, Horse spleen apoferritin: N-terminal and C-terminal residues, Arch. Biochem. Biophys. 125:975-980 (1968).
12. A. Huberman and E. Barahona, Primary structure of rat liver apoferritin: The amino end, Biochim. Biophys. Acta 533:51-56 (1978).
13. G. W. Richter, Comparison of ferritins from neoplastic and non-neoplastic human cells, Nature (Lond.) 207:616-618 (1965).
14. Y. Makino and K. Konno, A comparison of ferritin from normal and tumor bearing animals, J. Biochem. (Tokyo) 65:471-473 (1969).
15. J. W. Drysdale, Microheterogeneity in ferritin molecules, Biochim. Biophys. Acta 207:256-258 (1970).
16. J. W. Drysdale, Heterogeneity in tissue ferritins displayed by gel electrofocusing, Biochem. J. 141:627-632 (1974).
17. I. Urushizaki, Y. Niitsu, K. Ishitani, M. Matsuda, and M. Fukuda, Microheterogeneity of horse spleen ferritin and apoferritin, Biochim. Biophys. Acta 243:187-192 (1971).
18. B. K. van Kreel, V. Eijk, and B. Leijnse, The isoelectric fractionation of rabbit ferritin, Acta Haematol. 47:59-64 (1972).
19. L. W. Powell, E. Alpert, K. J. Isselbacher, and J. W. Drysdale, Abnormality in tissue isoferritin distribution in idiopathic haemochromatosis, Nature (Lond.) 250:333-335 (1974).
20. L. W. Powell, E. Alpert, K. J. Isselbacher, and J. W. Drysdale, Human isoferritins: Organ specific iron and apoferritin distribution, Br. J. Haematol. 30:47-55 (1975).
21. I. Urushizaki, K. Ishitani, and Y. Niitsu, Microheterogeneity of rat liver ferritin: Comparison of electrofocusing and chromatographic fractions, Biochim. Biophys. Acta 328:95-100 (1973).
22. L. W. Powell, L. V. McKeering, and J. W. Halliday, Alterations in tissue ferritins in iron storage disorders, Gut 16:909-912 (1975).
23. I. Urushizaki, K. Ishitani, M. Natoni, M. Yokota, M. Mitago, and Y. Niitsu, Heterogeneity of ferritin from 3'-methyl-4-(dimethylamino)azobenzene, Gann 64:237-246 (1973).
24. J. R. Hazard, M. Yokota, and J. W. Drysdale, Immunological differences in human isoferritins: Implications for immunological quantitation of serum ferritin, Blood 49:139-146 (1977).

25. T. G. Adelman, P. Arosio, and J. W. Drysdale, Multiple subunits in human ferritins: Evidence for hybrid molecules, Biochem. Biophys. Res. Commun. 63:1056-1062 (1975).
26. E. Alpert, Characterization and subunit analysis of ferritin isolated from normal and malignant human liver, Cancer Res. 35:1505-1509 (1975).
27. D. J. Lavroie, K. Ishikaawa, and I. Listowsky, Correlations between subunit distribution, microheterogeneity, and iron content of human liver ferritin, Biochemistry 17:5448-5454 (1978).
28. C. F. A. Bryce and R. R. Crichton, Microheterogeneity in apoferritin molecules—An artifact, Hoppe-Seyler's Z. Physiol. Chem. 354:344-346 (1973).
29. S. Shinjo, H. Abe, and M. Masuda, Carbohydrate composition of horse spleen ferritin, Biochim. Biophys. Acta 411:165-167 (1975).
30. J. W. Drysdale, Ferritin phenotypes: Structure and metabolism, Ciba Foundation Symposium 51:41-67 (1977).
31. J. W. Halliday, L. V. McKeering, R. Tweedale, and L. W. Powell, Serum ferritin in haemochromatosis—Changes in the isoferritin composition during venesection therapy, Bi. J. Haematol. 36:395-404 (1977).
32. J. W. Drysdale and R. M. Singer, Carcinofetal isoferritins in placenta and Hela cells, Cancer Res. 34:3352-3354 (1974).
33. J. C. K. Lee and G. W. Richter, Distinctive properties of ferritin from the Reuber H-35 rat hepatoma, Cancer Res. 31:566-572 (1971).
34. M. C. Linder, J. R. Moor, H. N. Munro, and H. P. Morris, Structural differences in ferritins from normal and malignant rat tissues, Biochim. Biophys. Acta 386:409-421 (1975).
35. D. Buffe, C. Rimbaut, and P. Burtin, Presence d'un ferroproteine d'origine tissulaire, l'α_2H globuline dans le serum des sujets atteints d'affections lignes, Int. J. Cancer 3:850-856 (1968).
36. D. Buffe, C. Rimbaut, J. Lemeale, O. Schweioguth, and P. Burtin, Presence d'une ferroproteine d'origine tissulaire, l'α_2H globuline dans le serum d'enfants porteurs de tumeurs, Int. J. Cancer 5:85-87 (1970).
37. D. Buffe and C. Rimbaut, α_2-H-globulin: A hepatic glycoferroprotein: Characterization and clinical significance, Ann. N.Y. Acad. Sci. 259:417-426 (1975).
38. E. Alpert, K. J. Isselbacher, and J. W. Drysdale, Beta-fetoprotein: Identification as normal liver ferritin, Lancet 1:44 (1973).
39. D. Marcus and N. Zinberg, Isolation of ferritin from human mammary and pancreatic carcinomas by means of antibody immunoadsorbents, Arch. Biochem. Biophys. 162:493-501 (1974).
40. D. Snyder and E. Alpert, Human liver ferritin: Tumor-specific alteration in structure and antigenicity, Gastroenterology 72:1189 (1977).

41. E. Alpert, D. S. Snyder, L. Davenport, and A. Quaroni, Human liver ferritin: Alterations in primary structure and antigenicity in a malignant liver cell line, in Carcinoembryonic Proteins (F. G. Lehmann, ed.), Elsevier, Amsterdam, 1979, pp. 261-271.
42. P. Arosio, M. Yokota, and J. W. Drysdale, Structural and immunological relationships of isoferritins in normal and malignant cells, Cancer Res. 36:1735-1739 (1976).
43. M. Worwood, B. M. Jones, and A. Jacobs, The reactivity of isoferritins in a labelled antibody assay, Immunochemistry 13:477 (1976).
44. K. R. Reissman and M. R. Dietrich, On the presence of ferritin in peripheral blood of patients with hepatocellular disease, J. Clin. Invest. 35:588-595 (1956).
45. G. M. Addison, M. R. Beamish, C. N. Hales, M. Hodgkins, A. Jacobs, and P. Clewellin, An immunoradiometric assay for ferritin in the serum of normal subjects and patients with iron overload, J. Clin. Pathol. 25:326-329 (1972).
46. A. Jacobs and M. Worwood, Ferritin in serum: Clinical and biochemical implications, N. Engl. J. Med. 292:951-956 (1975).
47. D. A. Lipschitz, J. D. Cook, and C. A. Finch, A clinical evaluation of serum ferritin as an index of iron stores, N. Engl. J. Med. 290:1312-1316 (1974).
48. J. W. Drysdale and H. N. Munro, Regulation of synthesis and turnover of ferritin in rat liver, J. Biol. Chem. 241:3630-3637 (1966).
49. A. Unger and C. Hershko, Hepatocellular uptake of ferritin in the rat, Br. J. Haematol. 28:169-180 (1974).
50. D. A. Lipschitz, A. Pollack, M. A. Savin, and J. D. Cook, The kinetics of serum ferritin, Clin. Res. 24:571a (1976).
51. P. Arosio, M. Yokota, and J. W. Drysdale, Characterization of serum ferritin in iron overload, Br. J. Haematol. 36:199-207 (1977).
52. M. Worwood, S. Dawkins, H. Wagstaff, and A. Jacobs, The purification and properties of ferritin from human serum, Biochem. J. 157:97-103 (1976).
53. M. Worwood, S. J. Cragg, M. Wagstaff, and A. Jacobs, Binding of human serum ferritin to conconavalin A, Clin. Sci. 56:83-87 (1979).
54. A. Yachi, A. Yoshikozua, T. Akira, and W. Takeo, Clinical significance of α H_2 globulin, Ann. N.Y. Acad. Sci. 249:435-445 (1975).
55. A. Jacobs, A. Slater, J. A. Whittaker, G. Canellos, and P. Wiernik, Serum ferritin concentration in untreated Hodgkin's disease, Br. J. Cancer 34:162-166 (1976).
56. D. H. Parry, M. Worwood, and A. Jacobs, Serum ferritin in acute leukaemia at presentation and during remission, Br..Med. J. 1:245-247 (1975).
57. G. P. White, M. Worwood, D. H. Parry, and A. Jacobs, Ferritin synthesis in normal and leukemic leukocytes, Nature (Lond.) 250:584-585 (1974).

58. E. J. Sarcione, J. R. Smalley, M. J. Lerna, and L. Stutzman, Increased ferritin synthesis and release by Hodgkin's disease peripheral blood lymphocyte, Int. J. Cancer 20:339-346 (1977).
59. M. C. Kerr, J. D. Torrence, D. Derman, M. Simon, G. M. Mac Nab, R. W. Charlton, and T. M. Bothwell, Serum and tumor ferritins in primary liver cancer, Gut 19:294-299 (1978).
60. Y. Niitsu, Y. Kohgo, S. Ohtsuka, N. Watanabe, J. Koseki, K. Shibata, K. Ishitani, T. Nagai, Y. Gocho, and I. Urushizaki, Ferritins in serum and tissue and their implication in malignancy, in Onco-Developmental Gene Expression (W. H. Fishman and S. Sell, eds.), Academic Press, New York, pp. 757-763 (1976).
61. B. Wahren, E. Alpert, and P. Esposti, Multiple antigens as marker substances in germinal tumors of the testis, J. Natl. Cancer Inst. 58:489-505 (1977).
62. D. M. Marcus and N. Zinberg, Measurement of serum ferritin by radioimmunoassay: Results in normal individuals and patients with breast cancer, J. Natl. Cancer Inst. 55:791-795 (1975).
63. A. Jacobs, B. Jones, C. Ricketts, J. L. Hayward, R. D. Bulbrook, and D. Y. Wang, Serum ferritin concentration in early breast cancer, Br. J. Cancer 34:286-290 (1976).
64. J. Prieto, M. Barry, and S. Sherlock, Serum ferritin in patients with iron overload and with acute and chronic liver disease, Gastroenterology 68:525-533 (1975).
65. L. W. Powell, J. W. Halliday, L. B. McKeering, and R. Tweedale, Alerations in serum and tissue isoferritins in disease states: Hemochromatosis and malignant disease, in Proteins of Iron Metabolism (E. B. Brown, P. Aisen, J. Fieldings, and R. R. Crichton, eds.), Grune & Stratton, New York, 1977, pp. 61-65.
66. M. Worwood, M. Wagstaff, B. M. Jones, S. Dawkins, and A. Jacobs, Biochemical properties of human isoferritins, in Proteins of Iron Metabolism (E. B. Brown, P. Aisen, J. Fielding, and R. R. Crichton, eds.), Grune & Stratton, New York, 1977, pp. 79-91.

10

POLYAMINES AS BIOCHEMICAL MARKERS
OF TUMOR GROWTH PARAMETERS

DIANE HADDOCK RUSSELL / Department of Pharmacology, University of Arizona Health Sciences Center, Tucson, Arizona

- I. Introduction   241
- II. Polyamine Accumulation in Tumor Cells   243
- III. Intra- and Extracellular Polyamines During Animal Tumor Regression   245
  - A. Polyamines During Mammary Tumor Growth and Regression   245
  - B. Polyamines in Rats with 3924A Hepatomas After Chemotherapy   246
  - C. Polyamines in Rats with 3924A Hepatomas After Local Radiation   247
- IV. Extracellular Polyamine Levels in Diagnosed Cancer Patients   248
  - A. Age and Sex Differences in Polyamine Excretion   251
  - B. Dietary and Diurnal Effects on Polyamine Excretion   251
  - C. Physiological Variations in Polyamine Levels   252
- V. Polyamines as Markers of Response to Cancer Therapy   252
- VI. Urinary Polyamine Values and Disease Activity   259
  - A. Multiple Markers to Prescribe Disease Status   260
- VII. Summary   260
   References   262

I. INTRODUCTION

Discrete intracellular accumulation patterns of the polyamines, spermidine, and spermine, and their precursor, putrescine, have been substantiated during growth and differentiation in normal and neoplastic cells in culture and in animal tissues [1-4]. Increased ribosomal RNA in cells during growth and proliferation is paralleled by increased spermidine and preceded by an accumulation of putrescine [5-11]. In fact, the spermidine concentration could be used in place of protein, RNA, or DNA determinations to evaluate alterations in the specific activity of various components of cell growth.

A spermidine/spermine ratio of less than 1.0 is typical of a tissue with low biosynthetic activity or of differentiated tissues with a constant rate of RNA and protein synthesis. A spermidine/spermine ratio of greater than 2.0 is typical of a tissue undergoing hypertrophy and/or hyperplasia. The spermine concentration is higher in differentiated tissues, decreases during dedifferentiation processes such as liver regeneration, and is constant or increases with age.

Studies of polyamine profiles of hepatomas as a function of their growth rates indicate that the total concentrations of polyamines in a tissue cannot be interpreted with any reliability in relation to growth kinetics without information about turnover rates in that tissue, as well as measurements of the activities of the enzymes responsible for their synthesis, particularly ornithine decarboxylase, the rate-limiting enzyme in putrescine synthesis, and S-adenosyl-L-methionine decarboxylase, the rate-limiting enzyme in spermidine synthesis [12, 13].

The generalizations summarized so far indicate that (1) increased accumulation of intracellular putrescine in tissues indicates initiation of a growth process; (2) accumulation of spermidine concomitant with ribosomal RNA prescribes increased mass or hypertrophy, which may or may not be followed by DNA synthesis and cell division; and (3) increased spermine appears to be a marker of differentiation, whereas decreased spermine parallels dedifferentiation processes in many cell regulatory systems such as lymphocyte mitogenesis and regenerating rat liver.

Since alterations in intracellular polyamine accumulation patterns are indicators of growth, proliferation, and differentiation states, it is not surprising that there are altered extracellular levels in various pathological disorders [4]. The phenomenon of polyamine excretion has been well-characterized in bacterial systems and usually occurs in response to a reduction in growth rate [1]. It also was reported that during early cleavage stages of sea urchin development, all three amines decreased markedly in concentration at mitosis [14].

The rapid declines in the concentrations of these compounds suggested (1) that there were enzymes present for their selective degradation; or (2) that the compounds were secreted in the surrounding medium at that time. The rapidity of decline suggests that the second alternative is more likely. In mammalian systems, there is little evidence that polyamines are metabolized intracellularly and substantial evidence that they are excreted in response to growth arrest. In cell cycle progression, intracellular polyamine concentrations increase as cells leave the G_1 phase and progress through S and G_2 phases [15, 16].

When Chinese hamster ovary cells are exposed to 43°C in the G_1 phase, there is an immediate depletion of normal intracellular spermidine and spermine content, followed by a delay in DNA replication kinetics [17]. The amines can be recovered from the culture medium, suggesting that they are excreted in response to heat. As spermidine and spermine reaccumulate, DNA replication proceeds at a reduced rate until near normal S phase levels of these polyamines are attained. Hyperthermic treatment of the same cells after they had entered S phase resulted in a specific reduction of the spermidine level and a reduction in the rate of DNA replication [17].

Recently it was reported that nonconfluent monolayers of BHK cells (when transferred from 10% serum to 1% serum, which leads to a significant reduction in the growth rate) excrete polyamines, particularly spermidine, which results in dramatic decreases in the intracellular concentration of spermidine [18]. In rats and humans, results obtained after the injection of tracer amounts of [^{14}C]putrescine, [^{14}C]spermidine, or [^{14}C]spermine indicated that there was not a significant uptake of the amines by tissues [19, 20]. Rather, there is a rapid conjugation of [^{14}C]-putrescine and [^{14}C]spermidine and excretion. [^{14}C]Spermine is excreted without prior conjugation. In humans, [^{14}C]putrescine and [^{14}C]spermidine were almost completely conjugated within 4 to 5 min of injection [20].

The lack of tissue uptake of extracellular polyamines, and their rapid conjugation and excretion, as well as the significant recovery rates of injected amines in humans [20] make extracellular polyamine levels excellent markers of alterations in cell kinetics. Since other comprehensive reviews exist [4, 21], this review will attempt to outline the main concepts related to the major usefulness of polyamines as markers of alterations in tumor kinetics in response to therapy.

II. POLYAMINE ACCUMULATION IN TUMOR CELLS

Polyamine biosynthesis and accumulation has been most extensively studied in mouse L1210 leukemia [22-26]. Solid tumors can be obtained by injecting the tumor cells subcutaneously, or ascites tumors can be obtained by intraperitoneal injection. Because of its rapid growth characteristics (the lifespan of a mouse injected with 10^5 or 10^6 tumor cells is 7 to 10 days), it is a good system in which to study the chain of events occurring in response to tumor inoculation.

In solid tumors, at the earliest time that enough tissue is available to assay, both ornithine decarboxylase activity and S-adenosyl-L-methionine decarboxylase activity are markedly elevated above the level in normal tissues known to have substantial polyamine biosynthetic activity [22]. In line with previous comments, both putrescine and spermidine concentrations were nearly twice the control concentration within 6 days of tumor inoculation, whereas the spermine concentration was lower than control values. Therefore the spermidine-to-spermine ratio in the tumor was greater than 2:0, and in other tissues, such as liver, it was less than 1:0.

Drugs known to increase the survival time of L1210 tumor-bearing mice decreased the ability of the tumor to accumulate polyamines in direct

TABLE 1 Polyamine Concentrations in the Mouse L1210 Tumor as Affected by Drugs

Drug treatment of L1210 tumor-bearing mice	Increase in survival time of control (%)	Putrescine	Spermidine (nmol/g)	Spermine
Cytosine arabinoside	110	233	1315	871
NSC 82196	150	151	1019	760
NSC 45388 (DTIC)	200	168	800	518
None	100	345	1518	709

Four groups of 5-10 DBA/2 mice received subcutaneous injections of 10^5 L1210 ascites cells and various drugs on the following schedules: group 1, cytosine arabinoside 10 mg/kg i.p. daily × 6; group 2, NSC 45388 12.5 mg/kg i.p. twice daily × 6; groups 3, NSC 82196, one s.c. injection (70 mg/kg) on day 3; group 4, no drug. All drug-injected groups and a control group were killed 8 days after the initial injection of ascites cells; the tumors and control tissues were then excised, chilled on ice, and homogenized in 4 vol 0.1 N HCl. The tissues were assayed for endogenous levels of putrescine, spermidine, and spermine by extraction of amines into alkaline 1-butanol, separation of amines by high-voltage electrophoresis, and quantification after staining with an acid ninhydrin solution. Each value represents the mean of at least five determinations. Survival time was assessed in controls by recording the number of days survived after tumor injection in a group of 10 mice.
Source: Data reproduced from Ref. 22.

proportion to their ability to increase lifespan (Table 1) [22]. This was the first demonstration in tumors that the inhibition of putrescine and spermidine accumulations paralleled the attenuation of the growth rate of the tumor, a fact whose profound importance has only recently begun to be explored in terms of the development of drugs which block the enzymes in the biosynthetic pathway as possible antitumor agents [27]. More extensive studies of the effects of drugs on polyamine accumulation patterns were conducted on L1210 ascites tumors grown in BDF/1 mice [24, 25]. These studies corroborated the findings in the L1210 solid tumor system, that is, an increase in survival time in response to various antineoplastic agents was concomitant with a decrease in the spermidine-to-spermine ratio [24].

III. INTRA- AND EXTRACELLULAR POLYAMINES DURING ANIMAL TUMOR REGRESSION

Russell reported in <u>Nature</u> in 1971 [28] that patients with metastatic cancer excreted increased levels of polyamines in their urine. In a patient with a large ovarian teratoma, polyamine excretion decreased toward normal value after resection of the bulk of the tumor. Other patients with solid and hematological tumors exhibited some pattern of enhanced polyamine concentrations in their urine. Further, this early report indicated that leukemia patients in remission had only slightly increased concentrations of polyamines. It was suggested that increased concentrations of polyamines in the urine might provide a diagnostic tool to evaluate tumor activity [29]. These data suggested that studies of animal tumor models in relation to polyamine metabolism would be useful in substantiating polyamine excretion as a tumor-specific event.

A. Polyamines During Mammary Tumor Growth and Regression

Therefore it became important to study alterations in intra- and extracellular polyamines in tumors during growth and regression. The rat MTW9 mammary carcinoma first produced by Kim and Furth [30] appeared ideal for studies of polyamines because its growth is hormone-dependent and the course of regression, which is rapid, can be studied after removal of the source of the mammotrophic hormone [31]. The tumor is inoculated into female rats at the same time that an MTTW10 pituitary tumor that secretes mammotrophins is also established in the same rat. Castration, along with removal of the pituitary tumor, initiates the process of tumor regression. Similar to the pattern seen in the growing L1210 tumor, substantial levels of putrescine and spermidine were present in the growing tumor with a spermidine-to-spermine ratio of 4:2 [32]. Tumor-bearing rats were then castrated, the hormonal source was ablated, and alterations in polyamine levels in tumors and in serum were monitored for 72 hr.

Within an average of 60 hr after initiation of tumor regression, the tumor volume is reduced to one-half its original volume and therefore constitutes the time of most rapid tumor regression. Within 24 hr, the spermidine-to-spermine ratio had decreased to 2:7, within 48 hr to 2:9, and at 72 hr to 1:8 (Table 2). Spermidine was at its highest level in serum at 48 hr and was nearly threefold higher than the value in the growing tumor. Therefore we postulated that intracellular spermidine levels that increase during tumor growth are lowered by excretion during tumor regression, and we suggested that increased spermidine levels in serum reflected tumor cell death.

This concept was further substantiated by sampling tumor interstitial fluid and measuring polyamine levels before and after initiation of regression.

TABLE 2 Polyamines in MTW9 Mammary Tumors and Sera of Rats After Initiation of Tumor Regression

Status	Putrescine	Spermidine	Spermine	Spermidine-to-spermine ratio
Growing				
Tumor[a]	77 ± 2.9	1570 ± 26	372 ± 14	4.2
Serum[b]	n.d.[c]	3.8 ± 0.9	n.d.	
24-hr regressing				
Tumor	45 ± 0.75	1480 ± 30	546 ± 20	2.7
Serum	1.7 ± 0.4	7.3 ± 1.8	n.d.	
48-hr regressing				
Tumor	32 ± 3.1	1040 ± 100	483 ± 95	2.1
Serum	1.5 ± 0.14	11.2 ± 2.2	0.18 ± 0.02	
72-hr regressing				
Tumor	39 ± 2.5	980 ± 14	554 ± 24	1.8
Serum	1.9 ± 0.2	7.2 ± 0.8	0.18 ± 0.03	

[a] Concentration expressed as nmol/g wet weight.
[b] Concentration expressed as nmol/ml.
[c] Amine was not detectable.
Polyamine concentrations were measured with a Beckman Model 121 automatic amino acid analyzer. The sensitivity of this method is no greater than 1 nmol. Data represent the mean ± SE of samples from four rats.
Source: Data reproduced from Ref. 32.

Compared to growing tumors, the spermidine concentration was increased at 24 and 48 hr after initiation of regression, similar to the increases in serum levels. Although there was considerable variation in the levels of spermidine detectable in the tumor interstitial fluid of rats with growing and regressing tumors, in all cases the concentration of spermidine in the interstitial fluid was higher in the animals with regressing tumors. This suggested that at least a portion of the elevated spermidine in serum came from the tumor as it regressed.

B. Polyamines in Rats with 3924A Hepatomas After Chemotherapy

Rapidly growing tumors are known to respond favorably to chemotherapy, and the 3924A hepatoma with a doubling time of approximately 24 hr has been extensively studied as a model of tumor growth. Before any therapy, the 3924A hepatoma is comprised of approximately 51% hepatoma cells, 26% connective tissue, 18% necrotic tissue, and 5% blood [33, 34]. Within 48 hr of treatment with 5-fluorouracil (5-FU), these percentages change markedly. The tumor is then composed of approximately 27% hepatoma cells, 40% connective tissue, and 30% necrotic tissue. Within 24 hr of treatment, the largest decrease is in the percentage of hepatoma cells, dropping from 51 to 30%. Spermidine concentration in the serum of tumor-bearing rats after a single administration of 5-FU is more than doubled within 24 hr, concomitant with the large decrease in tumor cellularity [35].

The level remains elevated at 48 hr and drops below the initial level detected in the serum at 72 hr. In rats without tumors that are administered 5-FU, there is only a slight increase in the spermidine level at 24 hr, the response being approximately 25% of the amount of the response in tumor-bearing rats. This suggested that 5-FU was killing some normal cells but mainly was targeted toward the tumor cells. Spermidine concentration in the 3924A tumor tissue dropped to approximately 65% of control within 48 hr, paralleling the decrease in the number of tumor cells, although tumor volume was constant during this period.

C. Polyamines in Rats with 3924A Hepatomas
   After Local Radiation

In studies of the response of 3924A rat hepatomas to chemotherapy, it could not be totally ruled out that the elevated extracellular spermidine level might be derived from host tissues rather than tumor tissue. Therefore we studied the effects of local radiation on the 3924A rat hepatoma [36].

Under these conditions, any increases in polyamine concentrations extracellularly could be attributed only to release from the tumor tissue, the only site of tissue damage. Spermidine concentration in the serum increased from 3.8 to 9.6 nmol/ml serum within 24 hr in the radiation-treated tumor rats. Further, putrescine increased from 22 nmol/ml serum to 91 nmol/ml serum within the same time. A rapid decrease in the tumor level of spermidine was detected after chemotherapy. There was an early transitory increase in putrescine concentration, suggesting that its synthesis was stimulated in response to radiation. Radiation is known to synchronize cells, and extensive human studies of polyamine levels in extracellular fluids (to be discussed later) have indicated that increased extracellular levels of putrescine are related to the number of cells in cell cycle within the tumor [37-39].

In summary, alterations in intracellular and extracellular levels of putrescine and spermidine in response to chemotherapy and radiation therapy provided the first solid evidence that these polyamines were markers of tumor kinetics and not of tumor burden, as had been assumed in certain earlier studies.

IV. EXTRACELLULAR POLYAMINE LEVELS IN DIAGNOSED CANCER PATIENTS

Patients with both hematological and solid tumors have significant elevations in urinary levels of putrescine, spermidine, and spermine [38] (Table 3). The urinary levels of all three polyamines were higher in patients with hematological tumors compared to patients with solid tumors. For spermine, there was no overlap in the range of values in patients with cancer versus normal volunteers, whereas for putrescine and spermidine, the range overlapped but was extended in all tumor categories with significant elevations in the mean and median values for these polyamines.

In an extensive study, Dreyfus et al. [40] compared polyamine excretion in patients with malignant and nonmalignant diseases, and the data are summarized in Table 4. They reported that 78% of patients in the cancer group excreted an elevated amount of urinary spermidine. They also found elevated spermine specific for female cancer patients. The case was less obvious in males since seven of the male controls excreted increased amounts of spermine. They postulated that this elevation might be of prostatic origin, since seminal fluid is known to contain a high concentration of spermine. These workers also found that 52% of the patients with malignant diseases excreted elevated levels of putrescine contrasted to 26% of the patients with nonmalignant disease. Of patients in the control group who excreted an elevated level of putrescine, most had infectious or inflammatory diseases or were undergoing regenerative processes such as fracture repairs or recovery from megaloblastic anemia or blood loss [40]. Therefore the abnormal amount of putrescine was probably derived from bacteria in the case of infectious diseases and derived from normal tissues in the case of regenerative processes (discussed later under putrescine as a marker of growth kinetics). They concluded that elevated levels of all three amines, putrescine, spermidine, and spermine, were exclusive for patients with cancer (Table 4).

Extensive studies of elevated urinary polyamine levels have been reported in Burkitt's lymphomas [41], in prostate carcinomas [42], in breast carcinomas [43], in genitourinary tract malignancies [44], and in many studies of leukemia (for review see Ref. 4). Polyamine levels also are elevated in the plasma and serum of patients with malignancies [38, 45, 46]. Further studies are necessary to determine whether plasma or urine samples are more useful in determining disease status.

In cerebrospinal fluid, putrescine and spermidine and sometimes spermine were elevated in most patients with central nervous system tumors, particularly those with glioblastomas or meduloblastomas [47]. Patients with other central nervous system disorders, such as seizures and hydrocephaly, had polyamine levels in the normal range, suggesting that elevated polyamines are not common to all central nervous system pathologies.

Serum elevations of polyamines in patients with solid tumors were reported by Nishioka and Romsdahl [45]. They found that putrescine values

TABLE 3 Pretreatment Levels of Polyamines in the Urine of Patients with Cancer and of Normal Control Subjects

	Putrescine (μg/mg creatinine)				Spermidine (μg/mg creatinine)				Spermine (μg/mg creatinine)			
	No. of subjects	Mean ± SD	MDN[a]	Range	No. of subjects	Mean ± SD	MDN	Range	No. of subjects	Mean ± SD	MDN	Range
Hematological tumors	68	4.4 ± 0.66[b]	4.2	0.45–38	68	3.7 ± 0.79[b]	2.7	0.58–25	62	0.8 ± 0.32[b]	1.0	0.25–8.0
Solid tumors	55	3.7 ± 0.39[b]	3.8	0.29–10	56	2.7 ± 54[b]	1.8	1.09–8.1	56	0.6 ± 0.28[b]	0.55	0.12–3.7
Normals	16	2.1 ± 0.62	1.0	0.40–2.1	16	1.2 ± 0.18	0.8	0.40–2.1	16	0.04 ± 0.007	0.04	0.02–0.3

[a] Median.
[b] The difference from values for normal control subjects is significant ($p < 0.001$).
Polyamines were assayed in 24-hr urine collections by means of an amino acid analyzer technique.
Source: Data reproduced from Ref. 38.

TABLE 4 Comparison of Polyamine Excretion in Patients with Malignant and Nonmalignant Diseases

	No. of patients with	
	Malignant	Nonmalignant
Total cases	42	54[a]
Elevated excretion of any one polyamine	37 (88%)	24 (44%)
Normal total polyamine excretion	5	30
Individual polyamine excretion[b]		
Putrescine $\geqslant 5$ mg	22 (52%)	14 (26%)
Cadaverine[c]		
$\geqslant 5$ mg	3	3
$\geqslant 2$ mg	7	7
Spermidine $\geqslant 2$ mg	33 (78%)	13 (24%)
Spermine $\geqslant 0.5$ mg	16 (38%)	7 (13%)[d]
Polyamine excretion patterns		
Putrescine $\geqslant 5$ mg + cadaverine $\geqslant 5$ mg	2	2
Spermidine $\geqslant 2$ mg + spermine $\geqslant 0.5$ mg	5	4[d]
Putrescine $\geqslant 5$ mg + spermine $\geqslant 0.5$ mg	1	0
Putrescine $\geqslant 5$ mg + spermidine $\geqslant 2$ mg, with or without other elevated polyamines	19 (45%)	6 (11%)
Putrescine $\geqslant 5$ mg + spermidine $\geqslant 2$ mg, with normal spermine excretion	10	5
Putrescine $\geqslant 5$ mg + spermidine $\geqslant 2$ mg + spermine $\geqslant 0.5$ mg	9	1

[a] Fourteen patients in this group had infectious or inflammatory disease, nine (64%) of whom had elevated total polyamine excretion. Of the 40 patients with various other nonmalignant diseases, 15 (37%) had elevated total polyamine excretion.
[b] As an isolated finding or combined with other elevated polyamine excretion.
[c] Only 19 patients with malignant disease and 42 with nonmalignant disease were examined for cadaverine.
[d] All these patients were males.
Percentages are of the total number of patients in each group.
Source: Data reproduced from Ref. 40.

from sera of cancer patients were considerably higher than the corresponding values from normal subjects, and that spermidine was higher than normal in 75% of sera examined while spermine values were only occasionally elevated. On the basis of their studies, they felt that putrescine was the most reliable marker of neoplastic growth.

The same investigators studied serum polyamine levels of 53 patients with colorectal carcinoma and 28 normals or patients with benign bowel disease [48]. One or more of the polyamines were elevated in the sera of 66% of the patients with preoperative colorectal carcinoma; nine patients with benign bowel disease had normal serum polyamine levels, and one patient with a villous adenoma had elevated serum polyamine levels.

A. Age and Sex Differences in Polyamine Excretion

Age-related changes in tissue and extracellular levels of polyamines have been noted in animals [49, 50]. For example, a decrease in tissue spermidine levels with age was reported in rats [49]. However, in analyses of human tissues obtained from young adults and children [51], no marked age or sex differences in polyamines were found. Waalkes et al. [52], measuring urinary polyamines, found that children excreted somewhat greater amounts than adults based on weight. The same workers also noted that females excreted substantially more urinary putrescine than males, and that the average level of putrescine decreased with age in females. Although this age-related difference was not found in males, the males excreted more spermine overall than females. The mean values for spermidine, on the other hand, were essentially the same for males and females. Other less well-defined changes related to the menstrual cycle have been reported, but they may not be pertinent to extracellular changes since they involved whole blood polyamine levels [53].

B. Dietary and Diurnal Effects on Polyamine Excretion

Waalkes et al. [52] assessed the effect of diet and diurnal changes on urinary putrescine levels. There was no difference in urinary polyamine levels between patients on a low-purine diet and those eating added meat portions for 3 days (Table 5). It thus seems that major fluctuations in urine levels must be determined by endogenous effects, as already noted. Waalkes et al. [52] also noted no major diurnal variations. Other workers, including Russell and Russell [46], observed some diurnal variation in extracellular polyamines. Extracellular diurnal changes are rather small when compared to whole blood analysis of male-female variations and changes in women through the menstrual cycle which reflect intracellular pools of polyamines [53].

TABLE 5 Effect of Diet on Polyamine Urine Levels

Subject	Diet[a]	Concentration (μmol/kg/24 hr)		
		Putrescine	Spermidine	Spermine
A	1	0.133	0.051	0.003
	2	0.127	0.057	0.009
B	1	0.248	0.096	0.011
	2	0.225	0.092	0.012
C	1	0.274	0.172	0.015
	2	0.299	0.115	0.019
D	1	0.180	0.066	0.011
	2	0.298	0.091	0.011
E	1	0.281	0.134	0.026
	2	0.151	0.071	0.019

[a] Diet 1 = no meat for 3 days (low-purine diet); diet 2 = regular diet with added meat portions for 3 days. Urine was collected for analysis on days 3 and 6.
Source: Data reproduced from Ref. 52.

C. Physiological Variations in Polyamine Levels

Besides these effects—particularly of sex and to some extent age—on polyamine levels, variations have been noted in a number of physiological circumstances. For example, any situation in which there is rapid regeneration or regrowth of a tissue is associated with elevated polyamines. Thus wound healing after accidental or surgical trauma is associated with increased levels of urinary polyamines, particularly putrescine [38]. Liver regeneration after acute hepatitis also has been associated with increased levels of polyamines [40]. These types of fluctuations must be noted carefully for comparison with fluctuations to be used as clinical indicators in individual patients.

V. POLYAMINES AS MARKERS OF RESPONSE TO CANCER THERAPY

Figure 1 illustrates the concept of the spontaneous cell loss factor of a tumor being expressed as an elevated spermidine excretion pattern before chemotherapy. Note that for a patient with acute lymphocytic leukemia, spermidine excretion in terms of milligrams per day before chemotherapy was approximately 4.2 mg, compared with an upper limit of normal of 2.5 mg per day. This concept was based on two tenets derived from the animal tumor model studies discussed earlier, as well as on the elevation of polyamines, particularly spermidine, in patients with diagnosed cancer. The tenets are (1) during either normal or neoplastic growth there are increased cellular concentrations of polyamines, particularly spermidine [12, 28, 29, 54]; and (2) the importance of cell loss factor as the cause of the elevations of extracellular polyamines, that is, tumor growth rate is best described not by the mitotic rate but by cell loss factor [55].

Certain tumors have cell loss factors as high as 70-80%. The animal tumor models which we have discussed indicated that the elevation of spermidine in serum was associated with tumor cell death; therefore we predicted that those tumors with a high cell loss factor would be associated with the highest levels of spermidine in serum and urine. This prediction has held remarkably true in studies of human cancer patients, even though it was published in 1974 [35], before extensive human studies. A recent study of extracellular polyamines in Ehrlich ascites tumor-bearing mice suggests they are released from dead or dying tumor cells [56].

FIGURE 1 Urinary spermidine and number of circulating blast cells before chemotherapy, after chemotherapy, and during remission in a patient with acute lymphocytic leukemia.

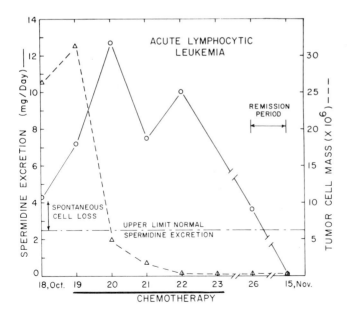

TABLE 6 Polyamines as Biochemical Markers of Tumor Cell Growth and Tumor Cell Death

Tumor growth
1. Intracellular <u>and</u> extracellular concentration of putrescine is proportional to the <u>number</u> of cells in the cell cycle, i.e., the growth fraction of the tumor.
2. Intracellular spermidine concentration is related to the amount of ribosomal RNA in the tumor. Extracellular spermidine concentration is proportional to the spontaneous cell loss factor of the tumor.
Tumor cell kill (chemotherapy and radiation therapy)
1. Extracellular concentration of spermidine is proportional to tumor cell kill, and thus is a reliable index of the effectiveness of the therapy.
2. Extracellular concentration of putrescine is proportional to both the growth fraction of the tumor, suggesting that putrescine may be excreted from cells traversing cell cycle, as well as to the tumor cell kill (release of total intracellular pool of putrescine).

Further predictions were made at that time on the basis of the two assumptions of the model: (1) After chemotherapy or radiation therapy, there should be increases in polyamine levels in serum and urine. The extent of the elevation should be related to the efficacy of the therapy; for example, effective therapy regimens that result in high cell kill also should result in the largest manifestations of polyamines in serum and urine. (2) Surgical ablation of large portions of the tumor should result in an immediate reduction in the level of polyamines excreted without any intervening fluctuations. (3) False positives would be expected in patients with other pathological conditions that are expressed through a high cell loss factor. Again, these predictions have held remarkably true in human trials and have been expanded. Therefore in Figure 1 the increase in spermidine seen immediately after initiation of chemotherapy is related to tumor cell kill. The dashed line represents the tumor cell mass, which drops dramatically concomitant with an increase in spermidine excretion. Note that within 24 hr there was over a threefold increase in the spermidine excretion and, as the patient went into remission, the level of spermidine excretion dropped into the normal range. Table 6 summarizes our current model of polyamines as biochemical markers of tumor cell growth and tumor cell death.

The relation of intra- and extracellular spermine and tumor kinetics is not well-defined. It was previously stated that extracellular spermidine and putrescine were rapidly conjugated before excretion [20]. Spermine does not appear to be conjugated; it is found in higher concentrations in differentiated tissues and is known to be intraconvertible to spermidine [3]. Note in Table 3 that the urinary spermine level in normal volunteers was only 3% that of normal spermidine excretion, and in patients with hematological tumors, the spermine level excreted was approximately 20% that of spermidine. A similar relationship was found for spermine excretion in patients with solid tumors. It should be emphasized, however, that an increased spermine excretion is characteristic of patients with cancer. The low level of urinary spermine compared to putrescine and spermidine suggests that some of the early methods to detect polyamines (which were not sensitive to the low picomole range) would have tended to underestimate spermine excretion [3].

To substantiate that post- to pretreatment ratios of spermidine and putrescine are important in evaluating the response to chemotherapy, we studied 22 patients with no response to chemotherapy, 16 patients with a partial response, and 18 patients with a complete response (Table 7) [38]. Of these patients, 36 had hematological tumors and 20 had solid tumors. Note that when there was no response to chemotherapy as judged by later clinical criteria, the post- to pretreatment increase in the spermidine ratio was 1:2. Those patients with either a partial or complete response had a mean increase in this ratio of 3:7 and 3:6, respectively.

Putrescine concentration usually increased after chemotherapy, but was not correlated with the response. In fact, the nonresponders tended to show elevations of putrescine without major increases in spermidine in line with concepts derived from animal tumor model studies; i.e., intracellular and extracellular putrescine concentrations increase as cells traverse cell cycle, and putrescine is also excreted from the cell during the cell death process in response to chemotherapy.

In the same table, we have plotted the post- to pretreatment spermidine ratio over the post- to preputrescine ratio. Complete responders had a ratio of 1:4, partial responders, 1:2, and nonresponders, 0:4. These data underline the importance of alterations in the spermidine concentration extracellularly as a marker of tumor cell loss and therefore of response to chemotherapy.

Figure 2 illustrates the mean spermidine levels in three groups of patients having complete, partial, and no response to a given course of combination chemotherapy. There is a highly significant early rise in spermidine levels associated with either a complete or partial response to therapy. Other investigators have substantiated the usefulness of elevations in spermidine after multimodality therapy as an index of treatment efficacy [40, 48, 52, 57, 58].

TABLE 7 Response to Chemotherapy and Urinary Spermidine and Putrescine Excretion in 56 Patients Studied Serially

Type of response	No. of cases	A Post-treatment/[a] pretreatment putrescine ratio	B Post-treatment/ pretreatment spermidine ratio	B/A ratio
None	22	2.7 ± 3.5[b]	1.2 ± 0.5	0.4
Partial	16	3.0 ± 3.1	3.7 ± 2.1[c]	1.2
Complete	18	2.5 ± 1.2	3.6 ± 1.3[c]	1.4
Combined, hematological tumors	36	2.6 ± 2.2	3.0 ± 2.0[d]	1.2
Combined, solid tumors	20	2.8 ± 3.4	2.0 ± 1.3	1.4

[a] Post-treatment indicates highest value within 72 hr of initiation of therapy.
[b] Mean ± SD.
[c] Difference from values for patients with no response is significant ($p < 0.0001$).
[d] Difference from values for patients with solid tumors is significant ($p > 0.05$).
Source: Data reproduced from Ref. 38.

Figure 3 shows the mean putrescine excretion in patients with complete response, partial response, or no response to chemotherapy. In both the complete and the partial response, there is an early increase in putrescine excretion, suggesting that those patients who respond have a significant number of tumor cells in cell cycle; thus they would contain considerable amounts of putrescine. Those with no response to chemotherapy have a late response in terms of putrescine excretion. This increase may be due to recruitment of cells into cell cycle, which is known to occur [59]. The latter excretion would occur during cell cycle traverse and would not be attributable to cell loss.

Putrescine is a marker of the number of cells in cell cycle in studies of multiple myeloma patients before chemotherapy. Studies of urinary putrescine levels in patients with multiple myeloma as a function of [^{3}H]thymidine labeling index indicated a direct correlation between apparent tumor growth fraction and urinary putrescine levels [4] (Fig. 4). This initial observation was subsequently extended to include patients with other hematological malignancies, including myelogenous leukemias and lymphomas. Therefore urinary putrescine has the capacity to be used as a marker of tumor kinetics, and this could prove particularly valuable in the design of chemotherapy schedules.

There has remained considerable speculation [60] over how important the knowledge of tumor kinetics is in the design of the therapy; however, a number of recent studies have confirmed the prognostic importance of pretreatment kinetic data. The pretreatment labeling index clearly correlates with survival in patients with high cell mass multiple myeloma [4]. In addition, tumor kinetics clearly influence response to treatment and survival at the relapse phase of disease. Thus pretreatment kinetics can affect drug selection and may influence intervals between courses of therapy. The correlations between putrescine and the labeling index indicate that measurements of putrescine will have the same type of prognostic value.

FIGURE 2 Urinary spermidine levels in patients receiving various cancer chemotherapeutic regimens. Time zero was before chemotherapy, and day 1 was the start of chemotherapy. Each point is the mean from 16 to 22 different patients with no response, partial response, or a complete response to therapy as judged by later standard clinical parameters. (From D. H. Russell and B. G. M. Durie, Ref. 4, <u>Polyamines as Biochemical Markers of Normal and Malignant Growth</u>, 1978, reproduced by permission of Raven Press, New York.)

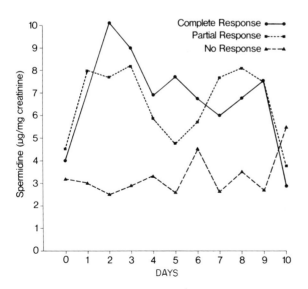

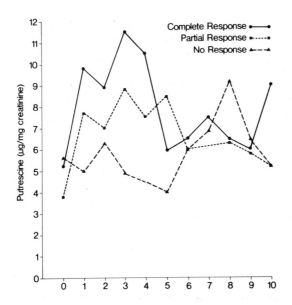

FIGURE 3 Urinary putrescine level in patients receiving various cancer chemotherapeutic regimens. Time zero was before chemotherapy, and day 1 was the start of chemotherapy. Each point is the mean from 16 to 22 different patients with no response, partial response, or complete response to therapy as judged by later standard clinical parameters.

FIGURE 4 Urinary putrescine levels in patients with multiple myeloma as a function of the [^{3}H]thymidine labeling index. These patients had not undergone any chemotherapy within several weeks. (From D. H. Russell and B. G. M. Durie, Ref. 4, <u>Polyamines as Biochemical Markers of Normal and Malignant Growth</u>, 1978, reproduced by permission of Raven Press, New York.)

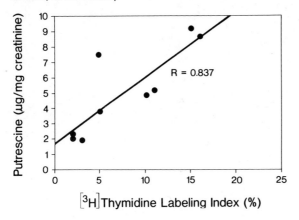

VI. URINARY POLYAMINE VALUES AND DISEASE ACTIVITY

Table 8 [38] shows that high levels of both putrescine and spermidine are related to progressive disease, whereas lower values occur in the same disease state that is slowly progressive. Therefore longitudinal studies of polyamine levels in patients could be invaluable in assessing the time at which the disease is shifting from a slowly progressive to a rapidly progressive state. There are no other markers that allow the clinician to rapidly assess alterations in tumor kinetics. This type of information is important in timing the reinitiation of chemotherapy before a major relapse.

Urinary putrescine levels as a function of remission or relapse in multiple myeloma patients indicate a greater than sixfold increase in the mean urinary putrescine excretion in patients in relapse versus remission (Table 9) [38]. Further, there is no overlap of the range of putrescine excretion in remission versus relapse. A number of other studies [61, 62] have now shown significant associations of elevated polyamines related to relapse. In a study by Nishioka and Romsdahl [62], serum levels of

TABLE 8 Baseline Urinary Polyamine Values and Disease Activity in Hematological and Solid Cancers

Disease	No. of patients	Concentration (μg/mg creatinine)	
		Putrescine	Spermidine
Acute myelogenous leukemia	12	11.4 ± 6.5[a]	9.7 ± 7.1
Smoldering leukemia	4	3.0 ± 1.0	1.8 ± 0.6
Chronic myelogenous leukemia			
Blast phase	2	27.6 ± 14.6	25.8 ± 12.5
Chronic phase	4	5.6 ± 1.7	6.8 ± 6.5
Disseminated soft tissue sarcomas			
Rapidly progressive	2	18.0 ± 9.6	9.9 ± 2.6
Slowly progressive	4	3.2 ± 1.0	2.1 ± 0.7
Metastatic colon carcinoma			
Rapidly progressive	5	12.7 ± 6.7	8.0 ± 6.4
Slowly progressive	6	4.0 ± 1.5	2.1 ± 1.6

[a] Mean ± SD.

Source: From D. H. Russell and B. G. M. Durie, Ref. 4, Polyamines as Biochemical Markers of Normal and Malignant Growth, 1978, reproduced by permission of Raven Press, New York.

TABLE 9 Urinary Putrescine Levels in Patients with Multiple Myeloma as a Function of Remission or Relapse

Condition	No. of patients	Urinary putrescine (µg/mg creatinine)	
		Mean	Range
Remission	13	0.97	0.4-1.88
Relapse	20	6.4	2.3-13.21

Source: From D. H. Russell and B. G. M. Durie, Ref. 4, Polyamines as Biochemical Markers of Normal and Malignant Growth, 1978, reproduced by permission of Raven Press, New York.

putrescine, spermidine, and spermine were all significantly elevated as a function of recurrent disease.

A. Multiple Markers to Prescribe Disease Status

Because extracellular elevations of polyamines really prescribe pathological cell growth and cell loss, it is important for specificity to utilize a series of markers that will be more specific for cancer as opposed to non-cancer pathology. Woo et al. [63] performed multiple correlations of various marker combinations with disease status of breast cancer patients and found an $R = 0.891$ based on urinary polyamines, nucleosides, and CEA (Fig. 5). Polyamines and nucleosides together were excellent markers of disease status ($R = 0.843$). Alone, polyamines have an $R = 0.594$, nucleosides an $R = 0.658$, and CEA an $R = 0.259$. Therefore the specificity lacking in either nucleosides alone or polyamines alone is compensated for by measuring both series of markers.

VII. SUMMARY

The polyamines, putrescine, spermidine, and spermine have been established as biochemical markers of normal and malignant growth. In malignancy, extracellular urinary and plasma levels of spermidine are related to tumor cell loss, and extracellular levels of putrescine are related to both the number of tumor cells in cell cycle and to tumor cell loss. A greater than twofold increase in urinary spermidine within 24 to 48 hr of chemotherapy predicts a complete or a partial response with a high degree of accuracy. Putrescine may be valuable not only in assessing the

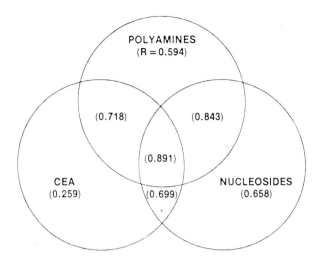

Multiple Correlations (R) of Various Marker Combinations

with Disease Status of Breast Cancer Patients: (Patient Number = 40).

FIGURE 5 Venn diagram for comparision of various marker combinations in predicting the disease status of patients with breast carcinoma. Polyamines: POLY, POLYR, POLYDT, POLYDTR. Nucleosides: NUC, NUCR, NUCDT, NUCDTR. CEA: CEA, CEADT. (From Ref. 63.)

early response to therapy, but also in determining whether the chemotherapy promotes a later burst of cell proliferation.

Therefore there is no doubt at this time that polyamine determinations in extracellular fluids could assist the clinician in (1) diagnosis, in combination with other markers, (2) early assessment of response to multimodality therapy, (3) tumor staging, and (4) assessment of disease activity including long-term monitoring of polyamine levels to pinpoint remission and relapse in adjuvant patients.

Table 10 reflects the major areas for future evaluation of the clinical usefulness of polyamines [4]. These areas can be developed utilizing current methodology (separation, quantitation using amino acid analyzers) [4, 46], but routine tests that are useful to the clinician and have a rapid turnaround time demand the development of simple, reliable, inexpensive assays for polyamines in biological fluids. The most promising approach seems to be the development of radioimmunoassays specific for the extracellular putrescine and spermidine conjugates. With the availability of such rapid tests as radioimmunoassays, monitoring of polyamine levels

TABLE 10 Major Areas for Further Evaluation of the Clinical Usefulness of Polyamines

1. Use of plasma spermidine to evaluate cell kill. Correlation with (a) clinical response and (b) in vitro drug sensitivity.
2. Use of urine (+ plasma) putrescine to evaluate cell kinetics. Correlations with autoradiography of marrow and tumor cells.
3. Long-term monitoring of extracellular polyamines regarding remission and relapse, e.g., adjuvant patients.
4. Evaluation of tumor status and burden.
5. Development and application of radioimmunoassays in clinical studies.

Source: From D. H. Russell and B. G. M. Durie, Ref. 4, Polyamines as Biochemical Markers of Normal and Malignant Growth, 1978, reproduced by permission of Raven Press, New York.

could assist the clinician in the rapid, accurate assessment of tumor activity and its response to current therapy. This information should be invaluable in prolonging the survival time of patients, as well as in designing the most effective therapy regimens.

ACKNOWLEDGMENT

This work was supported by USPHS Grant CA-14783 from the National Cancer Institute. Dr. Russell is the recipient of Research Career Development Award CA-00072 from the National Cancer Institute.

REFERENCES

1. S. S. Cohen, Introduction to the Polyamines, Prentice-Hall, Englewood Cliffs, New Jersey, 1971.
2. D. H. Russell (ed.), Polyamines in Normal and Neoplastic Growth, Raven Press, New York, 1973.
3. U. Bachrach, Function of the Naturally Occurring Polyamines, Academic Press, New York, 1973.
4. D. H. Russell and B. G. M. Durie, Polyamines as Biochemical Markers of Normal and Malignant Growth, Raven Press, New York, 1978.
5. W. G. Dykstra, Jr. and E. J. Herbst, Science 149:428 (1965).
6. J. Jänne, Acta Physiol. Scand. (Suppl.) 300:7 (1967).
7. A. Raina and J. Jänne, Fed. Proc. 29:1568 (1970).

8. A. Raina, J. Jänne, and M. Siimes, Biochim. Biophys. Acta 123:197 (1966).
9. D. H. Russell and J. Lombardini, Biochim. Biophys. Acta 240:273 (1971).
10. D. H. Russell and T. A. McVicker, Biochem. J. 130:71 (1972).
11. D. H. Russell, V. J. Medina, and S. H. Snyder, J. Biol. Chem. 245:6732 (1970).
12. D. H. Russell, in Polyamines in Normal and Neoplastic Growth (D. H. Russell, ed.), Raven Press, New York, 1973, p. 1.
13. H. G. Williams-Ashman, G. L. Coppoc, and G. Weber, Cancer Res. 32:1924 (1972).
14. C. A. Manen and D. H. Russell, J. Embryol. Exp. Morphol. 30:243 (1973).
15. D. H. Russell and P. J. Stambrook, Proc. Natl. Acad. Sci. USA 72:1482 (1975).
16. D. J. M. Fuller, E. W. Gerner, and D. H. Russell, J. Cell. Physiol. 93:81 (1977).
17. E. W. Gerner and D. H. Russell, Cancer Res. 37:482 (1977).
18. M. A. Melvin and H. M. Keir, Exp. Cell Res. 111:231 (1978).
19. M. G. Rosenblum and D. H. Russell, Cancer Res. 37:47 (1977).
20. M. G. Rosenblum, B. G. M. Durie, S. E. Salmon, and D. H. Russell, Metabolism of [^{14}C]spermidine and [^{14}C]putrescine in normal volunteers and in cancer patients, Cancer Res. 38:3161 (1978).
21. U. Bachrach, Ital. J. Biochem. 25:77 (1976).
22. D. H. Russell and C. C. Levy, Cancer Res. 31:248 (1971).
23. D. H. Russell, Cancer Res. 32:2459 (1972).
24. O. Heby and D. H. Russell, in Polyamines in Normal and Neoplastic Growth (D. H. Russell, ed.), Raven Press, New York, 1973, p. 221.
25. O. Heby and D. H. Russell, Cancer Res. 33:159 (1973).
26. O. Heby, S. Sauter, and D. H. Russell, Biochem. J. 136:1121 (1973).
27. H. G. Williams-Ashman, A. Corti, and B. Tadolini, Ital. J. Biochem. 25:5 (1976).
28. D. H. Russell, Nature 233:144 (1971).
29. D. H. Russell, C. C. Levy, S. C. Schimpff, and I. A. Hawk, Cancer Res. 31:1555 (1971).
30. U. Kim and J. Furth, Proc. Soc. Exp. Biol. Med. 103:643 (1960).
31. P. M. Gullino, F. H. Grantham, I. Losonczy, and B. Berghoffer, J. Natl. Cancer Inst. 49:1333 (1973).
32. D. H. Russell, P. M. Gullino, L. J. Marton, and S. M. LeGendre, Cancer Res. 34:2378 (1974).
33. C. J. Kovacs, H. A. Hopkins, R. M. Simon, and W. B. Looney, Br. J. Cancer 32:42 (1975).
34. W. B. Looney, A. A. Mayo, C. J. Kovacs, H. A. Hopkins, R. Simon, and H. P. Morris, Life Sci. 18:377 (1976).

35. D. H. Russell, W. B. Looney, C. J. Kovacs, H. A. Hopkins, L. J. Marton, S. M. LeGendre, and H. P. Morris, Cancer Res. 34:2382 (1974).
36. D. H. Russell, W. B. Looney, C. J. Kovacs, H. A. Hopkins, J. W. Dattilo, and H. P. Morris, Cancer Res. 36:420 (1976).
37. D. H. Russell, B. G. M. Durie, and S. E. Salmon, Lancet 2:797 (1975).
38. B. G. M. Durie, S. E. Salmon, and D. H. Russell, Cancer Res. 37:214 (1977).
39. D. H. Russell, Clin. Chem. 23:22 (1977).
40. F. Dreyfuss, R. Chayen, G. Dreyfuss, R. Dvir, and J. Ratan, Isr. J. Med. Sci. 11:785 (1975).
41. T. P. Waalkes, C. W. Gehrke, W. A. Bleyer, R. W. Zumwalt, C. L. M. Olweny, K. C. Kuo, D. B. Lakings, and S. A. Jacobs, Cancer Chemother. Rep. 59:721 (1975).
42. W. R. Fair, N. Wehner, and U. Brorsson, J. Urol. 114:88 (1975).
43. D. C. Tormey, T. P. Waalkes, D. Ahmann, C. W. Gehrke, R. W. Zumwalt, J. Snyder, and H. Hansen, Cancer 35:1095 (1975).
44. E. J. Sanford, J. R. Drago, R. J. Rohner, G. F. Kessler, L L. Sheehan, and A. Lipton, J. Urol. 113:218 (1975).
45. K. Nishioka and M. Romsdahl, Clin. Chim. Acta 57:155 (1974).
46. D. H. Russell and S. D. Russell, Clin. Chem. 21:860 (1975).
47. L. J. Marton, O. Heby, and C. B. Wilson, Int. J. Cancer 14:731 (1974).
48. K. Nishioka, M. M. Romsdahl, and M. J. McMurtrey, J. Surg. Oncol. 9:555 (1977).
49. J. Jänne, A. Raina, and M. Siimes, Acta Physiol. Scand. 62:352 (1964).
50. M. E. Feroli and R. Comolli, Exp. Gerontol. 10:13 (1975).
51. M. Siimes, Abstracts of 3rd Meeting Fed. Eur. Biochem. Soc., p. 10, Warsaw (1966).
52. T. P. Waalkes, C. W. Gehrke, D. C. Tormey, R. E. Zumwalt, J. N. Hueser, K. C. Kuo, D. B. Lakings, D. L. Ahmann, and C. G. Moertel, Cancer Chemother. Rep. 59:1103 (1975).
53. D. W. Lundgren, P. M. Farrell, L. F. Cohen, and J. Hankins, Proc. Soc. Exp. Biol. Med. 152:81 (1976).
54. D. H. Russell, Life Sci. 13:1635 (1973).
55. G. G. Steel, Eur. J. Cancer 3:381 (1967).
56. O. Heby and G. Andersson, Acta Pathol. Microbiol. Scand. (A) 86:17 (1978).
57. M. Nitta, Nagoya Med. J. 21:95 (1976).
58. L. J. Marton, O. Heby, V. A. Levin, W. P. Lubich, D. C. Crafts, and D. B. Wilson, Cancer Res. 36:973 (1976).
59. S. E. Salmon, Blood 45:119 (1975).
60. L. M. VanPutten, Cell Tissue Kinet. 7:493 (1974).

61. T. D. Miale, O. M. Rennert, D. L. Lawson, J. B. Shukla, and J. L. Frias, Med. Pediatr. Oncol. 3:209 (1977).
62. K. Nishioka and M. M. Romsdahl, Cancer Lett. 3:197 (1977).
63. K. B. Wood, T. P. Waalkes, D. L. Ahmann, D. C. Tormey, C. W. Gehrke, and V. T. Oliverio, Cancer 41:1685 (1978).

11

ECTOPIC HORMONE PRODUCTION BY TUMORS

A. MUNRO NEVILLE / Ludwig Institute for Cancer Research (London Branch), Royal Marsden Hospital, Sutton, Surrey, United Kingdom

I. Introduction   267
II. Definition of Ectopic Hormone Production   268
III. Evidence Required to Establish Ectopic Hormone Production by a Tumor   268
   A. Arteriovenous Hormone Gradients Across the Tumor Bed   268
   B. In Vitro Production and Release   269
   C. Demonstration of Ectopic Hormones in Tumors   271
   D. Structure of Ectopic Hormones   277
IV. Range of Ectopic Hormonal Manifestations   280
   A. Clinical Syndromes   280
   B. Incidence   281
   C. Multiple Ectopic Hormone Production   283
   D. Ectopic Releasing Factor-Like Activity   283
V. Clinicopathological Applications   284
   A. Clinical Roles   284
   B. Pathological Implications   285
VI. Biological Significance   287
VII. Summary   290
   References   291

I. INTRODUCTION

The ectopic production and release of polypeptide hormones by so-called nonendocrine tumors is now a well-recognized phenomenon. The frequency of its detection appears to be rising, possibly as a result of increasing clinical awareness and access to a protean range of hormonal immunoassay methods.

The subject has been reviewed recently on several occasions from clinical, pathological, biochemical, and biological standpoints [1-15]. The present treatise will concentrate on certain aspects, and in particular will review the criteria necessary to establish that a tumor is in fact producing ectopic hormones. An assessment of their present clinical, pathological, and biological implications will be attempted and, in the process, some of the problems still awaiting solution will be highlighted.

II. DEFINITION OF ECTOPIC HORMONE PRODUCTION

A hormone is defined as ectopic when it is produced by a tumor that has arisen from, and is composed of, cells not normally regarded as being engaged in the production of the hormone [16]. This implies that we are aware of all the known sites of hormone manufacture, which is almost certainly untrue. The recent recognition of novel gastrointestinal hormones and their cellular distribution, and of the peptidergic nervous system serve amply to temper the above definition [17, 18].

Ectopic hormone production, however, does not appear to be a random process. It is confined to protein and polypeptide hormones, apparently some more frequently than others; it does not involve steroid hormones, thyroid hormones, or the catecholamines [2]. It exhibits a high degree of association with particular histological types of tumors and sites (vide infra). Consequently, as will become apparent in the ensuing discussion, further studies using sensitive cell localization methods are still needed to outline the "truly normal" cellular distribution of the various protein and peptide hormones which at this time are regarded as ectopic when of tumor origin. Such data are essential before the biological significance of their release from tumors can be appreciated.

III. EVIDENCE REQUIRED TO ESTABLISH ECTOPIC HORMONE PRODUCTION BY A TUMOR

In recent years, several groups of workers have outlined criteria which must be met prior to acceptance of a particular tumor's ability to release an ectopic hormone (Table 1). Some are more definitive and important than others. Probably no one criterion is adequate. The first four, in particular, are highly relevant and justify amplification.

A. Arteriovenous Hormone Gradients Across the Tumor Bed

This is a most important criterion. However, it is possible that the cells (including the stromal cells) of a tumor could bind a hormone specifically or nonspecifically, and thereafter release and return the hormone into the circulation [19]. Thus, on its own, it is only highly suggestive and needs

TABLE 1 Evidence Required to Establish Ectopic Hormone Production by a Tumor

1. Arteriovenous hormone gradient across the tumor bed
2. In vitro production and release by the tumor
3. Demonstration of the hormone in the tumor and at higher levels than adjacent uninvolved tissues
4. Quantitative or qualitative hormone differences of the hormone with respect to the hormone in question when of nontumor origin
5. Association of the tumor with a clinical endocrine syndrome
6. Association of the tumor with elevated or inappropriate hormone levels in the blood
7. Fall in the blood hormone levels with clinical improvement on removal or therapeutic regression of the tumor
8. No change in the blood hormone levels on removal of the gland usually associated with production of the hormone in question

Source: Modified from Rees and Ratcliffe [2] and Blackman et al. [12].

confirmation by, for example, demonstrating the in vitro synthesis and release of the hormone in question.

Unfortunately, arteriovenous gradients across the tumor bed have been demonstrated infrequently (Table 2), possibly because of technical difficulties at the time of operation, a long plasma half-life of the hormone, or low levels of hormone production [2] coupled with assay imprecision or insensitivity. Moreover, ectopic hormone production and/or release may be episodic [20] or masked by a rise in the level of the hormone under study which is derived from the eutopic organ as a result of surgical stress, etc. The technique, however, has been employed successfully to localize tumors in vivo before surgery [21] (vide infra) and may become more important with the development of more advanced blood sampling techniques.

B. In Vitro Production and Release

This is probably the most reliable criterion providing potentially irrefutable proof. It is subject to limitations such as difficulty in maintaining hormone secretion in vitro, and the release of low hormone concentrations below the limits of assay sensitivity [22].

Although organ cultures of tumors may be easier to establish, their use to demonstrate ectopic hormone production and release should ideally

TABLE 2   Some Examples of Arteriovenous Differences Demonstrated Across Tumors*

Location	Tumor type	Hormone
Bronchus	Carcinoma	FSH, LH
	Carcinoid	ACTH
	Oat cell carcinoma	Calcitonin[†]
	Hypernephroma	PTH
Kidney	Hypernephroma	PTH
Adrenal	Pheochromocytoma	ACTH[‡]
	Pheochromocytoma	ACTH, βMSH

*Source: From Rees and Ratcliffe [2].
[†]Source: From Silva et al. [110, 111].
[‡]Source: From Jeffcoate and Rees [21].

involve an intrinsic radioactive labeling technique due to the inherent problem of in vivo hormone absorption and subsequent release in vitro. Intrinsic labeling of ectopic hormones has been reported on only a few occasions (Table 3), possibly due to the low biosynthetic level compared with the levels which would be derived from culture of the appropriate eutopic source (e.g., pituitary cultures). A further drawback is that the viable life expectancy of organ cultures is relatively short [22].

Monolayer cell cultures provide a more suitable and reliable alternative technique. Several studies have provided firm evidence of ectopic hormone synthesis and release over quite long periods of time (one month or so). More valuable, however, are functioning cell lines if they can be derived from such primary monolayer cultures. This approach has yielded several interesting cell lines capable of synthesizing a variety of ectopic hormones, including ACTH, vasopressin, HCG and its α and β subunits, and calcitonin [23-32]. Such lines provide accessible and reproducible sources of ectopic hormones, while also being of value as probes to study factors modulating hormone production. In addition, they represent a useful approach to study the chemical nature of ectopic hormonal products [8].

The implantation of human tumors into immune-deprived animals and the subsequent development of functioning transplantable lines provides an in vivo approach which may complement or be an alternative to the monolayer system.

The exciting new developments in molecular biology may also have an important future in providing a fuller understanding of many aspects of ectopic hormone production [33, 34]. To date, mRNA has been drived from eutopic hormonal and related product sources, cDNA has been prepared by the use of reverse transcriptase, and the nucleotide sequence has been determined. In this manner, for example, the corresponding amino acid sequence for the bovine ACTH-β lipotropin molecule has been derived [35].

C. Demonstration of Ectopic Hormones in Tumors

Many of the original studies aimed at proving ectopic hormone synthesis by a particular tumor were based on tumor extraction data. The finding of higher hormone levels in a nonendocrine tumor compared with related normal uninvolved tissue was taken as evidence of ectopic hormone production [16]. While valuable results were obtained in this manner, they are by themselves inadequate proof of ectopic hormone production, for the reasons stated above.

A recent study of human breast carcinomas by Ratcliffe [36] is pertinent in this context. By extraction of the tumors, he found higher levels of ACTH, growth hormone, and prolactin than in the associated normal breast tissue. From studies of the molecular nature of the hormones, he concluded that these hormones were probably not ectopic products, but were eutopic in origin and present due to receptors for them in the tumor tissues.

Not infrequently, extracts of tumors associated with a clinically overt hormonal syndrome appear to contain amounts similar to those present in tumors of the same type and site unassociated with a clinically overt hormonal syndrome [10, 37-39]. Such apparently inconclusive findings may be due to a variety of factors, including hormone structural heterogeneity including bioinert forms, assay specificities, and a low hormonal storage capacity. Rarely is no hormone corresponding with the clinical syndrome detected in the tumor.

TABLE 3 Intrinsic Radiolabeling of Ectopic Hormones

Location	Type of carcinoma	Hormone	Reference
Lung	Oat cell	Vasopressin	Klein et al. [112]
	Undifferentiated	Vasopressin	George et al. [113]
	Undifferentiated	Growth hormone	Greenberg et al. [114]
	Undifferentiated	HCG, HPL	Rabson et al. [24]
	Epidermoid	Calcitonin	Lumsden et al. [115]
Uterus	Anaplastic	Vasopressin	Martin et al. [116]
Kidney	Adenocarcinoma	PTH	Greenberg et al. [117]

During the past few years, immunocytochemical methods have become more sophisticated. In particular, the immunoperoxidase technique has been applied to demonstrate many hormones and antigens at a cellular level [40]. This method gives excellent localization and permanent preparations in which good morphological detail is also preserved (Fig. 1). It has yet to

FIGURE 1 An immunoperoxidase stain for βHCG is illustrated in an example of an embryonal carcinoma. The βHCG is located in the cytoplasm of cells forming a syncitium around a blood vessel in the tumor (× 300).

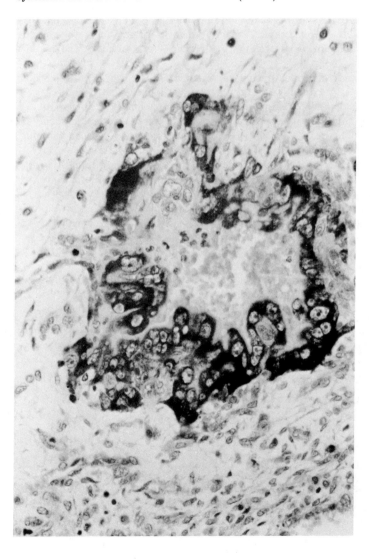

be applied successfully to demonstrate ectopic products, but it is hoped that modifications of the technique will prove sensitive enough to demonstrate the small amounts of such ectopic hormones and to give data on their cellular distribution. This approach may yield results that will have major biological and pathological significance in improving our understanding of the ectopic hormonal syndromes. It is most important in this context to establish for certain whether such hormonal products are truly ectopic to the cells from which the tumor arises.

Throughout the body, there is a series of cells, referred to as the Amine Precursor Uptake and Decarboxylation (APUD) cells, which derive their name from their ability to take up amine acid precursors and to decarboxylate them [41, 42]. In addition to amine storage and production, they can elaborate peptide hormones.

Immunocytochemical techniques may also be applied at the ultrastructural level. Electron-dense granules are present in the cytoplasm of APUD cells. These granules are considered, but not yet proved, to be the storage site of the hormones. Not surprisingly, tumors producing ectopic hormones of the type described above may contain similar granules (Figs. 2 and 3). Their frequency varies from presence in most or many cells of carcinoid tumors, to presence in very few cells of oat cell tumors [43-48]. The cells of the pituitary producing ACTH, the calcitonin producing C cells of the thyroid, the bronchial Kulchitsky cells, the endocrine cells of the gastrointestinal tract, the islets of Langarhans, and the pheochromocytes of the adrenal medulla are characteristic examples.

In view of the presence of APUD cells in the lung, for example, it has been suggested that these cells give rise to tumors that produce APUD-like hormones (e.g., ACTH, ADH, calcitonin), such as carcinoid tumors and oat cell carcinomas, and that their hormone production is a normal appropriate, and not an inappropriate, function. This would argue, therefore, that the tumor cells producing such ectopic hormones have not changed their differentiated characteristics, but simply are an amplification of a small number of bronchial cells which produce that hormone in amounts which have not yet proved detectable under normal conditions.

Insufficiently sensitive analytical methods make this hypothesis difficult to disprove. However, Ellison [8] argued that the weight of existing evidence suggests that this could not be a general explanation. Since such a wide range of hormones is represented by bronchial tumors, it would be necessary to postulate a large number of minor populations of endocrine cells. In addition the presence of Kulchitsky cells in the lung would not account for the production by bronchial tumors of hormones which are not linked with such pronounced APUD cell characteristics, for example, parathyroid hormone (PTH) and gonadotropins. Furthermore, this possible explanation for lung-tumor-derived hormones could not directly be applied to the wider phenomenon of ectopic hormone production by tumors at other sites where there are apparently no normal APUD cell components.

On the other hand, the undeniably high incidence of ectopic hormones, and especially some of the APUD-type hormones in lung carcinomas, suggests that the presence of normal endocrine cells may indeed be highly

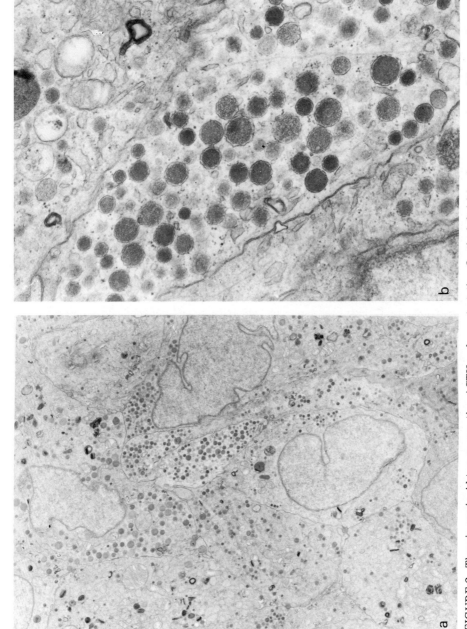

FIGURE 2 Thymic carcinoid tumor secreting ACTH and somatostatin. On the left (a) is a low-power view of the granule-containing tumor cells (× 5000), while on the right (b) granule heterogeneity and irregularity can be observed (× 28,700). The relative numbers of granules vary considerably from cell to cell.

11 / Ectopic Hormone Production

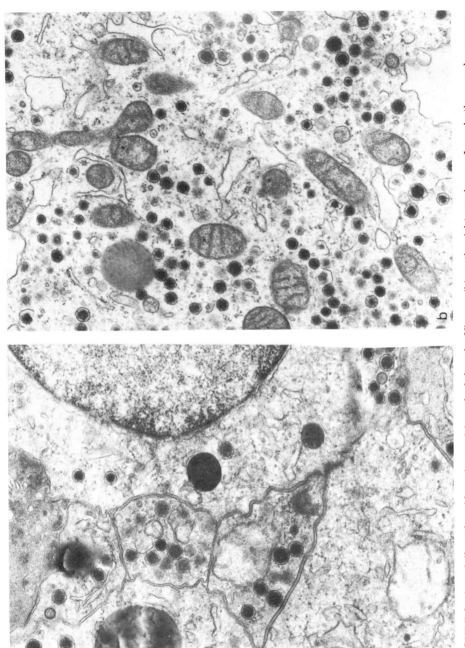

FIGURE 3 A medullary thyroid carcinoma (a) on the left and a Vipoma (b) on the right are shown. In each, the granules appear more regular and have denser cores ($\times$ 20,000).

TABLE 4 Some of the Clinical Syndromes and Most Characteristic Tumors Associated with Ectopic Hormone Production

Hormone	Most characteristic associated neoplasms	Syndrome
ACTH and related moeities	Oat cell bronchial carcinoma Islet cell carcinoma Thymic carcinoma Pheochromocytoma and other neuroectodermal tumors Carcinoid tumor Medullary thyroid carcinoma	Cushing's syndrome
Arginine vasopressin	Bronchial carcinoma (oat cell) Carcinoid tumors	Inappropriate antidiuresis
PTH and other osteolytic agents	Squamous cell carcinoma of lung Hepatoma Hypernephroma Mammary carcinoma	Hypercalcemia
Gonadotropins	Large cell carcinoma of lung Pancreatic carcinoma Gastric carcinoma Hepatoblastoma Hypernephroma	Gynecomastia (adult males) Precocious puberty
HPL	Bronchial carcinoma	Gynecomastia
Growth hormone	Bronchial carcinoma Carcinoid tumor	Hypertrophic pulmonary osteoarthropathy
Prolactin	Hypernephroma Bronchial carcinoma	Galactorrhoea
Calcitonin	Oat cell bronchial carcinoma	–
Somatomedin (NSILA)	Hepatoma Mesenchymal tumor Adrenocortical carcinoma	Hypoglycemia
Vasoactive intestinal peptide (VIP)	Bronchial carcinoma	Watery diarrhea and hypokalemic alkalosis (WDHA)
Enteroglucagon	Hypernephroma	Malabsorption Constipation
Erythropoietin	Uterus	Erythrocytosis

relevant to the expression of the phenomenon in lung cancer. The observed nonrandom association of hormones produced with the histological type and, possibly, cell of origin of the tumor affords further evidence of such a link (see Table 4).

Multiple ectopic hormone production by a particular tumor is now well established (see Table 5). It will be interesting to ascertain whether one tumor cell can produce all of the ectopic hormones noted for a single tumor (Fig. 2), or whether this is a property residing in different tumor cells with one cell, one hormonal product being the rule. If the same cell manufactures all the hormones, then it will be interesting to discover whether this is a property of a single granule or whether one hormone only is contained in a particular granule. In addition, the sites of synthesis of ectopic hormones not normally stored in granules such as parathyroid hormone and the gonadotropins may be detected in this manner.

D. Structure of Ectopic Hormones

The structure and chemical nature of ectopic hormones has been a question of importance to which many workers have addressed themselves in recent years. The presently available evidence, derived primarily from physiochemical and immunochemical studies rather than from amino acid sequence data, suggests that ectopic hormones are the same as, or very similar to, normal hormones, their physiological precursors, or metabolic fragments.

Hormonally active peptides would appear to be derived by a series of progressive degradative steps from a bioinert higher molecular weight precursor [35, 49-54]. Moreover, the bioactive peptides may be subject to further degradation. This peptide cascade, which occurs in the endocrine glands normally and which is subject to variation in ectopic secreting tumors, lies at the basis of ectopic hormone structural heterogeneity.

Thus larger molecular weight forms of PTH, ACTH, and calcitonin have been extracted from ectopic tumor sources and/or shown to be released in vivo or in vitro by such tumors [10, 26, 48, 52, 55, 56]. High molecular ("big") ACTH has been shown to contain not only $ACTH_{1-39}$ but also β-lipotrophic hormone (βLPH), a 91 amino acid peptide [35]. This may account for the high frequency with which both are associated as ectopic tumor products. In patients with lung tumors and no evidence of hypercortisolism, it has been suggested that these tumors produce "big" bioinert ACTH, whereas when a clinical syndrome is apparent both "big" ACTH and the bioactive $ACTH_{1-39}$ form may be released [10].

Similar findings have been recorded for the inappropriate production of calcitonin. While monomeric calcitonin, a 32 amino acid peptide, may be released into the blood or in vitro by lung tumors, higher molecular weight forms tend to predominate [26, 56, 57]. Interestingly, medullary thyroid carcinoma, the eutopic tumor source of calcitonin, also produces a range of higher molecular weight forms in addition to the monomeric form [58, 59].

TABLE 5 Some Reported Examples of Multiple Ectopic Hormone Production

Organ	Histology	Hormones
	Oat cell	ACTH, PTH
	Oat cell	ACTH, AVP
	Carcinoid	ACTH, AVP, neurophysin
Lung	Oat cell	ACTH, βMSH, CLIP, AVP, insulin, oxytocin, neurophysin, prolactin
	Oat cell	Oxytocin, AVP, neurophysin
		ACTH, βMSH, CT
	Oat cell	ACTH, βMSH, ADH
	Oat cell	ACTH, βMSH, ADH, CT
	Oat cell	ACTH, βMSH, CT
	Adenocarcinoma	ACTH, βMSH, CT
Liver	Carcinoid	ACTH, βMSH, CT
Thymus	Carcinoid	ACTH, βMSH, CT
	Carcinoid	ACTH, somatostatin, gonadotropin-releasing activity
Adrenal	Pheochromocytoma (2)	ACTH, βMSH, CT
Stomach	Carcinoid	ACTH, βMSH, CT
	Carcinoid	PTH, CT
	Carcinoid	ACTH, βMSH
Esophagus	Anaplastic small cell	ACTH, βMSH, CT
	Islet cell	ACTH, βMSH
Pancreas	Islet cell	PTH, CT

Source: From the data of Liddle et al. [119], Cullen and Tomlinson [120], O'Neal et al. [121], Bailey [20], Hamilton et al. [122], Rees et al. [88], Rees and Ratcliffe [2], Deftos et al. [123], Gomi et al. [124], Hirata et al. [125, 126], Abe et al. [127], Coscia et al. [128], Himsworth et al. [95], Yamaguchi et al. [129], and Shalet et al. [91].

However, not all ectopic hormones, for example, arginine vasopressin, are secreted in higher molecular forms [60].

In certain ectopic hormone situations, it has been found that the hormone present is mainly of a smaller molecular weight and represents a fragment of the normal product. A study of ACTH will illustrate this point. Tumor extracts contain an excess of C- over N-terminal immunoreactive ACTH material [2]. This has been shown to be due to the presence of a peptide with the sequence of 18-39 ACTH. The peptide in question is referred to as corticotropin-like intermediate lobe peptide (CLIP), and it is also detectable in the intermediate lobe of animal pituitaries [61].

CLIP may be produced as a by-product of $ACTH_{1-39}$ in the breakdown of $ACTH_{1-39}$ to yield α-MSH. A 15 amino acid peptide corresponding to the N-terminal fragment of ACTH has also been found ectopically in tumors [62], and Lowry et al. [63] recently extracted from a thymic carcinoid tumor a peptide consisting of the 2-38 amino acid sequence of authentic $ACTH_{1-39}$.

The glycoprotein hormones provide some of the best evidence for fragment or subunit release by tumors. The glycoprotein hormones (FSH, LH, TSH, and HCG) exist as heterodimers composed of noncovalently bound dissimilar α and β subunits [64]. The α subunits are immunologically indistinguishable, whereas the β subunits are distinct and confer biochemical specificity, immunological specificity, and biospecificity. Differences in the ratio of subunits released have been found in their ectopic and eutopic production, as well as in their ability to recombine [25, 28, 65]. These differences may be related to the release of higher molecular weight forms of the α subunit. This α and β subunit discordance suggests that their synthesis is controlled by different genes and unique mRNAs.

In summary, ectopic hormones may differ in dominant molecular type from their eutopic normal source. However, such moieties are formed in the normal sequence of evolution or represent breakdown products derived from the normal hormone after its formation. The mechanistic reasons for the molecular heterogeneity of ectopic hormones are unknown. It is apparent that a peptide cascade occurs normally, and it is unlikely that ectopic tumors will be understood better until we learn more about the controlling factors and mechanisms of their normal production.

As mentioned above, many ectopic hormone-producing tumors contain cytoplasmic granules in which the hormones are thought to be stored (Fig. 2). Similar granules occur in neuroectodermal tumors such as pheochromocytomas, and abnormalities in their mode of catecholamine storage have been documented [66]. Consequently, abnormalities in hormone packaging and storage in the granules, together with altered release mechanisms, could account for some of the observed hormonal structural heterogeneity. The heterogeneity of granule structure is well documented

[67, 68], and unlike the normal situation, such "ectopic" granules may be released intact from the tumor cells [69].

It is also conceivable that such tumors may express inappropriate enzymes or fail to possess the enzymes necessary for normal synthesis of such hormones. In addition, the sites of enzyme packaging in the tumor cells may be abnormal.

It is obvious, therefore, that many further studies of the intercellular location of the hormones, their precursors, and fragments, together with an appreciation of their synthetic mechanisms, are necessary and will almost certainly be rewarding.

IV. RANGE OF ECTOPIC HORMONAL MANIFESTATIONS

A. Clinical Syndromes

There is a wide variety of clinical syndromes associated with ectopic hormone production. Some of the more common examples, together with their most characteristic associated tumors, are listed in Table 4. Their clinical manifestations, diagnosis, and therapy have been discussed extensively elsewhere [3, 6, 11-14, 40, 60, 70, 71]. A direct causative association of syndrome and ectopic hormone has been proved for most of these syndromes; however, the etiology of others, including gynecomastia and hypertrophic pulmonary osteoarthropathy, remains to be clarified.

In addition, the WDHA has recently been reported in association with tumors in the absence of raised VIP (vasoactive intestinal peptide) levels [72].

It is interesting to note the paucity of ectopic hormonal syndromes in association with mammary carcinoma (Table 4). The hypercalcemia of breast cancer is usually associated with osteolytic metastases, and an increased local production of prostaglandins may play an etiological role [73].

The ectopic production of gonadotropins by breast tumors has been reported by several workers [74, 75], and their cytochemical demonstration in tumor cells has been suggested as a poor prognostic indicator [76, 77]. A recent study by Monteiro et al. [78], however, failed to confirm (cytologically or by assay of plasma) any significant content or production of gonadotropins and other placental proteins in a large series of breast cancer subjects. Further studies are therefore needed.

Raised plasma calcitonin levels in breast cancer have been reported by Coombes et al. [56] and shown to be ectopic in origin in one tumor by passage and growth in nude mice. However, more recent studies tend to support the notion that such calcitonin elevations are more usually eutopic than ectopic in source [79].

Some tumors may be associated with the release of several ectopic hormones; generally the physiological effects of only one of the hormones tend to dominate the clinical picture [80]. In addition to the tumor types listed, others more rarely may cause one or more of these syndromes. For example, the ectopic production of ACTH by neoplasms has been reported at such diverse sites as the gallbladder [81], prostate [82, 83], salivary gland [84], and cervix [85]. Many, but not all, such tumors exhibited morphological similarities to tumors known to arise from APUD cells [15] (Figs. 2 and 3). Recent data have also shown that some of the syndromes may result not from ectopic but from eutopic production of the causative hormone whose increased release was stimulated by inappropriate tumor formation of releasing factor-like activity (vide infra). This may be the principal reason why some tumors do not contain increased hormone levels yet are associated with a particular syndrome due to its excess in plasma.

There are many other syndromes associated with cancers which may be due to hormonal factors, but these syndromes await characterization and clarification. These include cachexia, hepatic dysfunction, pyrexia, and neuromuscular and dermatological manifestations. Each represents a future fascinating area for study. Of interest in this context is the recent description of a human hypernephroma xenograft model which caused cachexia in a patient before its removal and continues to do so as a tumor xenograft growing in immune-deprived mice [86].

B. Incidence

The precise incidence of ectopic hormone production is unknown and will be extremely difficult to ascertain. The factors which may operate to obscure the accurate acquisition of this information have been discussed by Rees [4, 5] and include inadequate clinical follow-up, production of apparently bioinert hormones (e.g., calcitonin) or precursors and fragments, the availability of assays, and investigator bias. It will also depend on whether overt clinical or solely biochemical evidence is employed to assess hormone production.

Table 6 lists the reported range of overt hormonal syndromes in association with the different forms of bronchial carcinoma. However, if biochemical evidence is used, the incidence is much higher.

Gewirtz and Yalow [38] found that almost all lung tumors contained elevated levels of "big" ACTH. Odell et al. [10] more recently found that 95% of tumors contained immunoassayable "big" ACTH, αLPH, α subunit, HCG, and vasopressin in amounts greater than blood. Moreover, the "big" ACTH, βLPH, and vasopressin levels were higher than in appropriate control tissues. In a study of oat-cell carcinomas, Ratcliffe [36] found that 90% of the tumors had elevated levels of ACTH, βLPH, and calcitonin. In 50% of the tumors, growth hormone was also elevated.

Elevated blood levels of ACTH have been reported in between 67 and 72% of patients with lung cancer [10, 38]. Table 7 lists comparative data

TABLE 6 Incidence of Overt Hormonal Syndromes in Lung Cancer

Hormone	Syndrome	Incidence (%)
ACTH and related moeities	Cushing's syndrome	3-22
ADH	Inappropriate Antidiuresis	8-35
PTH and other osteolysins	Hypercalcemia	6-16
Gonadotropin	Gynecomastia	2
Calcitonin	–	70

Source: From data of Rees and Ratcliffe [2], Rees [4, 5], Coombes et al. [96], Odell and Wolfsen [13], and Singer et al. [118].

for other hormones and also for patients with colorectal cancers. However, Ratcliffe [36] failed to confirm these findings in a series of unselected lung cancer patients when plasma ACTH and βLPH levels were compared with age- and sex-matched patients with non-neoplastic lung disease.

These discrepancies may be related to assay specificity. Immune complexes containing higher molecular weight forms of immunoassayable ACTH have been described in the blood of patients with overt ectopic ACTH

TABLE 7 Percentage of Patients with Elevated Plasma Peptide Levels in Carcinoma of Lung and Colon

Peptide	Lung	Colon
ACTH-proACTH	72	58
βLPH	61	12
α subunit (men)	30	26
AVP	41	43
HCG	7	6

Source: From the data of Odell et al. [10].

syndromes [87]. Such complexes may influence incidence assessment and assay results.

Finally, it cannot be overemphasized that evidence derived solely by tumor extraction is inadequate. As mentioned before, Ratcliffe [36] found elevated amounts of ACTH, βMSH, and growth hormone in a series of mammary tumors. The tumor ACTH and βMSH were chemically similar to their normal pituitary counterparts and were present probably because breast cells contained appropriate receptors for them.

Thus the true prevalence of ectopic hormone production is far from being gauged and will probably depend on a series of experiments using unselected tumors and in which tumor extraction, assay of blood levels, and in vitro synthesis and release are correlated.

C. Multiple Ectopic Hormone Production

There are numerous documented reports of the ectopic tumor production of multiple hormones (Table 5). In many instances, such as the association between ACTH, βMSH and βLPH, and vasopressin (AVP) and neurophysin, the multiplicity for each group is explicable on the basis of a common precursor molecule. However, there are other examples where this cannot be the sole explanation. These include the association of ACTH, calcitonin, and growth hormone in oat cell carcinomas, ACTH (and related moeities), AVP (and related moeities), insulin, and prolactin in an oat cell carcinoma [88] and calcitonin and ACTH in medullary thyroid carcinoma [9]. The implications of multiple hormone production are discussed below in the section on biological significance.

D. Ectopic Releasing Factor-Like Activity

During the past few years, some studies have suggested that some humoral and metabolic syndromes may be caused in part or whole by the ectopic production of releasing factor-like activity.

There are several interesting reports of endocrine syndromes, such as Cushing's syndrome or acromegaly [89] in association with various neoplasms in which plasma ACTH or growth hormone levels were raised but their ectopic production by the tumor could not be demonstrated [90, 91]. Moreover, there was no tumor or primary abnormality of the pituitary gland, and removal of the tumors resulted in amelioration of the syndromes. These observations drew attention to the possibility that these tumors might be manufacturing hypothalamic-type releasing factors to act on the pituitary, and in this way cause the various syndromes.

Acromegaly has been reported in association with lung tumors [89]. While the production of growth hormone by the tumor may explain some of the syndromes, others could not be due to such hormone production because there was no evidence of growth hormone in the tumor. This raised the possibility of the production of growth-hormone-releasing activity. This proposition has been given a firm foundation by the recent studies of Shalet et al. [91], who found growth-hormone-releasing activity in a bronchial carcinoid tumor by using a flow culture system whereby the media from the tumor cells in culture were passed over pituitary cells also in culture when a stimulation of growth hormone release was observed. The nature of the releasing activity was not determined, but neither dopamine nor 5-hydroxytryptamine was involved. This particular tumor was also found, in this manner, to be producing corticotropin-releasing activity, as had already been demonstrated for some tumors [90, 92]. Other carcinoid tumors have also been shown to produce somatostatin and gonadotropin-releasing activity [91].

Of extreme interest and potential biological significance is the detection of ectopic tumor production of corticotropin-releasing activity in association with ACTH. This might suggest an alternative mechanism leading to "ectopic" ACTH production in which CRF-like activity is locally active and effective.

It may be, therefore, that this form of ectopic hormone production is more common than hitherto anticipated, and may represent another mechanism whereby various hormonal syndromes can arise.

V. CLINICOPATHOLOGICAL APPLICATIONS

Despite the time that has elapsed since the first recognition of ectopic hormone production and the present awareness of the diversity of factors that can be released by tumors, their importance and value in the clinical management of oncological diseases remains to be fully appreciated and exploited. This may be due in part to the relatively few laboratories equipped and able to assay an extensive array of products, and thus to enable their clinical use. However, overt hormonal effects are often only recognized late in the course of the disease.

A. Clinical Roles

1. Systemic Manifestations

The ectopic production by tumors of known hormones, as will be readily apparent from the preceding sections, can account for many of the general systemic manifestations associated with neoplasia (Table 4). The etiology of other features such as cachexia still awaits precise elucidation. However, such discoveries will be important for the future total care of the cancer patient and for the design of appropriate therapeutic schedules. It is nevertheless important to recognize and appreciate the significance of those various systemic changes, since they may predate the overt physical detection of the tumor itself.

2. Biochemical Markers: Diagnosis, Localization, and Monitoring

Many clinicians have looked to the production of ectopic hormones and their assay in plasma or urine as potentially valuable biochemical indices to facilitate the detection and monitoring of neoplastic disease. This is not unreasonable when one reflects upon the value of eutopic hormonal markers in these respects, e.g., HCG or its β subunit and gestational choriocarcinoma, AFP and hepatomas or yolk sac carcinomas, calcitonin and medullary thyroid carcinomas, and steroid hormones and adrenocortical tumors [93].

It was hoped that ectopic hormones might facilitate the earlier detection of both primary and metastatic neoplasia. Although there are anecdotal examples [5], in general these products fail to fulfill the criteria necessary to achieve such goals. Ectopic products are not unique to tumors; seldom are plasma levels raised in the presence of small tumor bulk, and the available evidence tends to indicate that the plasma levels are not stoichiometric with respect to the viable tumor cell population. Moreover, their production may be cyclical [20] and may exhibit dependence on the tumor cell density and cell cycle stage [80].

In the future the ability to measure precursor forms which tend to be associated more with ectopic than with eutopic release may provide a new inroad to achieve some of these clinical goals.

Despite these limitations, there are now several well-documented examples emanating in general from clinics with access to laboratories capable of assaying a wide range of hormones and where a series of sophisticated, selective venous catheterizations have been done [21]. Measurement of hormone levels at different venous levels has resulted in the localization of tumors not detectable by other clinical or physical diagnostic methods. However, in each case a clinical syndrome has been present to guide the laboratory to measure a particular hormone.

The sequential monitoring of ectopic hormone levels after primary diagnosis and treatment could be valuable in the earlier detection of metastases and in assessing responses to chemotherapy and radiotherapy, as has been found for the oncofetal antigens and eutopic hormones [93]. This area has not been fully exploited. While several examples of the use of plasma ectopic hormone assays in these respects have been recorded [5, 2, 94-96], their full value and potential remains to be outlined (Fig. 4).

B. Pathological Implications

One reason why ectopic hormones may not have been used often enough as tumor markers may be the absence of overt signs of hormone excess to

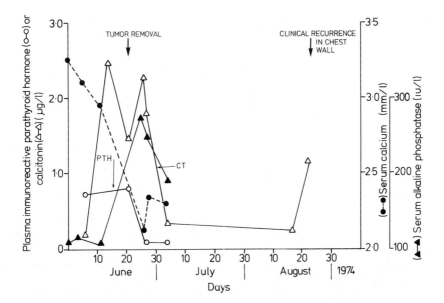

FIGURE 4 Ectopic production of immunoreactive PTH and calcitonin in a patient with a squamous cell bronchogenic carcinoma. PTH levels (normal <0.76 µg/liter) were high (for this serum calcium) preoperatively, as were calcitonin levels (normal <0.1 µg/liter). Following removal of the primary tumor both these hormone levels fell, and there was a concomitant rise in alkaline phosphatase. Calcitonin rose upon clinical recurrence of the tumor in the chest wall. The tumor contained significant amounts of immunoreactive PTH (6.44 µg/g tumor; 0.81 µg/g normal lung) and calcitonin (0.49 µg/g tumor; 0.016 µg/g normal lung).

guide the clinician on which hormone(s) to measure. In this context, immunocytochemical studies may have a future role if the results of the eutopic production of HCG and AFP by teratomas or CEA by various neoplasms can be applied to the ectopic situation [40].

It has been found that if a particular antigen or hormone can be demonstrated in the primary tumor by immunocytochemistry, raised levels will develop during the subsequent follow-up phase before or at the time when metastases are detected. Such technology remains to be attempted with ectopic tumors. It may be that present immunocytochemical methods are still not sensitive enough to demonstrate many of the ectopic hormones. Nonetheless, if antibody reagents to the precursor forms can be developed, success could be anticipated. Then the range of hormones associated with a particular lesion could be described and the clinician alerted to those hormones that may have value as plasma markers.

11 / Ectopic Hormone Production

This area of future research is a particularly exciting one for histopathologists. It may well enable their playing a more useful and dynamic role in patient management. Describing the functional properties of such tumors could lead to a new classification of histogenetic and prognostic significance. Finally, comparing the functional ectopic hormonal products in the cells of the primary and metastatic tumors may detect aspects of tumor cell selection and selective growth advantage.

VI. BIOLOGICAL SIGNIFICANCE

Quite apart from their potential practical clinical and pathological value as cell and plasma markers, inappropriate hormonal tumor products have fundamental importance as an expression of an aberration of cell differentiation linked with neoplasia. Clarification of the mechanisms leading to such functional abnormalities may contribute further to an understanding of events occurring during the change from a normal to a neoplastic cell.

Accordingly, it will be important to answer at least two questions. The first is ascertaining whether the ability to make apparently "new" products is the result of an actual change in gene structure or whether it is an epigenetic phenomenon involving the expression of normal but inappropriate genes [8]. The second question has been referred to above and concerns the biological effects which the tumor products may have, with particular respect to their ability to confer a selective advantage in favor of continued growth and progression of the tumor that produces them.

The present discussion will attempt to answer these questions by concentrating on bronchial tumors and will be based on certain facts which have emerged from the foregoing part of this treatise. There is no doubt that some tumors do in fact manufacture de novo and release hormones, and that this process is not a sponge effect as it was once considered [19]. Although conclusive data are still awaited, the bulk of morphological and biological evidence tends to indicate that the cells from which ectopic-hormone-producing tumors arise are probably not normally concerned with their synthesis. Except in rare instances, these ectopic hormones are remarkably similar, if not identical, to the product of the eutopic tissue. Under certain circumstances such cells are also capable of producing either precursor forms or fragments, as well as can the various ectopic-hormone-producing tumors. While this ability may be due in part to altered mechanisms of hormone storage, packaging, and/or release, the gene code for expressing the normal hormone must be present together with the enzymes which reduce it to its normal subconstituents.

The present data also allow the conclusion that appropriate hormone-producing cells and ectopic-hormone-producing cells may respond similarly both in vivo and in vitro to a variety of stimuli. Ellison and her colleagues have demonstrated this most elegantly for calcitonin production both by the C cells of the thyroid and by calcitonin-producing bronchial carcinomas [29, 30].

On these bases, it is now possible to attempt to explain the phenomenon and to use the APUD cell system as a model. It is clear that a random mutation of structural genes to produce "nonsense proteins" is inconsistent with the overall biological and pathological picture. It would seem that any proposed hypothesis must be based on the expression of a normal but different (ectopic, inappropriate) set of genes.

During early development, the cells destined to become endocrine APUD-type cells must differentiate from embryonic precursor cells. It is of little consequence whether such cells are of neurocetodermal origin migrating during fetal development to the endocrine glands or whether they differentiate from related endoderm [97]. If one postulates a similarity of structure due to their similar functions, involving a common ancestry becomes solely semantic. Rather one would propose that from the available genes (of which all cells carry a full set) those coding for APUD characteristics are selected in progenitor cells of whichever derivation, and are expressed during the final stages of differentiation in endocrine glands.

It is postulated that a number of stages of embryonic induction lead to a hypothetical pre-APUD cell which is, at that point, restricted to APUD-type differentiation but not to a specific APUD phenotype, that is, the precursor cell possesses the potential to produce a range of APUD-type hormones such as ACTH, vasopressin, calcitonin, and insulin. A further differentiating signal would then select one from among the many available sets of APUD-type genes, resulting finally in the expression of a specific hormone as the animal develops.

There is no evidence to suggest that subsequently in the normal life of the cell these phenotypes are interchangeable, but rather that the specific characteristics of each endocrine cell type are stable. Thus there is a built-in stability of the differentiated stage, although the different APUD cells are clearly related to one another and may be closely arranged spatially in the endocrine gland, as for example in the islets of Langerhans of the pancreas [98]. Indeed, there is evidence that at a point of heavy demand for new hormone-producing cells following alloxan-induced diabetes in rodents, it is ductal or acinar cells rather than other islet cells that have the capacity to form into beta cells [99].

The mechanism conferring the stability of differentiated characteristics is not fully understood. One possibility is that a product of the gene set being expressed in the cell acts back on the genome to ensure continued expression of the whole set in the manner of a positive feedback loop. If, then, at any point during the life of the cell, abnormal circumstances prevail and the feedback is interrupted, the cell may lose the ability to express its specific phenotype. (Such abnormal circumstances might be exposure to a carcinogenic stimulus, many of which may interact with components involved with gene expression [100].) If the interference has not been too severe or prolonged, however, the cell may still retain its hypothetical pre-APUD range of available phenotypes. In the absence of the specific selective signal present during its embryonic period, it may come at random to select any of these. If the alteration also involves a transformation to malignancy, an ectopic-hormone-producing tumor would then result.

This hypothetical mechanism could thus explain the relatively high incidence of ectopic-hormone-producing tumors arising in the lung, a site where normal APUD cells are found. It does not directly account for non-APUD ectopic hormones, but the same argument can be applied to events governing the stability of other differentiated cells, e.g., the metaplastic squamous bronchial epithelium. Low-level damage would be expected to produce small changes between closely related phenotypes, while greater changes at gene control level would produce more extensive alterations, such as the production of hormones by tumors in tissues unrelated to any endocrine cell.

This implies that within the whole phenomenon of ectopic hormone production, instances of hormone synthesis by tumors which arise from truly nonendocrine cells would be encountered more rarely than instances of ectopic or inappropriate hormone production by tumors at sites known to have an endocrine component. This prediction is borne out by available evidence (Table 4), especially with respect to the APUD-type hormones since classification of the hormone is clear and identification of APUD cells is relatively certain. Instances of ectopic APUD hormones produced by tumors of tissues having no normal APUD-cell component are relatively rare [47], but recently it has been demonstrated that APUD-type granules can develop in rodent liver parenchymal cells after the administration of carcinogens [101]. Evidence is less easy to assess for non-APUD hormones since the relationship between their normal producing cells and other cell types is not so well defined, and it is therefore more difficult to know whether the tumors which produce them ectopically arise from cells which truly have no endocrine relationships.

What then is the rationale behind the nonrandom distribution of ectopic hormone production? As suggested earlier, it may be related to the possibility that ectopic hormone production might confer a selective growth advantage to the tumor cell clone associated with its manufacture and release. In fact, if one of the hormones itself, or some other product within the same gene set, confers any selective advantage on the clone which expresses it, that hormone will be selected for and will appear with a higher incidence than hormones which carry no such advantage. The fact that ACTH [38] and calcitonin [96] are apparently present in a disproportionately high incidence in oat cell carcinomas of the bronchus suggests that some factors linked with either of these hormones may be advantageous to tumor growth and survival.

During the past few years, a series of growth factors of differing tissue origin and with effects on different cell types have been identified, for example, fibroblast growth factor, epidermal growth factor, nerve growth factor, and nonsuppressible insulin-like activity [102]. Like the hormonal peptides, however, one of their modes of action appears to be expressed via the agency of specific cell surface receptors [103, 104]. It is therefore possible to visualize their conferring selective growth advantage if tumor cell populations were able either to produce them ectopically and respond to them and/or to express inappropriate growth factor receptors and thereby respond to their normal physiological presence.

Recently accrued data suggest that both mechanisms may be operative in neoplasia. A human fibrosarcoma line has been reported to produce a growth factor similar to MSA (multiplication-stimulating activity) [105]. This moeity, normally of hepatic origin, is closely related to the group of hormones referred to as somatomedins which, like MSA, have nonsuppressible insulin-like activity. Non-islet-cell tumors associated with hypoglycemia contain and probably release high levels of somatomedins (Table 4) which could possibly be related to tumor growth and selective advantage [106, 107].

There is also evidence that tumor cells in the form of virus-transformed fibroblasts may express ectopic receptors capable of responding to various growth factors produced by the tumor cells themselves [108].

Neuroblastoma cells have nerve growth factor and have receptors for it on their cell surface [130] and pheochromocytomas in vitro also exhibit process formation and maturation in response to nerve growth factor [109].

Thus the fascinating possibility of cell autostimulation as a means of selective growth advantage or its modulation would seem a possibility. There are analogies in this vein with respect to ectopic hormones. Human bronchial carcinoma cells producing high molecular weight forms of calcitonin have receptors capable of binding the monomeric form of calcitonin [131]. The production of corticotropin-like releasing activity, often in association with ectopic ACTH secretion, may be a further facet of autostimulation.

Many of the ectopic hormones are produced in precursor forms. In the case of "big" ACTH, its components are known to break down to yield a series of biologically active moieties including bioactive ACTH, LPH, MSH, endorphins, and enkephalins. It would not seem unreasonable to assume and to test whether some of these moieties or those derived from other ectopic prohormones may confer growth and cell selective advantage.

VII. SUMMARY

Review of the present knowledge of ectopic hormone production by tumors reveals scientific facets and challenges which in many respects outweigh their clinical value. Such a comment must not detract from their clinical recognition and the administration of effective therapy where possible.

Nonetheless, the biological and pathological significance of ectopic hormones and their analogy to known growth factors remain future challenges.

Now that immunological reagents reacting against the chemically characterized prohormones which tend to be the dominant ectopic forms are becoming available, their mechanisms of production, the possible growth advantages they confer, and aspects of tumor cell selection with evolution of the neoplasm become potentially solvable research problems. Answers to these questions could have important applications in the future medical management of the cancer patient.

REFERENCES

1. A. Primack, The production of markers of bronchogenic carcinoma: A review, Semin. Oncol. 1:235-244 (1974).
2. L. H. Rees and J. G. Ratcliffe, Ectopic hormone production by non-endocrine tumours, Clin. Endocrinol. 3:263-299 (1974).
3. D. G. Bartuska, Humoral manifestations of neoplasms, Semin. Oncol. 2:405-409 (1975).
4. L. H. Rees, The biosynthesis of hormones by non-endocrine tumours— A review, J. Endocrinol. 67:143-175 (1975).
5. L. H. Rees, Concepts in ectopic hormone production, Clin. Endocrinol. 5:363s-372s (1976).
6. L. H. Rees, Hormone production by tumors, Antibiotics Chemother. 22:161-165 (1978).
7. W. M. Awad, Jr., P. J. A. Davies, and M. S. Wells, Biological relationships of ectopic hormone syndromes, in Hormones and Cancer (K. K. Charyulu and A. Sudarsanam, eds.), Grune & Stratton, New York, 1976, pp. 281-294.
8. M. L. Ellison, Cell differentiation and the biological significance of inappropriate tumour products, Proc. R. Soc. Med. 70:845-850 (1977).
9. G. Keusch, U. Binswanger, M. A. Dambacher, and J. A. Fischer, Ectopic ACTH syndrome and medullary thyroid carcinoma, Acta Endocrinol. 86:306-316 (1977).
10. W. Odell, A. Wolfsen, Y. Yoshimoto, R. Weitzman, D. Fisher, and F. Hirose, Ectopic peptide synthesis: A universal concomitant of neoplasia, Trans. Assoc. Am. Physicians 90:204-227 (1977).
11. G. Sufrin and G. P. Murphy, Humoral syndromes of renal adeno-carcinoma in man, Rev. Surg. 34:149-166 (1977).
12. M. R. Blackman, S. W. Rosen, and B. D. Weintraub, Ectopic hormones, Adv. Intern. Med. 23:85-113 (1978).
13. W. D. Odell and A. R. Wolfsen, Humoral syndromes associated with cancer, Ann. Rev. Med. 29:379-406 (1978).
14. H. Rochman, Tumor associated markers in clinical diagnosis, Ann. Clin. Lab. Sci. 8:167-175 (1978).
15. P. Skrabanek and D. Powell, Unifying concept of nonpituitary ACTH-secreting tumors: Evidence of common origin of neural-crest tumors, carcinoids, and oat-cell carcinomas, Cancer 42:1263-1269 (1978).

16. G. W. Liddle, W. E. Nicholson, D. P. Island, D. N. Orth, K. Abe, and S. C. Lowder, Clinical and laboratory studies of ectopic humoral syndromes, Recent Prog. Horm. Res. 25:233-314 (1969).
17. A. M. J. Buchan, J. M. Polak, E. Solcia, and A. G. E. Pearse, Localisation of intestinal gastrin in a distinct endocrine cell type, Nature 277:138-140 (1979).
18. J. Polak and S. R. Bloom, Peptidergic nerves of the gastrointestinal tract, Invest. Cell. Pathol. 1:301-326 (1978).
19. R. H. Unger, J. de V. Lochner, and A. M. Eisentraut, Identification of insulin and glucagon in a bronchogenic metastasis, J. Clin. Endocrinol. Metab. 24:823-831 (1964).
20. R. E. Bailey, Periodic hormonogenesis—A new phenomenon: Periodicity in function of a hormone-producing tumor in man, J. Clin. Endocrinol. Metab. 32:317-327 (1971).
21. W. J. Jeffcoate and L. H. Rees, Adrenocorticotropin and related peptides in non-endocrine tumors, in Current Topics in Experimental Endocrinology (L. Martini and V. H. T. James, eds.), vol. 3, Academic Press, New York and London, 1978, pp. 58-74.
22. M. J. O'Hare, M. L. Ellison, and A. M. Neville, Tissue culture in endocrine research: Perspectives, pitfalls and potentials, in Current Topics in Experimental Endocrinology (L. Martini and V. H. T. James, eds.), Academic Press, New York and London, 1978, pp. 2-56.
23. D. N. Orth, Establishment of human malignant melanoma clonal cell lines that secrete ectopic adrenocorticotropin, Nature (New Biol.) 242:26-28 (1973).
24. A. S. Rabson, S. W. Rosen, A. H. Tashjian, and B. D. Weintraub, Production of human chorionic gonadotrophin in vitro by a cell line derived from a carcinoma of the lung, J. Natl. Cancer Inst. 50:669-674 (1973).
25. A. H. Tashjian, B. D. Weintraub, N. J. Barowsky, A. S. Rabson, and S. W. Rosen, Subunits of human chorionic gonadotrophin: Unbalanced synthesis and secretion by clonal cell strains derived from a bronchogenic carcinoma, Proc. Natl. Acad. Sci. USA 70:1419-1422 (1973).
26. M. L. Ellison, D. Woodhouse, C. Hillyard, M. Dowsett, R. C. Coombes, E. D. Gilby, P. B. Greenberg, and A. M. Neville, Immunoreactive calcitonin production by human lung carcinoma cells in culture, Br. J. Cancer 32:373-379 (1975).
27. T. Kameya, H. Kuramoto, K. Suzuki, T. Kenjo, T. Oshikiri, H. Hayashi, and M. Itakura, A human gastric choriocarcinoma cell line with human chorionic gonadotropin and placental alkaline phosphatase production, Cancer Res. 35:2025-2032 (1975).
28. J. M. Lieblich, B. D. Weintraub, G. H. Krauth, P. O. Kohler, A. S. Rabson, and S. W. Rosen, Ectopic and eutopic secretion of

chorionic gonadotropin and its subunits in vitro: Comparison of clonal strains from carcinomas of lung and placenta, J. Natl. Cancer Inst. 56:911-917 (1976).
29. M. Ellison, C. J. Hillyard, R. C. Coombes, and A. M. Neville, Control of calcitonin release in vitro, J. Endocrinol. 71:85 (1976).
30. M. L. Ellison, C. J. Hillyard, G. A. Bloomfield, L. H. Rees, R. C. Coombes, and A. M. Neville, Ectopic hormone production by bronchial carcinomas in culture, Clin. Endocrinol. 5:397s-406s (1976).
31. O. S. Pettengill, C. S. Faulkner, D. H. Wurster-Hill, L. H. Maurer, G. D. Sorenson, A. G. Robinson, and E. A. Zimmerman, Isolation and characterization of a hormone-producing cell line from human small cell anaplastic carcinoma of the lung, J. Natl. Cancer Inst. 58:511-518 (1977).
32. J. Kanabus, G. D. Braunstein, P. K. Emry, P. J. DiSaia, and M. E. Wade, Kinetics of growth and ectopic production of human chorionic gonadotropin by an ovarian cystadenocarcinoma cell line maintained in vitro, Cancer Res. 38:765-770 (1978).
33. M. J. Gait, Synthetic genes for human insulin, Nature 277:429-431 (1979).
34. L. Hall, R. K. Craig, and P. N. Campbell, mRNA species directing synthesis of milk proteins in normal and tumor tissue from human mammary gland, Nature 277:54-56 (1979).
35. S. Nakanishi, A. Inoue, T. Kita, M. Nakamura, A. C. Y. Chang, S. N. Cohen, and S. Numa, Nucleotide sequence of cloned cDNA for bovine corticotropin-β-lipotropin precursor, Nature 278:423-427 (1979).
36. J. G. Ratcliffe, Ectopic hormone production by lung tumours, Abstr. 157 in Proceedings of the VIIth Meeting of the International Society for Oncodevelopmental Biology and Medicine, 1979.
37. J. G. Ratcliffe, R. A. Knight, G. M. Besser, J. Landon, and A. G. Stansfeld, Tumour and plasma ACTH concentrations in patients with and without the ectopic ACTH syndrome, Clin. Endocrinol. 1:27-44 (1972).
38. G. Gewirtz and R. S. Yalow, Ectopic ACTH production in carcinoma of the lung, J. Clin. Invest. 53:1022-1032 (1974).
39. I. M. Holdaway, G. A. Bloomfield, J. G. Ratcliffe, K. W. F. Hinson, G. M. Rees, and L. H. Rees, Adrenocorticotrophin levels in normal and neoplastic lung tissue, in Endocrinology 1973 (S. Taylor, ed.), Heinemann, London, 1974.
40. A. M. Neville, K. M. Grigor, and E. Heyderman, Biological markers and human neoplasia, in Recent Advances in Histopathology (P. P. Anthony and N. Woolf, eds.), vol. 10, Churchill Livingstone, Edinburgh, London and New York, 1978, pp. 23-44.

41. A. G. E. Pearse, The cytochemistry and ultrastructure of polypeptide hormone-producing cells of the APUD series and the embryologic, physiologic and pathologic implications of the concept, J. Histochem. Cytochem. 17:303 (1969).
42. A. G. E. Pearse and T. T. Takor, Neuroendocrine embryology and the APUD concept, Clin. Endocrinol. 5:2295-2445 (1976).
43. S. Kay and M. A. Willson, Ultrastructural studies of an ACTH-secreting thymic tumor, Cancer 26:445-452 (1970).
44. B. Corrin and M. McMillan, Fine structure of an oat cell carcinoma of the lung associated with ectopic ACTH syndrome, Br. J. Cancer 24:755-758 (1970).
45. K. G. Bensch, B. Corrin, R. Pariente, and H. Spencer, Oat cell carcinoma of the lung, Cancer 22:1163-1172 (1968).
46. T. Horai, H. Nishihara, R. Tateishi, M. Matsuda, and S. Hattori, Oat cell carcinoma of the lung simultaneously producing ACTH and serotonin, J. Clin. Endocrinol. Metab. 37:212-219 (1973).
47. R. J. Levine and S. A. Metz, A classification of ectopic hormone-producing tumors, Ann. N.Y. Acad. Sci. 230:533-546 (1974).
48. G. A. Bloomfield, I. M. Holdaway, B. Corrin, J. G. Ratcliffe, G. M. Rees, M. Ellison, and L. H. Rees, Lung tumours and ACTH production, Clin. Endocrinol. 6:95-104 (1977).
49. D. F. Steiner, J. L. Clark, C. Nolan, A. H. Rubenstein, E. Margoliash, B. Aten, and P. E. Oyer, Proinsulin and the biosynthesis of insulin, Recent Prog. Horm. Res. 25:207-282 (1969).
50. R. S. Yalow and S. A. Berson, Size and charge distinctions between endogenous human plasma gastrin in peripheral blood and heptadecapeptide gastrins, Gastroenterology 58:609-615 (1970).
51. A. K. Tung and F. Zerega, Biosynthesis of glucagon in isolated pigeon islets, Biochem. Biophys. Res. Commun. 45:387-395 (1971).
52. R. S. Yalow and S. A. Berson, Size heterogeneity of immunoreactive human ACTH in plasma and in extracts of pituitary glands and ACTH-producing thymoma, Biochem. Biophys. Res. Commun. 44:439-445 (1971).
53. B. Kemper, J. F. Haberner, J. T. Potts, Jr., and A. Rich, Pro-parathyroid hormone: Identification of a biosynthetic precursor to parathyroid hormone, Proc. Natl. Acad. Sci. USA 69:643-647 (1972).
54. Y. Hirata, Synthesis and secretion of ACTH and β-MSH by ectopic ACTH-producing tumors, Kobe J. Med. Sci. 22:91-101 (1976).
55. R. C. Benson, B. L. Riggs, B. M. Pickard, and C. D. Arnaud, Immunoreactive forms of circulating parathyroid hormone in primary and ectopic hyperparathyroidism, J. Clin. Invest. 54:175-181 (1974).
56. R. C. Coombes, M. L. Ellison, G. C. Easty, C. J. Hillyard, R. James, L. Galante, S. Girgis, L. Heywood, I. MacIntyre, and

A. M. Neville, The ectopic secretion of calcitonin by lung and breast carcinomas, Clin. Endocrinol. 5:387s-396s (1976).
57. P. Franchimont, A. Reuter, J. C. Hendrick, and P. F. Zangerle, Circulating forms of ectopic hormones detected by radioimmunoassay, in Hormonal Receptors in Digestive Tract Physiology (Bonfils et al., eds.), Elsevier/North-Holland Biomedical Press, Amsterdam, 1977, pp. 95-98.
58. L. J. Deftos, B. A. Roos, D. Bronzert, and J. G. Parthemore, Immunochemical heterogeneity of calcitonin in plasma, J. Clin. Endocrinol. Metab. 40:409-412 (1975).
59. R. H. Snider, O. L. Silva, C. F. Moore, and K. L. Becker, Immunochemical heterogeneity of calcitonin in man: Effect on radioimmunoassay, Clin. Chim. Acta 76:1-14 (1977).
60. J. J. Morton, P. Kelly, and P. L. Padfield, Antidiuretic hormone in bronchogenic carcinoma, Clin. Endocrinol. 9:357-370 (1978).
61. A. P. Scott, J. G. Ratcliffe, L. H. Rees, J. Landon, H. P. J. Bennett, P. J. Lowry, and C. McMartin, Pituitary peptide, Nature (New Biol.) 244:65-67 (1973).
62. D. N. Orth, W. E. Nicholson, W. M. Mitchell, D. P. Island, and G. W. Liddle, Biologic and immunologic characterisation and physical separation of ACTH and ACTH fragments in the ectopic ACTH syndrome, J. Clin. Invest. 52:1756-1769 (1973).
63. P. J. Lowry, L. H. Rees, S. Tomlin, G. Gillies, and J. Landon, Chemical characterization of ectopic ACTH purified from a malignant thymic carcinoid tumour, J. Clin. Endocrinol. Metab. 43:831-835 (1976).
64. J. G. Pierce, The subunits of pituitary thyrotropin—Their relationship to other glycoprotein hormones, Endocrinology 89:1331-1344 (1971).
65. C. R. Kahn, S. W. Rosen, B. D. Weintraub, S. S. Fajans, and P. Gorden, Ectopic production of chorionic gonadotropin and its subunits by islet-cell tumors: A specific marker for malignancy, N. Engl. J. Med. 297:565-569 (1977).
66. L. Stjarne, U. S. Euler, and F. Lishajko, Catecholamines and nucleotides in phaeochromocytoma, Biochem. Pharmacol. 13:809-818 (1964).
67. R. A. DeLellis, L. May, A. H. Tashjian, and H. J. Wolfe, C-cell granule heterogeneity in man: An ultrastructural immunocytochemical study, Lab. Invest. 38:263-269 (1978).
68. S.-N. Huang and D. Goltzman, Electron and immunoelectron microscopic study of thyroidal medullary carcinoma, Cancer 41:2226-2235 (1978).
69. M. Ellison, Ectopic hormones: An epigenetic change expressed in neoplasia, Br. J. Cancer 41:664-665 (1980).

70. A. de Troyer and J. C. Demanet, Clinical, biological and pathogenic features of the syndrome of inappropriate secretion of antidiuretic hormone, Q. J. Med. 180:521-531 (1976).
71. A. Gomez-Uria and A. G. Pazianos, Syndromes resulting from ectopic hormone-producing tumors, in Symposium on Malignant Disease, Med. Clin. North Am. 59:431-440 (1975).
72. K. Luey and B. A. Scobie, Watery diarrhoea (WDHA) syndrome associated with carcinoma of the lung, Aust. N. Z. J. Med. 6:490-491 (1976).
73. T. J. Powles, S. A. Clark, D. M. Easty, G. C. Easty, and A. M. Neville, The inhibition by aspirin and indomethacin of osteolytic tumour deposits and hypercalcaemia in rats with Walker tumour, and its possible application to human breast cancer, Br. J. Cancer 28:316-321 (1973).
74. N. A. Sheth, J. N. Saruiya, K. J. Ranadive, and A. R. Sheth, Ectopic production of human chorionic gonadotrophin by human breast tumours, Br. J. Cancer 30:566-570 (1974).
75. N. A. Sheth, J. N. Suraiya, A. R. Sheth, K. J. Ranadive, and D. J. Jussawalia, Ectopic production of human placental lactogen by human breast tumors, Cancer 39:1693-1699 (1970).
76. C. H. W. Horne, I. N. Reed, and G. D. Milne, Prognostic significance of inappropriate production of pregnancy proteins by breast cancers, Lancet 2:279-282 (1976).
77. R. A. Walker, Significance of α-subunit HCG demonstrated in breast carcinomas by the immunoperoxidase technique, J. Clin. Pathol. 31:245-249 (1978).
78. J. C. M. P. Monteiro, K. M. Ferguson, S. Biswas, J. G. Grudzinskas, and A. M. Neville, Human chorionic gonadotrophin, human placental lactogen and pregnancy-specific β_1-glucoprotein in human breast cancer, Abstr. 119 in Proceedings of the VIIth Meeting of the International Society for Oncodevelopmental Biology and Medicine, 1979.
79. K. J. Olsen, C. Gadeberg, H. E. Nielsen, and A. Johannsen, Increased serum calcitonin in patients with mammary carcinoma, Acta Radiol. Oncol. 17:263-268 (1978).
80. L.-I. Larsson, L. Grimelius, R. Hakanson, J. F. Rehfeld, F. Stadil, J. Holst, L. Angervall, and F. Sundler, Mixed endocrine pancreatic tumors producing several peptide hormones, Am. J. Pathol. 79:271-284 (1975).
81. R. W. Spence and C. J. Burns-Cox, ACTH-secreting 'apudoma' of gallbladder, Gut 16:473-476 (1975).
82. W. J. Lovern, B. L. Fariss, J. N. Wettlaufer, and S. Hane, Ectopic ACTH production in disseminated prostatic adenocarcinoma, Urology 5:817-820 (1975).

83. R. E. Wenk, B. S. Bhagavan, R. Levy, D. Miller, and W. Weisburger, Ectopic ACTH, prostatic oat cell carcinoma, and marked hypernatremia, Cancer 40:773-778 (1977).
84. M. Sugawara and G. A. Hagen, Ectopic ACTH syndrome due to salivary gland adenoid cystic carcinoma: Response to metyrapone, Arch. Intern. Med. 137:102-105 (1977).
85. H. W. Jones, S. Plymate, F. B. Gluck, P. A. Miles, and J. F. Greene, Jr., Small cell nonkeratinizing carcinoma of the cervix associated with ACTH production, Cancer 38:1629-1635 (1976).
86. A. J. Strain, G. C. Easty, and A. M. Neville, A new experimental model of human cachexia, Invest. Cell. Pathol. 2:87-96 (1979).
87. K. Havemann, C. Gropp, A. Scheuer, T. Scherfe, and M. Gramse, ACTH-like activity in immune complexes of patients with oat-cell carcinoma of the lung, Br. J. Cancer 39:43-50 (1979).
88. L. H. Rees, G. A. Bloomfeld, G. M. Rees, B. Corrin, L. M. Franks, and J. G. Ratcliffe, Multiple hormones in a bronchial tumor, J. Clin. Endocrinol. Metab. 38:1090-1097 (1974).
89. P. H. Sonksen, A. B. Ayres, M. Braimbridge, B. Corrin, D. R. Davies, G. M. Jeremiah, S. W. Oaten, C. Lowy, and T. E. T. West, Acromegaly caused by pulmonary carcinoid tumours, Clin. Endocrinol. 5:503-513 (1976).
90. T. Suda, H. Demura, R. Demura, I. Wakabayashi, K. Nomura, E. Odagiri, and K. Shizume, Corticotropin-releasing factor-like activity in ACTH producing tumors, J. Clin. Endocrinol. Metab. 44:440-446 (1977).
91. S. M. Shalet, C. G. Beardwell, I. A. MacFarlane, M. L. Ellison, C. M. Norman, L. H. Rees, and M. Hughes, Acromegaly due to production of a growth hormone releasing factor by a bronchial carcinoid tumour, Clin. Endocrinol. 10:61-67 (1979).
92. H. Yamamoto, Y. Hirata, S. Matsukura, H. Imura, M. Nakamura, and A. Tanaka, Studies on ectopic ACTH-producing tumours. IV. CRF-like activity in tumour tissue, Acta Endocrinol. 82:183-192 (1976).
93. A. M. Neville and E. H. Cooper, Biochemical monitoring of cancer: A review, Ann. Clin. Biochem. 13:283-305 (1976).
94. F. M. Muggia, S. W. Rosen, B. D. Weintraub, and H. H. Hansen, Ectopic placental proteins in nontrophoblastic tumors: Serial measurements following chemotherapy, Cancer 36:1327-1337 (1975).
95. R. L. Himsworth, G. A. Bloomfield, R. C. Coombes, M. Ellison, J. J. H. Gilkes, P. J. Lowry, K. D. R. Setchell, G. Slavin, and L. H. Rees, 'Big ACTH' and calcitonin in an ectopic hormone secreting tumour of the liver, Clin. Endocrinol. 7:45-62 (1977).

96. R. C. Coombes, M. L. Ellison, and A. M. Neville, Biochemical markers in bronchogenic carcinoma, Br. J. Dis. Chest 72:263-287 (1978).
97. A. Andrews, Further evidence that enterochromaffin cells are not derived from the neural crest, J. Embryol. Exp. Morphol. 31:589-598 (1974).
98. L. Orci, The microanatomy of the islets of Langerhans, Metabolism 25(Suppl.):1303-1313 (1976).
99. R. N. Melmed, C. J. Benitez, and S. J. Holt, Intermediate cells of the pancreas. I. Ultrastructural characterization, J. Cell Sci. 11:449-475 (1972).
100. H. C. Pitot and C. Heidelberger, Metabolic regulatory circuits and carcinogenesis, Cancer Res. 23:1694-1700 (1963).
101. Y. Yoshida, A. Kaneko, N. Chisaka, and T. Onoe, Appearance of intestinal type of tumor cells in hepatoma tissue induced by 3'-methyl-4-dimethylaminoazobenzene, Cancer Res. 38:2753-2758 (1978).
102. D. Gospodarowicz and J. S. Moran, Growth factors in mammalian cell culture, Ann. Rev. Biochem. 45:531-558 (1976).
103. P. S. Rudland, Hormones and cell culture, Nature 276:113-114 (1978).
104. P. S. Rudland and L. J. de Asua, Action of growth factors in the cell cycle, Biochim. Biophys. Acta 560:91-133 (1979).
105. J. E. De Larco and G. J. Todaro, A human fibrosarcoma cell line producing multiplication stimulating activity (MSA)-related peptides, Nature 272:358 (1978).
106. T. Hyodo, K. Megyesi, and C. R. Kahn, Adrenocortical carcinoma and hypoglycemia: Evidence for production of nonsuppressible insulin-like activity by the tumor, J. Clin. Endocrinol. Metab. 44:1175-1184 (1977).
107. K. Megyesi, C. R. Kahn, J. Roth, and P. Gorden, Hypoglycemia in association with extrapancreatic tumors: Demonstration of elevated plasma NSILA's by a new radioreceptor assay, J. Clin. Endocrinol. Metab. 38:931-934 (1974).
108. G. J. Todaro, J. E. de Larco, S. P. Nissley, and M. M. Rechler, MSA and EGF receptors on sarcoma virus transformed cells and human fibrosarcoma cells in culture, Nature 267:526-528 (1977).
109. A. S. Tischler, M. A. Dichter, B. Biales, and L. A. Greene, Neuroendocrine neoplasms and their cells of origin, N. Engl. J. Med. 296:919-925.
110. O. L. Silva, K. L. Becker, A. Primack, J. Doppman, and R. H. Snider, Ectopic production of calcitonin, Lancet 2:317 (1973).
111. O. L. Silva, K. L. Becker, A. Primack, J. Doppman, and R. H. Snider, Ectopic secretion of calcitonin by oat-cell carcinoma, N. Engl. J. Med. 290:1122-1124 (1974).

112. L. A. Klein, A. S. Rabson, and J. Worksman, In vitro synthesis of vasopressin by lung tumour cells, Surg. Forum 20:231-233 (1969).
113. J. M. George, C. C. Capen, and A. S. Phillips, Biosynthesis of vasopressin in vitro and ultrastructure of a bronchogenic carcinoma, J. Clin. Invest. 51:141-148 (1972).
114. P. B. Greenberg, C. Beck, T. J. Martin, and H. G. Burger, Synthesis and release of human growth hormone from lung carcinoma in cell culture, Lancet 1:350-352 (1972).
115. J. Lumsden, J. Ham, and M. L. Ellison, Purification and partial characterization of high-molecular-weight forms of ectopic calcitonin from a human bronchial carcinoma cell line, Biochem. J. 191:239-246 (1980).
116. T. J. Martin, P. B. Greenberg, C. Beck, and C. I. Johnston, Peptide hormone synthesis by human tumours in cell culture, in Proceedings of the 4th International Congress of Endocrinology, Excerpta Medica, Amsterdam, 1972.
117. P. B. Greenberg, T. J. Martin, and H. F. Sutcliffe, Synthesis and release of parathyroid hormone by a renal carcinoma in cell culture, Clin. Sci. Mol. Med. 42:183-191 (1973).
118. W. Singer, K. Kovacs, N. Ryan, and E. Horvath, Ectopic ACTH syndrome: Clinicopathological correlations, J. Clin. Pathol. 31:591-598 (1978).
119. G. W. Liddle, J. R. Givens, W. E. Nicholson, and D. P. Island, The ectopic ACTH syndrome, Cancer Res. 35:1057-1061 (1965).
120. D. R. Cullen and B. E. Tomlinson, Carcinoma with multiple ectopic hormone secretion and associated myopathy, Postgrad. Med. J. 44:472-477 (1968).
121. L. W. O'Neal, D. M. Kipnis, S. A. Luse, P. E. Lacy, and L. Jarret, Secretion of various endocrine substances by ACTH-secreting tumors—Gastrin, melanotropin, norepinephrine, serotonin, parathormone, vasopressin, glucagon, Cancer 21:1219-1232 (1968).
122. B. P. M. Hamilton, G. V. Upton, and T. T. Amatruda, Jr., Evidence for the presence of neurophysin in tumors producing the syndrome of inappropriate antidiuresis, J. Clin. Endocrinol. Metab. 35:764-767 (1972).
123. L. J. Deftos, P. J. McMillan, G. P. Sartiano, J. Abuid, and A. G. Robinson, Simultaneous ectopic production of parathyroid hormone and calcitonin, Metabolism 25:543-550 (1976).
124. K. Gomi, T. Kameya, M. Tsumuraya, Y. Shimosato, F. Zeze, K. Abe, and T. Yoneyama, Ultrastructural, histochemical and biochemical studies of two cases with amylase, ACTH, and βMSH producing tumor, Cancer 38:1645-1654 (1976).
125. Y. Hirata, S. Matsukura, H. Imura, T. Yakura, S. Ihjima, C. Nagase, and M. Itoh, Two cases of multiple hormone-producing small cell carcinoma of the lung, Cancer 38:2575-2582 (1976).

126. Y. Hirata, N. Sakamoto, H. Yamamoto, S. Matsukura, H. Imura, and S. Okada, Gastric carcinoid with ectopic production of ACTH and β-MSH, Cancer 37:377-385 (1976).
127. K. Abe, I. Adachi, S. Miyakawa, M. Tanaka, K. Yamaguchi, N. Tanaka, T. Kameya, and Y. Shimosato, Production of calcitonin, adrenocorticotrophic hormone, and β-melanocyte-stimulating hormone in tumors derived from amine precursor uptake and decarboxylation cells, Cancer Res. 37:4190-4194 (1977).
128. M. Coscia, R. D. Brown, M. Miller, K. Tanaka, W. E. Nicholson, K. R. Parks, and D. N. Orth, Ectopic production of anti-diuretic hormone (ADH), adrenocorticotrophic hormone (ACTH) and beta-melanocyte stimulating hormone (beta-MSH) by an oat cell carcinoma of the lung, Am. J. Med. 62:303-307 (1977).
129. K. Yamaguchi, K. Abe, I. Adachi, N. Tanaka, M. Tanaka, S. Miyakawa, T. Kameya, and T. Kimura, A case of small cell carcinoma of the lung producing ADH, ACTH, MSH and calcitonin: Successful treatment of severe hyponatremia with furosemide and hypertonic saline, Jap. J. Clin. Oncol. 7:111-118 (1977).
130. R. A. Murphy, N. J. Pantazis, B. G. W. Arnason, and M. Young, Secretion of a nerve growth factory by mouse neuroblastoma cells in culture, Proc. Natl. Acad. Sci. U.S.A. 72(5):1895-1898 (1975).
131. J. Ham, M. L. Ellison, and J. Lumsden, Tumour calcitonin: Interaction with specific calcitonin receptors, J. Biochem. 190:545-550 (1980).

12

IMMUNOGLOBULINS AS CANCER MARKERS IN HUMANS

BEN K. SEON and DAVID PRESSMAN* / Department of Immunology Research, Roswell Park Memorial Institute, Buffalo, New York

I. Introduction   301
II. Bence Jones Protein (Free Light Chain) and Light-Chain Fragment   302
   A. Bence Jones Protein   302
   B. Light-Chain Fragment   308
III. Monoclonal and Biclonal Gammopathies   310
   A. Monoclonal Gammopathies   310
   B. Biclonal Gammopathies   312
IV. Heavy-Chain Disease Proteins and Atypical Immunoglobulins   314
   A. Heavy-Chain Disease (HCD) Proteins   314
   B. Atypical Immunoglobulins   315
V. Conclusion   316
   References   316

I. INTRODUCTION

The abnormal features of immunoglobulins in the urine and serum of patients with various malignant diseases such as multiple myeloma and Waldenström's macroglobulinemia have been well documented [1-9]. This abnormality is reflected in the presence of peculiar immunoglobulin (Ig) proteins such as Bence-Jones protein, monoclonal Ig, and heavy-chain disease protein.

    It is important to note that normal Ig within a single class and subclass is a heterogeneous mixture of chemically different molecules which, however, share the same overall structure. The patients with cancer involving lymphorecticular or plasma cells often produce a chemically homogeneous

---

*Dr. Pressman is deceased.

Ig, i.e., a monoclonal Ig which appears to be a product of cells derived from a single clone.

Thus the chemical homogeneity of monoclonal immunoglobulins of these cancer patients is the most important feature that distinguishes cancer Ig from normal Ig and can be used for diagnostic purposes for some types of cancer. Bence Jones proteins correspond to the light chains of monoclonal immunoglobulins [10, 11], although Bence Jones proteins frequently show heterogeneity in size due to the existence of different polymerized forms, mainly monomer and dimer [12, 13]. Heavy-chain disease (HCD) proteins are variants of Ig heavy chains in which a portion of the heavy chain is deleted [7, 14-17]. Details of the cancer-associated immunoglobulins and clinical significance of these proteins are described below in each section.

Five classes of human immunoglobulins are known. They are IgG, IgA, IgM, IgD, and IgE [18-20]. The Ig molecule of each class is composed of heavy and light chains, linked by disulfide bonds and by noncovalent interactions. The general properties of Ig molecules of different types are described in Table 1.

The differences between immunoglobulins of different classes reside in the different types of heavy chains of the Ig molecule, i.e., γ, α, μ, δ, and ϵ. Only two types of light chains, κ and λ, are known. Either type of light chain can combine with any type of heavy chain to make up a complete Ig molecule. Although the δ heavy chain of IgD is associated with the λ light chain more frequently than with the κ light chain, the biological significance of this preferential association is not known [29, 30].

Each heavy and each light chain is composed of a variable (V) region and a constant (C) region. The V region represents the amino-terminal portion of the polypeptide chain and differs among heavy chains or light chains of different Ig proteins in the amino acid sequence [31-36]. In contrast, the C region represents the carboxy-terminal portion of polypeptide chain and is constant in the amino acid sequence for a particular class (e.g., κ or λ light chain) or for a particular subclass (e.g., $\gamma 1$ or $\alpha 1$ heavy chain) of Ig proteins except for allotype-associated differences [31, 32, 37-42].

Subclasses of γ or α heavy chain have been identified by their antigenic determinants, i.e., $\gamma 1$, $\gamma 2$, $\gamma 3$, and $\gamma 4$ for γ chain, and $\alpha 1$ and $\alpha 2$ for α chain. Differences between the subclasses are reflected in the small but significant differences of amino acid sequences of the constant region (the carboxy-terminal portion) of the γ or α heavy chain [43, 44].

II. BENCE JONES PROTEIN (FREE LIGHT CHAIN) AND LIGHT-CHAIN FRAGMENT

A. Bence Jones Protein

Bence Jones (BJ) protein was found over a century ago because of its unusual solubility characteristics in the urine of a patient with a then

TABLE 1 Classes of Human Immunoglobulins

Class	IgG	IgA	IgM	IgD	IgE
Heavy chains					
Class	γ	α	μ	δ	ϵ
Subclasses	$\gamma1, \gamma2, \gamma3, \gamma4$	$\alpha1, \alpha2$	—	—	—
Mol wt $\times 10^{-4}$	5.2–5.4	5.2–5.8	6.5–7.0	6.0	7.0–7.3
Light chains					
Class	κ, λ	κ, λ	κ, λ	κ, λ	κ, λ
Mol wt $\times 10^{-4}$	2.2–2.4	2.2–2.4	2.2–2.4	2.2–2.4	2.2–2.4
Whole molecule					
Formula	$\kappa_2\gamma_2$ or $\lambda_2\gamma_2$	$(\kappa_2\alpha_2)_n$ or $(\lambda_2\alpha_2)_n$ n = 1, 2, 3 ...	$(\kappa_2\mu_2)_5$ or $(\lambda_2\mu_2)_5$	$\kappa_2\delta_2$ or $\lambda_2\delta_2$	$\kappa_2\epsilon_2$ or $\lambda_2\epsilon_2$
Mol wt $\times 10^{-5}$	1.5	1.6	9–10	1.7	1.8–1.9
Carbohydrate (%)	2.9	7.5	7.7–10.7	11–12	11–12
Concentration in normal human serum (mean mg/ml)	11.4	1.8	1.0	0.03	0.0003

Source: Data are from Refs. 9 and 21–26. The nomenclature used is that recommended by the World Health Organization [27] and by the Nomenclature Committee of the IUIS [28].

mysterious disease [1, 45]. Later the presence of BJ protein in the urine and/or serum has been used as one of the most characteristic signs of myelomatosis and the related lymphoproliferative disorders [4, 6, 8, 9]. After more than a century since BJ protein was first discovered, it was shown that BJ proteins correspond to the light chains of immunoglobulins [10, 11], and much progress has been made in determining its chemical and physicochemical properties.

The incidence of BJ proteinuria in plasma cell-lymphocytic neoplasia is high, e.g., approximately 50-70% of patients with multiple myeloma [46]. The first step in detecting BJ protein in the urine is to subject the urine specimen to paper or cellulose-acetate electrophoresis. Although the urine must be concentrated (e.g., 50-fold) beforehand in most cases of the urine tests, in some cases BJ protein can be detected in the unconcentrated original urine. For instance, a distinct protein peak corresponding to a BJ protein was observed in the γ region of the cellulose-acetate electrophoretogram of an unconcentrated 24-hr urine specimen of a patient (TSCH) with plasma cell leukemia (Fig. 1). The BJ protein in the unconcentrated urine could also be detected by immunoelectrophoresis (see below). It is more difficult to detect BJ protein in the serum by cellulose-acetate or paper electrophoresis since serum contains large quantities of many protein components. With either urine or serum, BJ protein can be definitely determined by immunoelectrophoretic analyses using appropriate antisera. The results of urine and serum analyses by immunoelectrophoresis are shown in Figures 2, 3, and 4. In Figure 2 the original (unconcentrated) urine collected during a 24-hr period from patient TSCH and the 20-fold concentrated urine were analyzed by immunoelectrophoresis using antisera individually specific to κ BJ protein, κ light chain associated with Fd fragment of Ig heavy chain, and γ heavy chain.

The antiserum to κ BJ protein (anti-κ BJP antiserum) had been prepared against a free κ BJ protein, whereas the antiserum to the associated

FIGURE 1 Cellulose-acetate electrophoresis of an original, unconcentrated 24-hr urine specimen of a patient (TSCH) with plasma cell leukemia. Protein bands of the electrophoretogram were scanned by a densitometer. A protein peak is observed in the γ region. The major component of the protein peak was determined to be a κ type BJ protein (see Fig. 2).

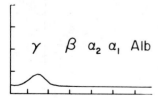

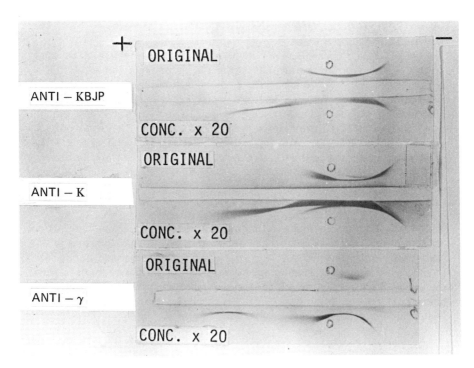

FIGURE 2 Immunoelectrophoretic analysis of a 24-hr urine specimen of patient TSCH with plasma cell leukemia. The urine was collected before the patient received chemotherapy, and part of the urine was concentrated 20-fold by use of an Amicon ultrafilter. Clinical features and various laboratory test results for patient TSCH at the time the urine was collected (October 4, 1974) were reported previously [48]. A κ type BJ protein and a monoclonal IgGκ were detected in the original, unconcentrated urine specimen. An anodic γ-chain fragment was detected in addition to the BJ protein and IgGκ in the 20-fold concentrated urine. No immunoprecipitin arcs were observed against antisera specific to λ BJ protein, λ chain, α chain, and μ chain. (Taken from B. K. Seon, S. Gailani, E. S. Henderson, and D. Pressman, reprinted with permission from Immunochemistry 14:703, Isolation and characterization of 7S IgG, a γ-fragment, β_2-microglobulin, and a Bence Jones protein in urine of a patient with plasma cell leukemia, copyright 1977, Pergamon Press, Ltd.)

κ light chain (anti-κ antiserum) was prepared by treatment of rabbit anti-Fab antiserum with soluble monoclonal IgGλ proteins and a λ BJ protein. The Fab fragment used was obtained by papain digestion of normal human IgG. The anti-κ antiserum is useful in detecting a κ chain which is associated with a heavy chain, since on some occasions anti-κ BJP antiserum reacts poorly with the associated κ chain, although it reacts well with free κ chain [47].

The original, unconcentrated urine showed a κ BJ protein as demonstrated by a precipitin arc against anti-κ BJP antiserum and anti-κ antiserum, and a small amount of IgGκ as demonstrated by a weak precipitin arc against anti-κ antiserum and anti-γ antiserum (the upper portion of each panel of Fig. 2). The 20-fold concentrated urine showed a γ fragment in addition to the κ BJ protein and the monoclonal IgG$_\kappa$ (the lower portion of each panel of Fig. 2). A single precipitin arc corresponding to a κ BJ protein was observed against anti-κ BJP, but an additional precipitin arc corresponding to IgG$_\kappa$ was observed against anti-κ antiserum. The anti-γ

FIGURE 3 Immunoelectrophoretic analysis of serum of a myeloma patient (RL) with light-chain disease. The serum was tested against antisera individually specific to the κ, λ, γ, α, μ, δ, and ϵ chains of immunoglobulins. Some of the results are illustrated in the figure. In each panel the serum of patient RL was placed in the upper well and the control normal serum was in the lower well. A monoclonal precipitin arc was observed only against anti-λ antiserum.

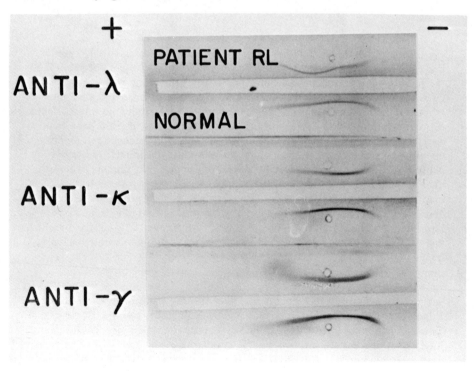

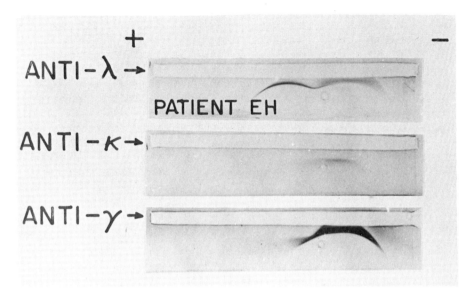

FIGURE 4 Serum of a myeloma patient (EH) showing both a monoclonal light chain and a monoclonal IgGλ. Immunoelectrophoretic analysis of the serum was carried out against antisera individually specific to the κ, λ, γ, α, μ, δ, and ε chains. The results with antisera to κ, λ, and γ chains are shown in the figure. The anodic arc and the cathodic arc in the top panel represent a monoclonal λ light chain and a monoclonal IgGλ, respectively.

antiserum showed an anodic arc corresponding to a γ fragment in addition to the cathodic IgG arc. No significant immunoprecipitin arc was observed for either the original or the 20-fold concentrated urine TSCH against antisera specific to λ BJ protein, λ chain associated with heavy chain, α chain, or μ chain [47]. The concentrations of the BJ protein, IgGκ, and the γ fragment in the urine varied in parallel with the clinical manifestations during the course of illness, i.e., before treatment, during remission after treatment, and after relapse [48].

Figures 3 and 4 illustrate the results showing the presence of serum light chain. Sera of two myeloma patients (RL and EH) were tested in the immunoelectrophoresis against antiserum to whole human serum and against antisera individually specific to κ, λ, γ, α, μ, δ, or ε chain of Ig. Serum RL presented a case of light chain disease and showed a monoclonal λ light chain (Fig. 3) without showing other Ig abnormalities. Serum EH, on the other hand, showed a monoclonal IgGλ in addition to a monoclonal light chain (Fig. 4).

The heavy and light chains of Ig are synthesized as separate polypeptide units on different polyribosomes in the same cell [49, 50]. In patient RL (light chain disease), the production of the heavy chain by the cells proliferated from a malignant clone is apparently blocked, whereas the production of the light chain by the same cell is not blocked. In patient EH the balanced production of the light chain and the heavy chain by cells proliferated from a malignant clone was disturbed, which resulted in the production of an excess of light chain. Myeloma cells which produce only light chains are apparently more anaplastic than the myeloma cells which continue to produce both light and heavy chains. Thus the light-chain myelomas grow faster, have more complications, and lead to an early demise [4].

B. Light-chain Fragment

There have been reports of the existence of light-chain fragments in the urine of some patients with myeloma [51-55]. These fragments are about one-half the size of light chain and appear to be catabolic degradation products of light chain rather than de novo synthetic products [56]. Similar light-chain fragments corresponding to the amino-terminal half fragment (V_L) or to the carboxy-terminal half fragment (C_L) can be obtained by in vitro digestion of light chain [57-59].

It is interesting that there is a strong difference in the susceptibility to enzymatic digestion between V_L and C_L at different temperatures [59]. The V_L was much more resistant to enzymatic digestion than the C_L at 37°C, whereas the situation was reversed at 55°C. The temperature differential permits the easy preparation and isolation of V_L and C_L since only the V_L and the remaining intact light chain are found in the digest after digestion at 37°C, and only the C_L is found after digestion at 55°C [60].

Some of the results of the digestion of a κ BJ protein (COL) at different temperatures are illustrated in Figure 5. The κ BJ protein obtained from a patient (COL) with myeloma was digested with pepsin at 37, 45, and 55°C, and the digests were analyzed by immunoelectrophoresis against anti-κ BJ protein (COL) antiserum and anti-κ BJ protein (ERI) antiserum. The former antiserum which had been prepared against the autologous BJ protein (COL) reacted with both V_L and C_L in addition to the intact BJ protein (COL). On the other hand the latter antiserum prepared against a different κ BJ protein (ERI) reacted only with the C_L and the intact BJ protein, but not with the V_L [59]. This happened because the amino acid sequence of the amino-terminal half of BJ protein varies among the different BJ proteins, while that of the carboxy-terminal half of the BJ proteins of the same

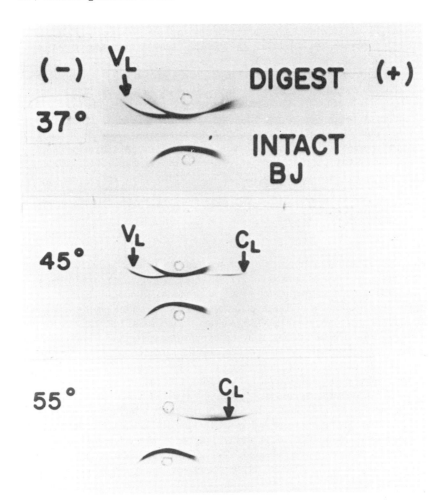

FIGURE 5 Immunoelectrophoretic analysis of peptic digests of the κ type BJ protein COL. The results shown are those obtained using an autologous antiserum (i.e., anti-BJ protein COL) against digests at 37, 45, and 55°C. In each panel the upper well contained the digest and the lower well contained the intact BJ protein (control). The 37°C digest shows V_L, the variable-half fragment, and the remaining intact BJ protein, while the 55°C digest shows only C_L, the constant-half fragment. The digest at the intermediate temperature, 45°C, shows both V_L and C_L in addition to the remaining intact BJ protein.

type (i.e., κ or λ type) is constant with only minor differences which are related to allotype [31, 32]. Only the results obtained with anti-κ BJ protein (COL) are shown in Figure 5.

In the bottom well of each panel of Figure 5 intact κ BJ protein (COL) was placed as a control. The 37°C digest showed V_L and the remaining intact BJ protein but no C_L, whereas the 45°C digest showed both V_L and C_L in addition to the remaining intact BJ protein. The 55°C digest showed only C_L. The V_L and the C_L could be isolated in high yields from the 37°C digest and the 55°C digest, respectively, by a simple gel filtration [60]. We have shown that pepsin selectively cleaved the peptide bond between phenylalanine 116 and isoleucine 117 of κ BJ protein to give V_L and C_L [61].

III. MONOCLONAL AND BICLONAL GAMMOPATHIES

A. Monoclonal Gammopathies

In monoclonal gammopathies, a single malignant clone of precursor cells of plasma cells appears to outgrow and overwhelm other clones resulting in the production of large quantities of a homogeneous (monoclonal) Ig. Any Ig of the five Ig classes of Table 1 can appear as a monoclonal Ig, although the incidence of monoclonal IgD [29, 30] and monoclonal IgE [62, 63] is low. The appearance of monoclonal Ig in the serum is usually readily detectable as a narrow protein band in an ordinary serum electrophoresis. The determination of the type of Ig, however, must be carried out by immunoelectrophoresis using appropriate, specific antisera.

In our laboratory, serum of any cancer patients which show a monoclonal Ig band in serum electrophoresis is routinely analyzed by immunoelectrophoresis using antisera individually specific to κ, λ, γ, α, μ, δ, and ϵ chains of immunoglobulins and antiserum to whole human serum. The procedures to prepare these specific antisera are described in Table 2. The specificity of each of these antisera was verified by immunoelectrophoretic analyses of the antisera against various purified Ig proteins and against whole human serum. The antisera specific to κ or λ chain had been prepared against a κ or λ chain associated with Fd fragment of γ chain.

It is important to use antisera directed to the associated κ or λ chain in the immunoelectrophoresis rather than antisera directed to free κ or λ chain. This advice is based on our observation that the κ or λ chain component of some Ig proteins reacts poorly with antisera prepared with free κ chain or free λ chain, whereas they react reasonably well with antiserum prepared with the associated κ or λ chain (e.g., see Fig. 2). This reaction pattern is most likely due to the concealment of some or many of the strong antigenic determinants on the free light chains when they are associated with the heavy chains.

Details of the immunoelectrophoretic patterns of sera of various types of monoclonal gammopathies have been reported previously [9, 66], and it

TABLE 2 Preparation of Antisera Specific to Each Ig Component Chain

Antisera specific to	Antigens used for immunization of animals [a]	Materials used for absorption of antiserum
κ	Fab [b]	IgGλ, λBJ protein
λ	Fab [b]	IgGκ, κBJ protein
γ	Normal IgG	Normal light chains [c]
α	Monoclonal IgA	Cord serum [d]
μ	Monoclonal IgM	Cord serum [e]
δ	Monoclonal IgD	Normal human serum
ε	Monoclonal IgE	Normal human serum

[a] Goats or rabbits.
[b] The Fab fragment was prepared by papain digestion of normal human IgG [64].
[c] The light chains were isolated from normal human IgG by mild reduction and alkylation [65].
[d] The human cord serum can be substituted for with IgA-deficient serum of patients.
[e] This cord serum can be substituted for with IgM-deficient serum of patients.

is relatively easy for an experienced investigator to determine the type of monoclonal gammopathies by immunoelectrophoresis, if the appropriate antisera are used.

In a conventional immunoelectrophoresis, it is difficult to determine monoclonal Ig at concentrations less than 0.1 mg/ml. We have developed a radioimmunoassay for monoclonal Ig with a sensitivity limit of 20 ng/ml [67]. Our radioimmunoassay involves the preparation of an antibody reagent specific for the idiotypic determinants of a particular monoclonal Ig protein. The use of such a sensitive and specific procedure will allow us to determine minute quantities of monoclonal Ig proteins in the serum and of the cells of patients who responded well to treatment.

Although the association of serum monoclonal Ig proteins with plasma cell-lymphocytic malignancies, especially with multiple myeloma and Waldenström's macroglobulinemia, is high, there are cases in which monoclonal Ig proteins have been found in the serum of patients with other diseases such as hepatoma and chronic infections [3, 4, 68]. In this respect, the term "plasma cell dyscrasia" was recommended to encompass the wide range of pathololological conditions involving unbalanced proliferative disorders of the cells that normally synthesize Ig proteins [3].

Rarely, monoclonal Ig is found in the serum of apparently healthy subjects. We recently found a monoclonal IgGκ in the serum of an apparently healthy young woman who came to Roswell Park Memorial Institute as a blood donor.* Extensive clinical examination of the subject did not show any sign of illness. This subject has been closely followed for about one year and the IgGκ in the serum has been gradually increasing, although no signs of malignancies have yet been detected. It is important to determine if the presence of the monoclonal IgGκ is an indication of premyelomatosis.

The molecular weights of the five classes of Ig proteins, i.e., IgG, IgA, IgM, IgD, and IgE, are larger than 100,000 (see Table 1), and these Ig proteins are retained by the glomerular filter of the kidney [69, 70]. Thus no significant or only a small amount of these Ig proteins is usually detected in normal urine. This is also true for the urine of many patients with myeloma and lymphoreticular malignancies, and relatively little was known about intact, whole Ig proteins in the urine. However, there have been reports indicating the presence of significant quantities of intact Ig proteins in the urine of some cancer patients with severe renal dysfunctions, although the detailed nature of these Ig proteins was not known [71, 72]. We have found large quantities of intact monoclonal IgGκ in the urine of a patient (TSCH) with plasma cell leukemia [47, 48]. Quantitative determination of IgG by radial immunodiffusion showed 610 mg of IgG in a 24-hr urine specimen, while 7.8 g of IgG was found in 100 ml of the serum of patient TSCH [47]. Both the urinary IgGκ and serum IgGκ were isolated in a pure form, and a detailed comparative study was undertaken. Both IgGκ proteins were intact 7S IgG molecules with a γ chain of γ1 subclass and with a κ chain of the VκIII subgroup [47]. The urinary IgGκ was indistinguishable from the serum IgGκ by chemical and immunological criteria. The quantities of these urinary and serum IgGκ proteins varied in parallel with the clinical manifestations during the course of the illness [48].

B. Biclonal Gammopathies

The simultaneous presence of two monoclonal immunoglobulins in a single patient is rare, and approximately 1% of the patients with myeloma have two monoclonal Ig proteins [73]. Relatively recently there have been several reports in which a comparative study was carried out for the immunological or chemical properties of two different monoclonal Ig proteins from a single patient [74-87].

---

*J. E. Fitzpatrick, S. Gailani, B. K. Seon, A. K. Bhargava, E. S. Henderson, and D. Pressman, unpublished observation.

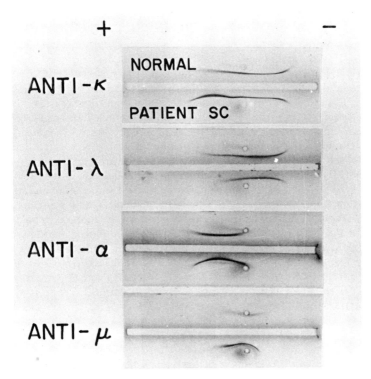

FIGURE 6 IgAκ and IgMκ biclonal gammopathies. Serum of a patient (SC) with lymphocytic lymphoma was tested against antisera individually specific to the κ, λ, γ, α, μ, δ, and ε chains of immunoglobulins, and some of the results are shown. In each panel the upper well contained normal serum as a control. A monoclonal IgAκ and a monoclonal IgMκ were detected in serum SC. These monoclonal proteins were isolated from the serum in pure form in a subsequent study.

An example of a biclonal gammopathy serum as revealed by immunoelectrophoretic analysis is shown in Figure 6. The serum was from a 78-year-old Caucasian male (SC) with lymphocytic lymphoma [88] and showed both a monoclonal IgAκ and a monoclonal IgMκ. These two monoclonal Ig proteins were isolated from the serum and comparative studies were carried out. The two proteins shared idiotypic determinants [77]. The light chains of these proteins were shown to be identical in urea disk electrophoresis, amino acid composition, amino-terminal amino acid sequence, and peptide map [78]. They carried the same InV allotype [78]. The amino-terminal amino acid sequence and peptide map studies suggested that the α and μ heavy chains of these monoclonal proteins share substantial portions of the common amino acid sequence in their variable regions (the amino-terminal portions) [79].

An extensive amino acid sequence study was carried out for a case of IgGκ and IgMκ biclonal gammopathies (Til) by Wang, Fudenberg, and their associates [75, 80, 87], and the variable regions of the γ heavy chain and the μ heavy chain of these two monoclonal proteins apparently had the same amino acid sequence. The clinical features of this patient closely resembled those of Waldenström's macroglobulinemia [89]. In many cases, biclonal gammopathy proteins in a single patient are Ig proteins of different classes and share a light chain of the same class (such as IgGκ and IgAκ, IgGλ and IgMλ, or IgAκ and IgMκ) [74, 76, 78, 81-83]. The two proteins share unique, individually specific antigenic determinants designated "idiotypic determinants" [75-77, 81, 82]. The light chains of the two proteins appear to be identical or very similar [74, 78, 83, 85]. The entire or a substantial portion of the variable portion of the heavy chains also appears to be the same in the two proteins [75, 79, 85, 87]. These two monoclonal Ig proteins appear to be produced by the cells derived from a single clone by a switch-over from one monoclonal Ig to the other monoclonal Ig.

IV. HEAVY-CHAIN DISEASE PROTEINS AND ATYPICAL IMMUNOGLOBULINS

A. Heavy-chain Disease (HCD) Proteins

HCD proteins are variant forms of the heavy chains of immunoglobulins and lack a portion of the amino acid sequence in the amino-terminal region of the heavy chains. HCD proteins are found free from light chains in the serum or in the urine, possibly because HCD proteins may be unable to associate effectively with light chains due to this defect.

Three different classes of HCD proteins are known: γHCD [7, 14], αHCD [15], and μHCD [16, 17]. Light-chain excretion in the form of a BJ protein has not been encountered in cases of γHCD and αHCD, whereas a BJ protein is frequently present in substantial amounts with μHCD [7, 90, 91]. Since serum and urine specimens of some γHCD or αHCD patients and of most μHCD patients show essentially normal paper (or cellulose-acetate) electrophoresis patterns, these HCD proteins can be detected only by immunoelectrophoresis. Biosynthetic [7, 91, 92] and structural studies (see Ref. 91 for a review) showed that most of the γ, α, and μHCD proteins are products of de novo synthesis and are not the products of postsynthetic degradation.

1. γHCD Proteins

Clinical features of patients with γHCD proteins are somewhat variable, but in general they resemble those of malignant lymphoma rather than multiple myeloma. The disease is most often seen in elderly males. Details of the clinical features of patients and the chemical characteristics of several different types of γHCD proteins have been described by Franklin and Frangione [7, 90, 91].

2. αHCD Proteins

Unlike γHCD, αHCD occurs most often in younger people [7, 90] and appears as a condition primarily affecting the secretory IgA system [93]. Although initially found only in residents of North Africa, the Middle East, and southern Europe, the disease has now been found also in northern Europe, South America, and North America, but with a much lower frequency [90]. Although αHCD was discovered more recently than γHCD, the former appears to be the most frequent disorder among the heavy-chain diseases.

3. μHCD Proteins

In most cases μHCD accompanies chronic lymphocytic leukemia (CLL), but there are cases without CLL [90]. Vacuolated plasma cells are seen in the bone marrow in most patients. Diagnosing μHCD is difficult because in most cases routine serum electrophoresis patterns are normal and the diagnosis can be proposed only on the basis of the presence of a free μ-chain component shown by immunoelectrophoresis.

B. Atypical Immunoglobulins

Atypical immunoglobulins reported are smaller than ordinary immunoglobulins because of a defect in either the light or the heavy chain or in both [94-105]. In some cases the defect appears to result in the formation of a half-molecule IgG [95, 98] or IgA [99, 101] which is composed of one light chain and one heavy chain instead of two light chains and two heavy chains as in ordinary IgG or IgA (see Table 1). The atypical immunoglobulins described differ from heavy-chain disease proteins in that the former contain both heavy- and light-chain components whereas the latter contain only heavy-chain components.

Diseases of patients with the reported atypical immunoglobulins are classified as plasmacytoma [95], plasma cell leukemia [98, 101], "mononucleosis" or "lymphoma-like" disease [97], multiple myeloma [99, 100], and plasma cell neoplasm (a tentative diagnosis) [94]. It is difficult to distinguish between ordinary monoclonal immunoglobulins and atypical immunoglobulins by routine serum electrophoresis and immunoelectrophoresis. Definite identification of the atypical immunoglobulins is only possible if the proteins are isolated from serum or urine and their chemical and immunological properties are determined.

V. CONCLUSION

The determination of immunoglobulins in the serum and urine from cancer patients has been used widely by clinical laboratories for diagnostic purposes. Methods are relatively simple, and in laboratories familiar with this determination the results can be quite accurate. It is well known that monoclonal immunoglobulins (M Ig) and Bence Jones protein corresponding to M Ig light chains are useful markers for the diagnosis and follow-up of patients with plasma cell neoplasms. More recently it was shown that M Ig are also useful markers for patients with some types of lymphoid cell neoplasms such as chronic lymphocytic leukemia and B-cell lymphoma. In these cases, however, the secretion of significant amounts of M Ig into the serum and urine is rather rare, so it is necessary to determine cellular M Ig. The monoclonal nature of Ig of tumor tissues or of isolated cells is shown either by the presence of only one type of light chain (κ or λ light chain), or more definitely by the presence of idiotypic determinants.

Determination of M Ig in cells is generally more laborious and difficult than in serum and urine. At the present, therefore, cellular M Ig are of limited use as cancer markers. However, in the future, the measurement of cellular Ig will be used increasingly, especially for the diagnosis and study of lymphoid cell neoplasms.

REFERENCES

1. H. Bence Jones, Lancet 2:88 (1847).
2. J. Waldenström, Acta Med. Scand. Suppl. 367:110 (1961).
3. T. Isobe and E. F. Osserman, Ann. N.Y. Acad. Sci. 190:507 (1971).
4. I. Snapper and A. Kahn, Myelomatosis, University Park Press, Baltimore, Maryland, 1971.
5. J. N. Buxbaum and E. C. Franklin, in Proteins in Normal and Pathological Urine (Y. Manuel, J. P. Revillard, and H. Betuel, eds.), University Park Press, Baltimore, Maryland, 1970.
6. R. Creyssel, in Proteins in Normal and Pathological Urine (Y. Manuel, J. P. Revillard, and H. Betuel, eds.), University Park Press, Baltimore, Maryland, 1970.
7. B. Frangione and E. C. Franklin, Semin. Hematol. 10:53 (1973).
8. A. Soloman, N. Engl. J. Med. 294:91 (1976).
9. R. A. Kyle, in Manual of Clinical Immunology (N. R. Rose and H. Friedman, eds.), American Society of Microbiology, Washington, D.C., 1976.
10. M. D. Poulik and G. M. Edelman, Nature 191:1 (1961).
11. G. M. Edelman and J. A. Gally, J. Exp. Med. 116:207 (1962).
12. G. M. Bernier and F. W. Putnam, Biochim. Biophys. Acta 86:295 (1964).

13. B. K. Seon, O. A. Roholt, and D. Pressman, Biochim. Biophys. Acta 194:397 (1969).
14. E. C. Franklin, J. Loewenstein, B. Bigelow, and M. Meltzer, Am. J. Med. 37:332 (1964).
15. M. Seligmann, F. Danon, D. Hurez, E. Mihaesco, and J. Pred'homme, Science 162:1396 (1968).
16. H. S. Ballard, L. M. Hamilton, A. J. Marcus, and C. H. Illes, N. Engl. J. Med. 282:1060 (1970).
17. F. A. Forte, F. Prelli, W. J. Yount, L. M. Jerry, S. Kochwa, E. C. Franklin, and H. G. Kunkel, Blood 36:137 (1970).
18. J. B. Fleischman, Ann. Rev. Biochem. 35:835 (1966).
19. D. S. Rowe and J. L. Fahey, J. Exp. Med. 121:171 (1965).
20. K. Ishizaka, T. Ishizaka, and M. M. Hornbrook, J. Immunol. 97:840 (1966).
21. G. M. Edelman and W. E. Gall, Ann. Rev. Biochem. 38:415 (1969).
22. H. Metzger, Ann. Rev. Biochem. 39:889 (1970).
23. K. J. Dorrington and C. Tanford, Adv. Immunol. 12:333 (1970).
24. H. M. Grey, C. A. Abel, and B. Zimmerman, Ann. N.Y. Acad. Sci. 190:37 (1971).
25. H. L. Spiegelberg, J. W. Prahl, and H. M. Grey, Biochemistry 9:2115 (1970).
26. S. Kochwa, W. D. Terry, J. D. Capra, and N. L. Yang, Ann. N.Y. Acad. Sci. 190:49 (1971).
27. Bull. World Health Org. 41:975 (1969).
28. J. Immunol. 108:1733 (1972).
29. J. L. Fahey, P. P. Carbone, D. S. Rowe, and R. Bachmann, Am. J. Med. 45:373 (1968).
30. Z. Jancelewicz, K. Takatsuki, S. Sugai, and W. Pruzanski, Arch. Intern. Med. 135:87 (1975).
31. K. Titani and F. W. Putnam, Science 147:1304 (1965).
32. N. Hilschmann and L. C. Craig, Proc. Natl. Acad. Sci. USA 53:1403 (1965).
33. B. A. Cunningham, M. N. Pflumm, U. Rutishauser, and G. M. Edelman, Proc. Natl. Acad. Sci. USA 64:997 (1969).
34. A. C. Wang, J. R. L. Pink, H. H. Fudenberg, and J. Ohms, Proc. Natl. Acad. Sci. USA 66:657 (1970).
35. G. P. Smith, L. Hood, and W. M. Fitch, Ann. Rev. Biochem. 40:969 (1971).
36. J. D. Capra and J. M. Kehoe, Proc. Natl. Acad. Sci. USA 71:845 (1974).
37. M. Wikler, K. Titani, T. Shinoda, and F. W. Putnam, J. Biol. Chem. 242:1668 (1967).
38. G. M. Edelman, B. A. Cunningham, W. E. Gall, P. D. Gottlieb, U. Rutishauser, and M. J. Waxdal, Proc. Natl. Acad. Sci. USA 63:78 (1969).

39. F. W. Putnam, G. Florent, C. Paul, T. Shinoda, and A. Shimizu, Science 182:287 (1973).
40. S. Watanabe, H. U. Barnikol, J. Horn, J. Bertram, and N. Hilschmann, Hoppe-Seyler's Z. Physiol. Chem. 354:1505 (1973).
41. H. Kratzin, P. Altevogt, E. Ruban, A. Kortt, K. Staroscik, and N. Hilschmann, Hoppe-Seyler's Z. Physiol. Chem. 356:1337 (1975).
42. H. Bennich and H. Bahr-Lindström, in Progress in Immunology II (L. Brent and J. Holborow, eds.), vol. 1, Elsevier-North Holland Publishing Company, Amsterdam, 1974.
43. J. R. L. Pink, S. H. Buttery, G. M. De Vries, and C. Milstein, Biochem. J. 117:33 (1970).
44. A. Torano and F. W. Putnam, Proc. Natl. Acad. Sci. USA 75:966 (1978).
45. W. MacIntyre, Med. Chir. Soc. Trans. 33:211 (1850).
46. R. A. Kyle, Mayo Clin. Proc. 50:29 (1975).
47. B. K. Seon, S. Gailani, E. S. Henderson, and D. Pressman, Immunochemistry 14:703 (1977).
48. S. Gailani, B. K. Seon, and E. S. Henderson, J. Med. 8:403 (1977).
49. A. L. Shapiro, M. D. Scharff, J. V. Maizel, Jr., and J. W. Uhr, Proc. Natl. Acad. Sci. USA 56:216 (1966).
50. B. A. Askonas and A. R. Williamson, Proc. R. Soc. Biol. Med. 166:232 (1966).
51. H. F. Deutsch, Science 141:435 (1963).
52. A. Solomon, J. Killander, H. M. Grey, and H. G. Kunkel, Science 151:1237 (1966).
53. R. C. Williams, Jr., S. R. Pinnell, and G. T. Bratt, J. Lab. Clin. Med. 68:81 (1966).
54. D. Cioli and C. Baglioni, J. Mol. Biol. 15:385 (1966).
55. M. Tan and W. Epstein, J. Immunol. 98:568 (1967).
56. D. Cioli and C. Baglioni, J. Exp. Med. 128:517 (1968).
57. A. Solomon and C. L. McLaughlin, J. Biol. Chem. 244:3393 (1969).
58. F. A. Karlsson, P. A. Peterson, and I. Berggard, Proc. Natl. Acad. Sci. USA 64:1257 (1969).
59. B. K. Seon, O. A. Roholt, and D. Pressman, J. Biol. Chem. 247:2151 (1972).
60. B. K. Seon, O. A. Roholt, and D. Pressman, J. Immunol. 109:1201 (1972).
61. B. K. Seon, A. L. Grossberg, O. A. Roholt, and D. Pressman, J. Immunol. 111:269 (1973).
62. S. G. O. Johansson and H. Bennich, Immunology 13:381 (1967).
63. M. Ogawa, S. Kochwa, C. Smith, K. Ishizaka, and O. R. McIntyre, N. Engl. J. Med. 281:1217 (1969).

64. E. C. Franklin and F. Prelli, J. Clin. Invest. 39:1933 (1960).
65. J. B. Fleischman, R. H. Pain, and R. R. Porter, Arch. Biochem. Biophys. Suppl. 1:174 (1962).
66. Immunoglobulin Abnormality Detection, Applications Monograph AM 310, Millipore Corporation, Bedford, Massachusetts, 1973.
67. S. Gailani, B. K. Seon, A. Nussbaum, E. S. Henderson, and D. Pressman, J. Natl. Cancer Inst. 58:1553 (1977).
68. U. Axelsson, R. Bachmann, and J. Hällén, Acta Med. Scand. 179:235 (1966).
69. I. Berggård, in Proteins in Normal and Pathological Urine (Y. Manuel, J. P. Revillard, and H. Betuel, eds.), University Park Press, Baltimore, Maryland, 1970.
70. W. Strober and T. A. Waldmann, Nephron 13:35 (1974).
71. W. Pruzanski, M. E. Platts, and M. A. Ogryzlo, Am. J. Med. 47:60 (1969).
72. B. S. Ooi, A. J. Pesce, V. E. Pollak, and N. Mandalenakis, Am. J. Med. 52:538 (1972).
73. R. Bihrer, R. Flury, and A. Morell, Schweiz. Med. Wochenschr. 104:39 (1974).
74. A.-C. Wang, I. Y. F. Wang, J. N. McCormick, and H. H. Fudenberg, Immunochemistry 6:451 (1969).
75. A.-C. Wang, S. K. Wilson, J. E. Hopper, H. H. Fudenberg, and A. Nisonoff, Proc. Natl. Acad. Sci. USA 66:337 (1970).
76. G. M. Penn, H. G. Kunkel, and H. M. Grey, Proc. Soc. Exp. Biol. Med. 135:660 (1970).
77. Y. Yagi and D. Pressman, J. Immunol. 110:335 (1973).
78. B. K. Seon, Y. Yagi, and D. Pressman, J. Immunol. 110:345 (1973).
79. B. K. Seon, Y. Yagi, and D. Pressman, J. Immunol. 111:1285 (1973).
80. A.-C. Wang, J. Gergely, and H. H. Fudenberg, Biochemistry 12:528 (1973).
81. R. A. Rudders, V. Yakulis, and P. Heller, Am. J. Med. 55:215 (1973).
82. D. S. Fair, R. G. Krueger, G. J. Gleich, and R. A. Kyle, J. Immunol. 112:201 (1974).
83. C. Wolfenstein-Todel, E. C. Franklin, and R. A. Rudders, J. Immunol. 112:871 (1974).
84. R. Oriol, J. Huerta, J. P. Bouvet, and P. Liacopoulos, Immunology 27:1081 (1974).
85. D. S. Fair, C. Sledge, R. G. Krueger, K. G. Mann, and L. E. Hood, Biochemistry 14:5561 (1975).
86. J. E. Hopper, C. Noyes, R. Heinrikson, and J. W. Kessel, J. Immunol. 116:743 (1976).
87. A.-C. Wang, I. Y. Wang, and H. H. Fudenberg, J. Biol. Chem. 252:7192 (1977).

88. D. C. Tormey, R. R. Ellison, and D. K. Hossfeld, Cancer 36:1321 (1975).
89. H. H. Fudenberg, A.-C. Wang, J. R. L. Pink, and A. S. Levin, Ann. N.Y. Acad. Sci. USA 190:501 (1971).
90. E. C. Franklin, N. Engl. J. Med. 294:531 (1976).
91. B. Fragione and E. C. Franklin, in Progress in Immunology III (T. E. Mandel, C. Cheers, C. S. Hosking, I. F. C. McKenzie, and G. J. V. Nossal, eds.), Australian Academy of Science, Canberra, 1977.
92. A. Alexander, D. Barritault, and J. Buxbaum, Proc. Natl. Acad. Sci. USA 75:4774 (1978).
93. M. Seligmann and J. C. Rambaud, Israel J. Med. Sci. 5:151 (1969).
94. A. F. Lewis, D. E. Bergsagel, A. Bruce-Robertson, R. K. Schachter, and G. E. Connell, Blood 32:189 (1968).
95. J. R. Hobbs and A. Jacobs, Clin. Exp. Immunol. 5:199 (1969).
96. H. F. Deutsch and T. Suzuki, Ann. N.Y. Acad. Sci. 190:472 (1971).
97. T. Isobe and E. F. Osserman, Blood 43:505 (1974).
98. H. L. Spiegelberg, V. C. Heath, and J. E. Lang, Blood 45:305 (1975).
99. H. L. Spiegelberg and B. G. Fishkin, J. Clin. Invest. 58:1259 (1976).
100. C. Rivat, C. Schiff, L. Rivat, C. Ropartz, and M. Fougereau, Eur. J. Immunol. 6:545 (1976).
101. G. M. Bernier, J. H. Berman, and M. W. Fanger, Ann. Intern. Med. 86:572 (1977).
102. D. M. Parr, M. E. Percy, and G. E. Connell, Immunochemistry 9:51 (1972).
103. J. W. Fett, H. F. Deutsch, and O. Smithies, Immunochemistry 10:115 (1973).
104. H. L. Spiegelberg, V. C. Heath, and J. E. Lang, Biochemistry 14:2157 (1975).
105. F. A. Garver, L. Chang, J. Mendicino, T. Isobe, and E. F. Osserman, Proc. Natl. Acad. Sci. USA 72:4559 (1975).

13

MULTIPLE MARKERS IN THE MANAGEMENT
OF CANCER PATIENTS

MARKKU SEPPÄLÄ* / Department of Obstetrics and Gynecology, University of Helsinki, Helsinki, Finland, and Department of Reproductive Physiology, St. Bartholomew's Hospital, London

EEVA-MARJA RUTANEN* and JAN LINDGREN / Department of Bacteriology and Immunology, University of Helsinki, Helsinki, Finland

TORSTEN WAHLSTRÖM / Department of Obstetrics and Gynecology and Department of Pathology, University of Helsinki, Helsinki, Finland

I. Introduction   321
II. Trophoblastic Tumors   322
    A. Chorionic Gonadotropin and the Alpha Subunit   322
    B. Pregnancy-Specific Beta-1-Glycoprotein   323
    C. Placental Protein Five (PP_5)   326
    D. Other Markers   328
III. Nontrophoblastic Tumors of the Reproductive System   331
    A. Carcinoembryonic Antigen   331
    B. Chorionic Gonadotropin and the Alpha Subunit   337
    C. Pregnancy-Specific Beta-1-Glycoprotein (PSBG)   340
    D. Alpha Fetoprotein (AFP)   345
IV. Summary   345
    References   347

I. INTRODUCTION

The potential use of multiple markers for the management of cancer patients has been explored by several investigators [1-3]. The most optimistic views are presented by Paul Franchimont and his colleagues at the University of Liége in Belgium [3]. Using a battery of assays including carcinoembryonic antigen (CEA), alpha fetoprotein (AFP), chorionic gonadotropin (hCG), the beta subunit of hCG, and kappa casein in tests on

---

*Present affiliation: Department of Obstetrics and Gynecology, University Central Hospital, Helsinki, Finland.

more than 1450 individuals, they found an elevated circulating level of one marker in 72% of cancer patients at the onset of clinical course. In metastatic disease as many as 85-95% of cases showed one marker, and 20% had two markers in the blood. The cancers they studied were located in breast, lung, gastrointestinal tract, female and male reproductive tracts, urinary system, blood, lymphatic system, and trophoblast. However, apart from malignant disease, abnormal circulating levels of one or several markers were also observed in 15% of patients without malignancy, most notably in patients with hepatitis and cirrhosis of the liver, conditions in which tissue regeneration produces some of the fetal proteins [4, 5].

Coombes and co-workers [2] used 19 biochemical parameters, most of which have been individually advocated as tumor index substances for breast cancer, for the follow-up of 42 patients with active breast cancer. Seven of these parameters were raised in more than half of the 17 patients with overt metastases; these were serum ferritin (88%), C-reactive protein (87%), CEA (81%), acid glycoprotein (75%), total alkaline phosphatase (64%), sialyl transferase (56%), and the urinary hydroxyproline/creatinine ratio (73%). Their results indicated that the incidence of biochemical abnormalities compared favorably with the results of physical methods of detecting metastases. They also suggest that biochemical tests could assist in monitoring metastatic disease and could indicate, at the time of mastectomy, patients who might benefit from immediate systemic therapy in addition to local treatment of their breast carcinomas.

We have concentrated on the study of tumor-associated markers in malignant disease of the female reproductive system, and present a review of recent work in this field.

II. TROPHOBLASTIC TUMORS

A. Chorionic Gonadotropin and the Alpha Subunit

The introduction of a more specific hCG assay utilizing antiserum against the beta subunit of hCG [6] has enabled the measurement of hCG in the presence of physiological concentrations of human luteinizing hormone (LH), which cross-reacts in the conventional hCG assay. hCG provides an excellent tumor marker for choriocarcinoma [7-9], and new marker candidates should undergo testing against the hCG secretion. As long as hCG is detected, there is undoubtedly active disease, provided that pregnancy is excluded. It has been calculated that a highly sensitive hCG assay would detect secretion by 10^5 trophoblastic cells, whereas 10^9-10^{12} cells are required before the tumor is detectable by physical clinical methods [8].

Free alpha subunit secretion by the normal placenta increases as pregnancy progresses. In late pregnancy the concentration of free alpha may be 30% of the hCG concentration [10, 11]. In trophoblastic disease the relative concentration of free alpha is negligible, and it has been suggested that in patients with trophoblastic disease, free alpha secretion is an unfavorable sign [12]. However, there are studies reporting isolated alpha secretion also in patients with hydatidiform mole with an excellent prognosis, but the relative concentration of free alpha has been extremely low [13, 14].

B. Pregnancy-Specific Beta-1-Glycoprotein

Pregnancy-specific beta-1-glycoprotein (TBG, SP_1, PAPP-C, PSBG) [15-18] has been demonstrated by immunohistochemical methods in the normal and malignant trophoblast [19-21]. Elevated PSBG levels have been found in the serum of patients with choriocarcinoma [22], and the levels follow clinical course [23, 24] (Figure 1). However, the correlation between PSBG and hCG values in patients with gestational trophoblastic tumors is not close [25]. A critical question regarding the clinical utility

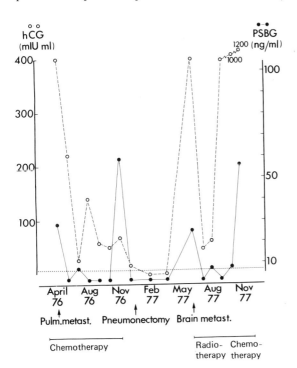

FIGURE 1 Circulating levels of hCG and PSBG in a choriocarcinoma patient with pulmonary and brain metastases. (From Ref. 24.)

of PSBG measurement in trophoblastic disease is whether PSBG is expressed in the absence of hCG, and if so, whether it is related to active disease.

By radioimmunoassay we studied the levels of circulating PSBG in 17 patients with choriocarcinoma whose serum samples came from a follow-up of 3 years [26]. As controls, patients with rheumatoid arthritis, liver disease, and infectious disease, and apparently healthy nonpregnant women were studied. The secretion of PSBG was compared with hCG secretion, and particular attention was paid to the frequency and magnitude of positive PSBG readings in hCG negative samples from patients with choriocarcinoma compared with healthy controls and patients with non-neoplastic disease [26].

Immunoreactive PSBG was observed in the serum of 4.7-5% of healthy adults or patients with non-neoplastic disease, the highest value being 4.8 µg/liter (Table 1). This was taken as the cutoff level for an elevated PSBG reading in trophoblastic disease.

Both hCG and PSBG were detected in the serum of all patients with trophoblastic disease at the early phase of illness. The decline of PSBG roughly paralleled that of hCG in most cases. However, 13 patients showed isolated PSBG or hCG after initial chemotherapy. PSBG concentrations over cutoff level were seen in 27 of 238 (11%) hCG-negative serum samples from these patients (Figure 2). After 3 months' absence of demonstrable hCG, PSBG was observed in 10% of hCG-negative samples from patients previously treated for choriocarcinoma (Table 2). Surprisingly, after 6 months remission PSBG was found in 12%, and after 1 year's remission PSBG was found in 17% of hCG-negative samples.

Patients who had been in remission for 12 months, but still occasionally had traces of circulating PSBG, are of particular interest. These patients are known to have an excellent prognosis: the recurrence rate is less than

TABLE 1 Circulating PSBG Levels in Controls

Subjects	Observed/ tested	Percent	Range (µg/liter)
Healthy adults	4/85	4.7	1.2-2.2
Septic infections	0/29	—	—
Rheumatoid arthritis	1/20	5.0	3.7
Liver disease	1/20	5.0	4.8
Total	6/154	3.9	1.2-4.8

Source: Data from Ref. 26.

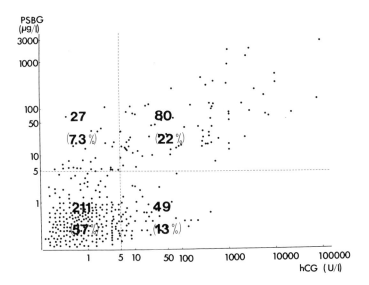

FIGURE 2   Distribution of serum hCG and PSBG levels in 367 serum samples from 17 patients with choriocarcinoma. (Data from Ref. 26.)

2% [27, 28]. Furthermore, Rosen and co-workers [29] have reported that some human fibroblast lines release PSBG in vitro. In light of these previous observations and our findings referred to above, it appears that although isolated PSBG peaks are sometimes seen in patients with choriocarcinoma after chemotherapy, the relationship of these peaks to an active disease is not clear. It may even be possible that chemotherapy causes release of a PSBG-like substance into the circulation.

At the present time there is no evidence to suggest that further treatment such as chemotherapy is indicated on the basis of an elevated PSBG level alone. One of our patients had had a brain metastasis and remained symptomless and hCG-negative for 2 years, until she experienced headache again. HCG was not demonstrated by routine immunochemical analysis of concentrated urine, and therefore additional treatment was withheld. We studied her serum samples in retrospect and found an increase and a subsequent decline of both hCG and PSBG, and the decline coincided with disappearance of the headache. Had we observed the elevation of trophoblastic markers in serum at the time the symptoms reappeared, additional chemotherapy would have been recommended. A likely explanation for the transient reappearance of PSBG and hCG may be reactivation and spontaneous regression of previously dormant tumor cells, since spontaneous regression of clinical choriocarcinoma has been well documented [30].

TABLE 2 Circulating PSBG in hCG-negative Patients with Choriocarcinoma

Clinical characteristics	No. of positive samples [a]	
	Observed/tested	Percent
After 3 months remission	11/112	10
After 6 months remission	9/76	12
After 12 months remission	7/42	17
All samples	27/238	11

[a] Cutoff level: 4.8 µg/liter.
Source: Data from Ref. 26.

Another interesting group included those patients in whom the circulating level of hCG was near the upper normal level (5 U/liter). This occurred in 35 samples from 9 patients (Table 3). In 15 samples from 7 patients the PSBG level was also elevated, providing further evidence for trophoblastic activity. This suggests that PSBG estimation may still have clinical significance in the management of those patients whose hCG value lies near the upper normal level.

C. Placental Protein Five (PP5)

PP5 is a unique protein isolated and characterized from the human placenta [31]. PP5 is a glycoprotein containing about 20% carbohydrate. It has a

TABLE 3 PSBG and hCG Levels in 367 Samples From 17 Patients with Choriocarcinoma

hCG (IU/liter)	PSBG (µg/liter)		
	<5	5-10	>10
<5	211	17	10
5-10	20	2	13
>10	29	3	62

Source: Data from Ref. 26.

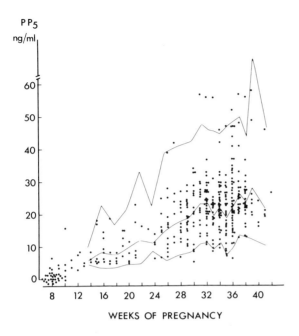

FIGURE 3 Serum PP5 levels in pregnancy. The solid lines are mean ± 2 SD. (From Ref. 33.)

molecular weight of 42,000, and one of its functions is to inhibit the activity of proteases such as trypsin and plasmin [32].

We studied the circulating levels of PP5 in pregnant women and patients with trophoblastic tumors, and compared the PP5 levels with those of hCG and PSBG [33]. With the immunoperoxidase technique we also studied the occurrence and localization of PP5 in normal placenta, hydatidiform mole, invasive mole, and choriocarcinoma.

All the serum samples from apparently healthy nonpregnant women were PP5-negative (detection limit 0.5 µg/liter). In normal pregnancy PP5 becomes detectable in serum around the eighth week after the last menstrual period. The levels rise as pregnancy progresses, peaking around the 37th to 39th weeks (Figure 3). In patients with hydatidiform mole only 2 of 19 serum samples from patients showed demonstrable PP5. The detectable levels were 0.6 and 2.0 µg/liter in the first pretreatment samples in which the hCG concentrations were 235 and 110,000 IU/liter and the PSBG concentrations 1300 and 4400 µg/liter, respectively. Seven serum samples from patients with invasive mole and 77 samples from patients with choriocarcinoma were all PP5-negative, even at the time the hCG and PSBG concentrations were high. One of these choriocarcinomas followed a term pregnancy, in which the PP5 levels are usually elevated between 30 and 40 μg/liter. Remarkably, PP5 was not detectable even in the first pretreatment sample of this patient, whose serum hCG concentration was 6000 IU/liter and PSBG concentration 185 μg/liter. Another deviation from the normal placental protein synthesis in this particular patient was a low alpha secretion compared with term pregnancy.

The disparity between the circulating levels of placental proteins in the normal and malignant trophoblast is intriguing. It may reflect differences in the circulating half-lives or in the synthesis and release of these proteins, or both. The half-life of PP5 is 15 min, whereas that of PSBG is 30-60 hr [34], and of hCG 30-40 hr [35]. The subunits of hCG also have a short half-life [35]. It is most likely that the half-life factor partly accounts for the differences in the circulating levels observed in trophoblastic disease.

Lee and co-workers [36] studied the concentrations of placental proteins in tissue and found striking changes at different stages of gestation for hPL, PSBG, and hCG. PSBG and hCG appear to have very similar clearance rates but very different blood/tissue ratios. Placental tissue levels of hPL and PSBG were higher in late pregnancy, hCG levels were lower, and PP5 showed no change.

We investigated the occurrence of PP5 in tissue by the immunoperoxidase technique [33]. Formalin-fixed, paraffin-embedded tissue sections were studied from normal placentae between 6 and 40 weeks of pregnancy, and from hydatidiform mole (10 cases), invasive mole (10 cases), and choriocarcinoma (6 cases). Adjacent control sections were prepared with anti-PP5 antiserum absorbed with PP5. PP5 was localized in the syncytiotrophoblast throughout gestation, whereas the cytotrophoblast was PP5-negative (Figure 4). All hydatidiform moles were PP5-positive (Figure 5), while one of 10 invasive moles took only a positive stain. All choriocarcinomata were PP5-negative by the immunoperoxidase method.

Our results indicate a relative absence of PP5 from the serum and malignant tissue of patients with choriocarcinoma compared with PP5 in the normal placenta and hydatidiform mole. Invasive moles lie between these two extremes. In light of the known biological role of PP5 as a protease inactivator [32], it is not surprising that the absence of PP5 is associated with trophoblastic invasiveness. These findings are important to tumor biology, and they may also have clinical significance: In normal pregnancy PP5, hCG, and PSBG are all detectable in serum by the 10th week of gestation, whereas in malignant trophoblastic disease hCG and PSBG are expressed in the absence of PP5.

D. Other Markers

Searching for other markers in trophoblastic disease, we measured the circulating levels of CEA, hPL, and AFP in patients with choriocarcinoma

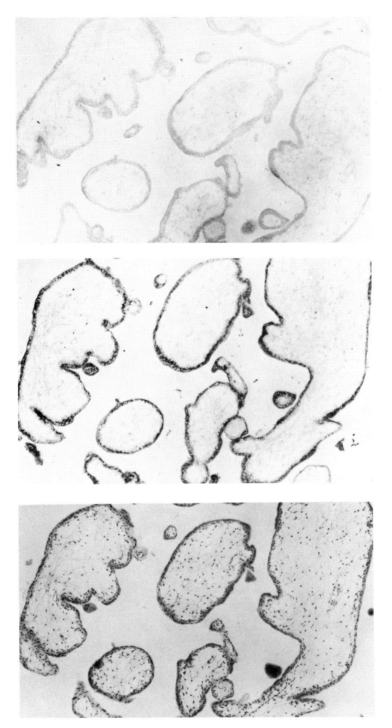

FIGURE 4 Immunoperoxidase localization of PP5 in the human placenta. Left: Hematoxylin–eosin (H-E) staining. Middle: Staining with anti-PP5 antiserum. Right: Adjacent control section stained with anti-PP5 antiserum absorbed with purified PP5. (From Ref. 33.)

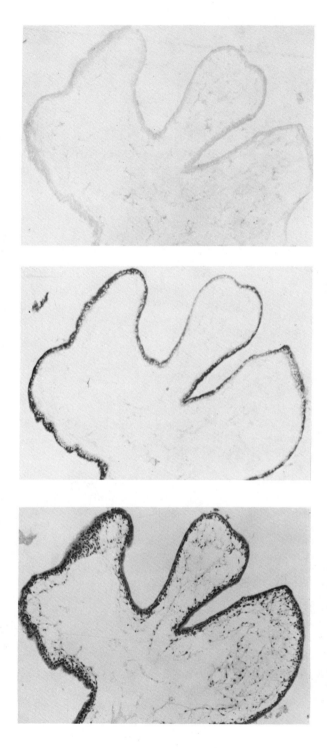

FIGURE 5 Immunoperoxidase demonstration of PP5 in hydatidiform mole. Left: H-E staining. Middle: Staining with anti-PP5 antiserum. Right: Adjacent control section stained with anti-PP5 antiserum completely absorbed with purified PP5. (From Ref. 33.)

whose urinary hCG excretion was negligible. The serum CEA concentration was slightly elevated in some patients, but it did not provide any additional information of trophoblastic activity when the highly sensitive radioimmunoassay of hCG was available [37]. The AFP level was normal in all cases. In normal trophoblastic tissue and hydatidiform mole, the AFP concentration is usually not elevated [38], although some studies have revealed an elevated AFP concentration in molar vesicular fluid [39]. We found no hPL in patients treated for choriocarcinoma. This was not unexpected, since the circulating levels of hPL are low, even in an untreated choriocarcinoma [40].

III. NONTROPHOBLASTIC CANCER OF THE REPRODUCTIVE SYSTEM

A. Carcinoembryonic Antigen

Apart from patients with malignant tumors of the gastrointestinal tract, lung, breast, and medullary thyroid gland, and with various nonmalignant conditions [see Ref. 41], several investigators have observed elevated CEA levels in the serum of patients with gynecologic malignancy [42-47]. The frequency of elevated circulating levels in gynecologic cancer has varied from 10 to 90% of cases, depending on the type of assay, cutoff level, and patients studied. Several immunological and physiocochemical criteria cannot distinguish the CEA of colorectal cancer from CEA isolated from mucinous cystadenocarcinomas or from ascitic fluid of patients with ovarian carcinoma, but a larger molecular weight has been described for CEA isolated from cervical carcinoma, and from the plasma and tumors of patients with ovarian carcinoma.

In gynecological cancer the occurrence of CEA in tissue varies according to the site and the spread of tumor. CEA is not detectable in normal tissues from the female reproductive tract by the immunoperoxidase technique, but it appears in some premalignant lesions of the uterine cervix (Figure 6) and in carcinoma in situ (Figure 7). The incidence of CEA positivity increases with advancing clinical spread (Figs. 8 and 9), so that in invasive disease 60-80% of carcinomas are CEA-positive. However, not all authors have found CEA in early premalignant lesions [48]. The immunoperoxidase method has a limited sensitivity requiring a CEA level in tissue of at least 3 μg/g for a positive reaction when applied to the study of formalin-fixed tissue [48].

1. Comparison Between Tissue and Serum CEA Concentrations

In cases in which the tumor tissue did not contain any demonstrable CEA by the immunoperoxidase technique, the serum CEA concentration was usually normal or slightly elevated. An elevated serum CEA level in the absence of CEA in tumor suggests that CEA is released outside the

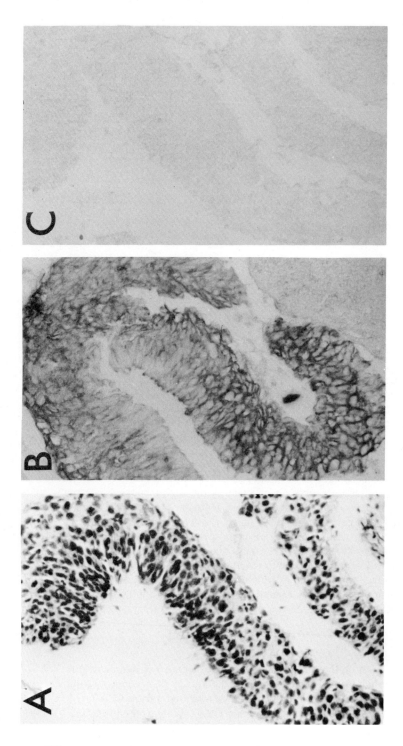

FIGURE 6 Immunoperoxidase staining of CEA in severe dysplasia of the uterine cervix. Left: H-E staining. Middle: Staining with anti-CEA antiserum. Right: Adjacent control section stained with anti-CEA antiserum after solid phase absorption with purified CEA. (From Ref. 49.)

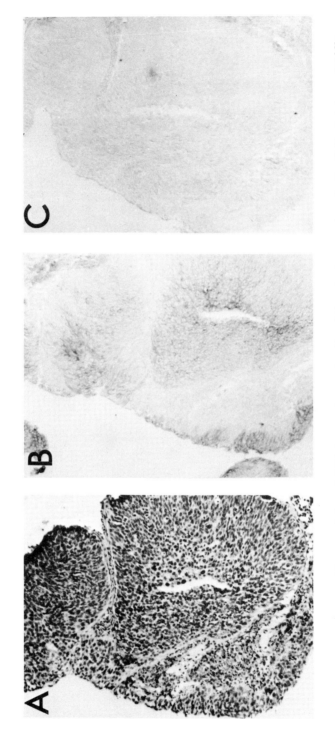

FIGURE 7 Immunoperoxidase staining of CEA in carcinoma in situ of the uterine cervix. Left: H-E staining, Middle: Anti-CEA antiserum. Right: Anti-CEA antiserum absorbed with CEA. (From Ref. 49.)

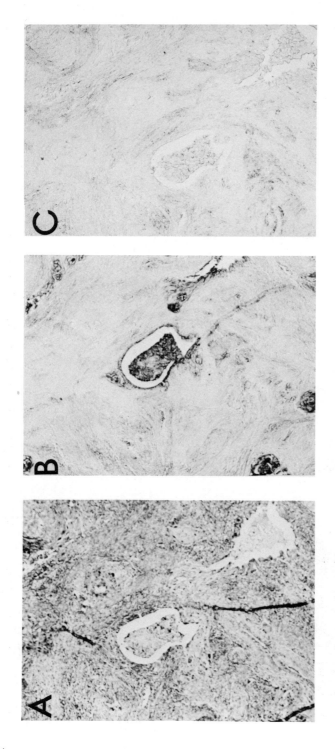

FIGURE 8 Immunoperoxidase staining of CEA in invasive epidermoid carcinoma of the uterine cervix. Left: H-E staining. Middle: Anti-CEA antiserum. Right: Anti-CEA antiserum absorbed with CEA. (From Ref. 49.)

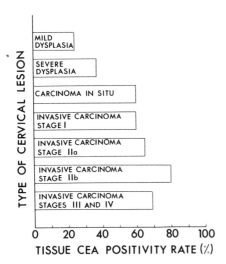

FIGURE 9 CEA positivity rates in premalignant and malignant epidermoid lesions of the uterine cervix. (Data from Ref. 49.)

tumor. In 19 patients whose tumor was CEA-positive, the circulating CEA concentration was elevated in 10 and normal in 9 (Figure 10). The finding that tissue CEA is more frequently positive than serum CEA is elevated is likely to reflect the great dilution of CEA released from the tumor into body fluids. Therefore immunoperoxidase staining of CEA in the tumor may be used to identify CEA-positive tumors for subsequent monitoring in those cases in which the circulating CEA level is not initially elevated.

Both the frequency and the magnitude of elevated serum CEA concentrations are low in patients with gynecological cancer compared with patients with colorectal cancer. The absence of recurrent tumor in many patients whose serum CEA concentration declined after surgery, and the maintenance of elevated CEA levels in those patients whose tumor had been incompletely removed, suggest that CEA measurement provides clinical information in a small proportion of patients. The CEA test is more likely to be helpful in patients with cervical and ovarian carcinomas, whereas those with endometrial carcinoma rarely express CEA in cancerous tissue, and in them the significance of elevated circulating levels is less clear.

The increasing tissue CEA positivity rate with advancing clinical spread of the tumor led us to investigate the prognostic significance of tumor CEA on patient survival. Cervical carcinoma is amenable to such an analysis, since CEA-positive and CEA-negative tumors can be distinguished by the immunoperoxidase method. Analyzing formalin-fixed paraffin-embedded

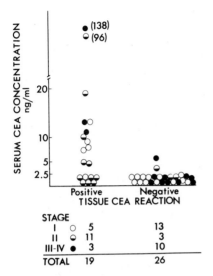

FIGURE 10 Circulating CEA concentrations relative to CEA in tissue by the immunoperoxidase method in 47 cancer patients classified according to clinical stage. (From Ref. 51.)

FIGURE 11 Ten-year survival rates of patients with CEA-positive and CEA-negative carcinomas of the uterine cervix treated by radical surgery. All cancers belonged to clinical stages Ib-IIa. (From Ref. 49.)

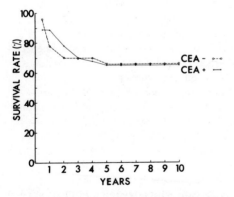

tissue blocks from patients treated 12 years ago, we found no difference in the survival rates between patients with CEA-positive and CEA-negative tumors (Figure 11). Thus, while the incidence of CEA positivity in the tumor increases with advancing clinical stage, the expression of CEA in tumors of the same clinical stage does not indicate more aggressive tumor

An important observation emerging from our recent studies is that endometrial adenocarcinoma is virtually always CEA-negative, whereas intracervical adenocarcinoma is CEA-positive. Clinical distinction between these two types of cancers is important because their treatments are different. However, the differential diagnosis is sometimes difficult because endocervical specimens obtained by curettage may be contaminated by endometrial tissue, and the histological distinction of these two tissues is not easy. In this situation we have found it most helpful to have an immunoperoxidase staining for CEA: A positive CEA reaction indicates that the adenocarcinomatous tissue derives from the endocervix [50].

2. Circulating CEA Levels

In our studies the circulating CEA levels were estimated in 328 patients with nontrophoblastic gynecological cancer and in 84 patients with benign tumors [51]. The serum CEA concentration was elevated (<5 µg/liter) in 9.8% of patients with malignant tumors and in 6% of those with benign tumors (Figure 12). While the incidence of elevated serum CEA levels was not strikingly different in malignant versus benign disease, the magnitude of CEA elevation was greater in malignant disease. In the benign group, elevated serum CEA levels were seen mostly in patients with mucinous tumors. In the malignant group the highest levels were seen in patients with ovarian carcinomas and the lowest in those with endometrial carcinoma.

After radical surgery the circulating CEA concentration declined below 2.5 µg/liter in 21 of 24 patients (Figure 13), and in 15 cases this decline was associated with uneventful recovery. However, in five patients the serum CEA concentration remained normal in spite of clinical recurrence. There were two patients in whom the removal of tumor had not been complete, and in both cases the elevated CEA level persisted.

B. Chorionic Gonadotropin and the Alpha Subunit

Ectopic production of hCG and/or its subunits has been reported to occur in a variety of malignant tumors [3, 52]. We focused our studies on gynecological tumors. Circulating hCG was found in 49 of 276 patients with gynecological cancer (18%) before treatment (Table 4). The elevated concentrations, varying between 7 and 60 IU/liter, were seen in patients with early as well as advanced disease [53]. No difference was observed in the frequency or magnitude of hCG levels in nondifferentiated versus differentiated cancers.

Surprisingly enough, low concentrations of hCG were also found in 15 of 104 patients (14%) with benign gynecological disease, but in none of 116 apparently healthy individuals tested. After radical surgery for cancer, the serum hCG concentration declined in 12 of 17 cases. But on the other hand some hCG-negative patients became hCG-positive after radical surgery

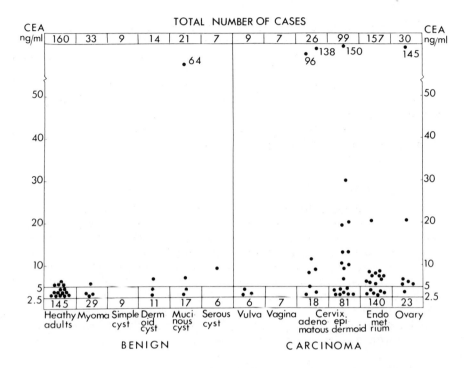

FIGURE 12 Circulating CEA concentrations in 160 apparently healthy adults and in 412 patients with malignant and nonmalignant gynecological disease before treatment. (From Ref. 51.)

including ovariectomy, which is known to result in elevation of the circulating LH level.

Recent findings suggest that there also is an hCG-like immunoreactive substance in the human pituitary that shares physiocochemical properties with hCG [54]. The potential interference of this material in the hCG assay has not been excluded in the patients who showed hCG for the first time after ovariectomy. Since removal of the ovaries is widely included in the surgical treatment of any type of gynecological cancer, and since elevated hCG levels may also be encountered in patients with nonmalignant tumors, the measurement of circulating hCG levels will hardly aid in the management of such patients.

Sodium-butyrate-stimulated HeLA cells produce free alpha subunits of glycoprotein hormones in vitro [55], and free alpha secretion has been observed in patients with pancreatic islet cell tumors [56]. We studied the circulating levels of immunoreactive alpha subunit by the hCG subunit radioimmunoassay in 101 patients with nontrophoblastic gynecological

TABLE 4 Occurrence of hCG in Serum of Patients with Nontrophoblastic Gynecological Tumors

	Elevated/total	Percent
Malignant tumors		
Vulva	2/8	25
Vagina	2/6	33
Cervix	23/11	21
Endometrium	17/125	14
Ovary	5/26	19
All	49/276	18
Benign conditions		
Endometriosis	0/15	0
Adenomyosis	1/6	17
Myoma	6/33	18
Simple cyst	1/9	11
Dermoid cyst	2/14	14
Serous cyst	2/7	29
Mucinous cyst	3/20	15
All	15/104	14

Source: From Ref. 53.

cancer and in 80 apparently healthy women, and found that in cancer patients the alpha subunit levels were not different from those of controls in the corresponding age groups [14].

1. Combined Measurement of CEA and hCG

Both CEA and hCG were estimated from serum samples of 312 patients with nontrophoblastic gynecological cancer. Both markers were elevated in 5 patients (1.6%), CEA alone was elevated in 27 (8.7%), and hCG alone in 40 cases (13%) (Table 5). The patients with the highest frequency of positive findings of either marker (33%) were those with ovarian cancer (Table 6). CEA and/or hCG was elevated in 30% of patients with cervical

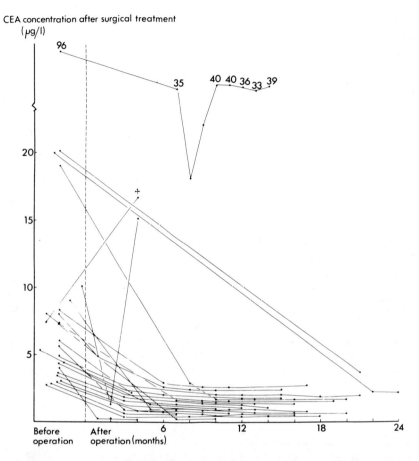

FIGURE 13 Circulating CEA concentrations before and after surgery for gynecological cancer. The highest levels came from a patient with stage IIb cervical epidermoid carcinoma in whom surgery was incomplete. (From Ref. 51.)

epidermoid carcinoma (Table 7), in 23% of patients with cervical adenocarcinoma (Table 8), and in 17% of patients with endometrial adenocarcinoma (Table 9).

C. Pregnancy-Specific Beta-1-Glycoprotein (PSBG)

PSBG has been identified in the serum of some patients with nontrophoblastic cancer [22, 57-59], but not all studies have done so [25]. Most notably, PSBG has been demonstrated in the tissue and serum of patients

TABLE 5 Occurrence of CEA (>5 ng/ml) and hCG (>7 mIU/ml) in Patients with Nontrophoblastic Gynecological Cancer

Marker	Observed/tested	Percent
CEA	32/312	10
hCG	45/312	14
CEA and/or hCG	72/312	23
CEA and hCG	5/312	1.6
CEA alone	27/312	8.7
hCG alone	40/312	13
Neither	240/312	77

Source: From Refs. 51 and 53.

with malignant teratomas [23, 60]. We have studied serum samples from patients with malignant teratoma and embryonal carcinoma, provided by Mrs. H. Orr at the Specific Protein Reference Unit, Putney Hospital, London, and from patients treated at the Department of Obstetrics and Gynecology, University of Helsinki. All 11 patients studied had an elevated serum AFP level. PSBG concentration was elevated in 2 patients (18%), hCG in 4 (36%), and PP5 in 3 (27%). The occurrence of placental proteins

TABLE 6 Occurrence of CEA (>5 ng/ml) and hCG (>7 mIU/ml) in Patients with Ovarian Carcinoma

Marker	Observed/tested	Percent
CEA	6/30	20
hCG	4/30	13
CEA and/or hCG	10/30	33
CEA and hCG	0/30	0
CEA alone	6/30	20
hCG alone	4/30	13
Neither	20/30	67

Source: Data from Fig. 12, which is taken from Ref. 51 and from Ref. 53.

TABLE 7 Occurrence of CEA (>5 ng/ml) and hCG (>7 mIU/ml) in Patients with Cervical Epidermoid Carcinoma

Marker	Observed/tested	Percent
CEA	10/99	10
hCG	20/99	20
CEA and/or hCG	30/99	30
CEA and hCG	0/99	0
CEA alone	10/99	10
hCG alone	20/99	20
Neither	69/99	70

Source: Data from Fig. 12, which is taken from Ref. 51 and from Ref. 53.

was occasional and, in keeping with the observations of Bagshawe and co-workers [25], PSBG and hCG levels were poorly correlated.

In some studies the circulating levels of PSBG have been followed during the course of treatment, and in many cases the levels have declined after removal of tumor [58]. We have seen a patient in whom PSBG, hCG, and CEA were all elevated at the onset of clinical disease, and in whom the levels declined after treatment (Figure 14). During the follow-up period some isolated PSBG peaks appeared without any relationship with clinical recurrence. Isolated PSBG peaks without evidence for clinical tumor have

TABLE 8 Occurrence of Elevated CEA (>5 ng/ml) and hCG (>7 mIU/ml) Levels in Patients with Cervical Adenocarcinoma

Marker	Observed/tested	Percent
CEA	5/26	19
hCG	4/26	15
CEA and/or hCG	6/26	23
CEA and hCG	3/26	12
CEA alone	2/26	7.7
hCG alone	1/26	3.8
Neither	20/26	77

Source: Data from Fig. 12, which is taken from Ref. 51 and from Ref. 53.

TABLE 9 Occurrence of Elevated CEA (>5 ng/ml) and hCG (>7 mIU/ml) in Patients with Endometrial Cancer

Marker	Observed/tested	Percent
CEA	11/157	7.0
hCG	17/157	11
CEA and/or hCG	26/157	17
CEA and hCG	2/157	1.3
CEA alone	9/157	5.7
hCG alone	15/157	9.6
Neither	131/157	83

Source: Data from Fig. 12, which is taken from Ref. 51 and from Ref. 53.

also been observed by other workers (F. Searle, personal communication, 1979). Rosen and colleagues [29] have observed that some human fibroblast lines release PSGB-like immunoreactive material into culture medium, and this material shares physicochemical properties with PSBG.

FIGURE 14 Circulating levels of PSBG, CEA, and hCG in a patient with invasive epidermoid carcinoma of the uterine cervix.

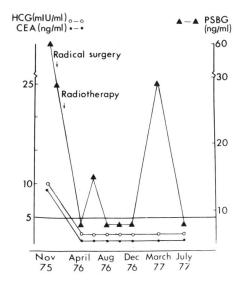

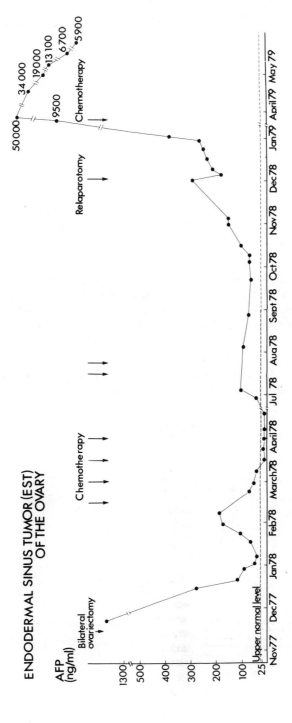

FIGURE 15 Circulating levels of AFP in a patient with endodermal sinus tumor. The patient occasionally showed low levels of hCG, PSBG, and PP5 in serum, but the levels of placental proteins bore no relationship to AFP levels.

Therefore the interpretation of PSBG values in patients with nontrophoblastic cancer is not clear, and more data are needed before the clinical significance of PSBG measurement is established.

D. Alpha Fetoprotein (AFP)

Since the discovery of AFP production by mouse hepatomas [61], the AFP assay has become widely used in the diagnosis and management of human primary liver cancer and also of embryonal teratocarcinomas. In these malignant conditions the AFP synthesis is compatible with the embryogenic sites of AFP production, i.e., the liver and the yolk sac [62]. In other types of cancer the serum AFP concentration is usually normal. Elevated serum AFP levels may occasionally be encountered in patients with liver metastases. We studied the serum AFP levels in 50 patients with advanced gynecological cancer and found an elevated AFP level in only one case [44].

Germ cell tumors of the gondas may contain yolk sac elements and produce AFP [63-68]. In published reports 36-86% of patients with germ cell tumors have had an elevated AFP level in serum. The great variation is likely to be due to different proportions of yolk sac tumors in these materials. Figure 15 shows representative example of the monitoring of treatment by the AFP assay. The initial decline of serum AFP concentration was prolonged to 7-8 days, indicating at an early stage of treatment that the tumor had not been completely removed. Following chemotherapy a normal AFP level was achieved, but the level increased again. The levels declined again after cis-platinum therapy. In spite of the fluctuations in AFP levels, the patient has remained asymptomatic.

Recent evidence suggests that AFPs synthesized by the yolk sac and the liver are glycosylated differently [69, 70]. Sera from patients with yolk sac tumors and amniotic fluid from early pregnancy contain a high proportion (15-45%) of AFP which does not bind to concanavalin A (Con A), whereas AFP in fetal and newborn sera, and in amniotic fluid from late pregnancy, contain less (2-6%) of this variant. The Con A binding characteristics do not affect the immunoreactivity of AFP and thus do not interfere with AFP estimation. In the patient described in Figure 15 as much as 75% of circulating AFP was of the Con A nonbinding type.

IV. SUMMARY

Summarizing our recent experience on the application of multiple markers for the management of patients with trophoblastic disease, it is obvious that information provided by hCG is overwhelming compared to that given by other placental proteins. It is not unexpected that in the treatment of tumors producing more than one marker dissociation may be observed between the response of different markers. PSBG has been identified in the absence of hCG in a number of choriocarcinoma patients, but the relationship of isolated PSBG secretion to active disease is less obvious. However, PSBG measurement may supplement the hCG test in those patients whose hCG value lies near the upper normal level. In general, it must be appreciated that failure to detect a tumor marker does not guarantee that there is in fact no tumor.

The relative absence of a placental protease inhibitor, PP5, from invasive trophoblastic disease is interesting insofar as it affects our understanding of the mechanisms related to tumor spread. It may be premature to make a statement on the prognostic significance of free alpha secretion in trophoblastic disease. Free alpha has been identified in cases with an unfavorable prognosis, but it has not been encountered in all those cases, and further evidence is required to establish whether free alpha secretion occurs before the unfavorable outcome is clinically evident.

In nontrophoblastic gynecological cancer there are several situations in which tumor markers can be applied to clinical practice. The presence or absence of CEA by the immunoperoxidase technique appears to be useful in the differential diagnosis between intracervical and endometrial adenocarcinomas. In split curettage, the endocervical specimen is often contaminated by endometrial tissue, and histopathological examination does not always distinguish between endocervical and endometrial adenocarcinomas. An accurate diagnosis is important, since the treatment of these two adenocarcinomas is different. Endocervical adenocarcinoma is usually CEA-positive and endometrial adenocarcinoma CEA-negative in immunoperoxidase staining.

In our experience the circulating CEA level is infrequently elevated in gynecological cancer, but the immunoperoxidase examination of the tumor offers a more effective means of identifying CEA-positive tumors. In the small number of patients in whom the preoperative serum CEA level is elevated, a fall in the level after surgery would indicate the completeness of surgical excision.

The application of hCG measurement to the management of nontrophoblastic gynecological cancer is more complex because treatment usually includes ovariectomy, after which supraphysiological LH concentrations may interfere in the assay. Elevated PSBG levels have occasionally been encountered in the serum of patients with nontrophoblastic gynecological cancer, but the relationship of isolated PSBG peaks to clinical recurrence is not clear.

Most would probably agree that none of the markers so far available are suitable for screening for cancer in a population. But when a tumor is identified the use of multiple markers provides an opportunity to find one or several for monitoring the course of treatment and detecting recurrent disease. Recently, hopes have been raised that radiolabeled antibodies to tumor-associated markers may eventually be used to localize metastases, but a prerequisite of this approach is the demonstration of a corresponding marker in tumor tissue.

ACKNOWLEDGMENT

The authors are grateful to Dr. R. E. Mansel, University Department of Surgery, Welsh National School of Medicine, Cardiff, Wales, for helpful comments and suggestions regarding the manuscript.

Our work was supported by grants from the Research Council for Medical Sciences, Academy of Finland (M.S.), the Medical Research Council, London (M.S.), the Finnish Cancer Society (E.-M.R.), and the National Institutes of Health, 1 RO 1 CA 23809-01 (T.W.).

REFERENCES

1. L. Stolbach, N. Inglish, C. Lin, R. N. Turksoy, W. Fishman, D. Marchant, and A. Rule, in Onco-Developmental Gene Expression (W. H. Fishman and S. Sell, eds.), Academic Press, New York, 1974, pp. 433-443.
2. R. C. Coombes, T. J. Powles, J. C. Gazet, H. T. Ford, J. P. Sloane, D. J. R. Laurence, and A. M. Neville, Lancet 1:132 (1977).
3. P. Franchimont, P. F. Zangerle, J. Nogerede, J. Bury, F. Molter, A. Reuter, J. C. Hendrick, and J. Collette, Cancer 38:2287 (1976).
4. G. I. Abelev, Transplant. Rev. 20:3 (1974).
5. E. Ruoslahti, H. Pihko, and M. Seppälä, Transplant. Rev. 20:38 (1974).
6. J. L. Vaitukaitis, G. D. Braunstein, and G. T. Ross, Am. J. Obstet. Gynecol. 113:751 (1972).
7. R. Hertz, J. Lewis, Jr., and M. B. Lipsett, Am. J. Obstet. Gynecol. 82:631 (1961).
8. K. D. Bagshawe, in Choriocarcinoma—The Clinical Biology of the Trophoblast and Its Tumours, Edward Arnold Ltd., London, 1969, p. 82.
9. D. P. Goldstein and T. S. Kosasa, in Progress in Gynecology (M. L. Taymor and T. H. Green, eds.), Vol. 6, Grune & Stratton, New York, 1975, p. 145.
10. P. Franchimont, U. Gaspard, A. Reuter, and G. Heynen, Clin. Endocrinol. 1:315 (1972).
11. J. L. Vaitukaitis, J. Clin. Endocrinol. Metab. 38:755 (1974).
12. J. L. Vaitukaitis and E. R. Ebersole, J. Clin. Endocrinol. Metab. 42:1048 (1976).
13. U. Gaspard, A. M. Reuter, J. C. Deville, Y. Gevaert-Wrindts, and P. Franchimont, Acta Endocrinol. (Kbh.), Suppl. 212:33 (1977).

14. E.-M. Rutanen, Int. J. Cancer 22:413 (1978).
15. Y. S. Tatarinov and V. N. Masyukevich, Byull, Eksp. Biol. Med. 69:66 (1970).
16. H. Bohn, Arch Gynak. 210:440 (1971).
17. T. M. Lin, S. P. Halbert, S. P. Kiefer, and W. N. Spellacy, Am. J. Obstet. Gynecol. 118:223 (1974).
18. C. H. W. Horne, G. D. Milne, and I. N. Reid, Lancet 2:279 (1976).
19. C. H. W. Horne, C. M. Towler, R. G. P. Pugh-Humphreys, A. W. Thomson, and H. Bohn, Experientia 32:1197 (1976).
20. T. M. Lin and S. P. Halbert, Science 193:1249 (1976).
21. Y. S. Tatarinov, D. M. Falaleeva, V. V. Kalashnikov, and B. O. Toloknov, Nature 260:263 (1976).
22. Y. S. Tatarinov and A. V. Sokolov, Int. J. Cancer 19:161 (1977).
23. F. Searle, B. A. Leake, K. D. Bagshawe, and J. Dent, Lancet 1:579 (1978).
24. M. Seppälä, E.-M. Rutanen, M. Heikinheimo, H. Jalanko, and E. Engvall, Int. J. Cancer 21:265 (1978).
25. K. D. Bagshawe, R. M. Lequin, P. Sizaert, and Y. S. Tatarinov, Eur. J. Cancer 14:1331 (1978).
26. E.-M. Rutanen and M. Seppälä, J. Clin. Endocrinol. Metab. 50:57 (1980).
27. C. B. Hammond, L. G. Borchert, L. Tyrey, W. T. Creasman, and R. T. Parker, Am. J. Obstet. Gynecol. 115:451 (1973).
28. G. T. Ross, D. P. Goldstein, R. Hertz, M. B. Lipsett, and W. D. Odell, Am. J. Obstet. Gynecol. 93:223 (1965).
29. S. W. Rosen, J. Kaminska, I. S. Calvert, and S. Aaronson, Am. J. Obstet. Gynecol. 134:734 (1979).
30. W. A. Bardawil and B. L. Toy, Ann. N.Y. Acad. Sci. 80:197 (1959).
31. H. Bohn, Arch. Gynak. 212:165 (1972).
32. H. Bohn and W. Winckler, Arch. Gynak. 223:179 (1977).
33. M. Seppälä, T. Wahlström, and H. Bohn, Int. J. Cancer 24:20 (1979).
34. P. Schultz-Larsen, Scand. J. Immunol. Suppl. 8:591 (1978).
35. J. L. Vaitukaitis, in Endocrinology (V. H. T. James, ed.), Vol. 2, Excerpta Medica, Amsterdam-Oxford, 1977, p. 104.
36. J. N. Lee, J. G. Grudzinskas, and T. Chard, Br. J. Obstet. Gynecol. 86:888 (1979).
37. M. Seppälä, E.-M. Rutanen, T. Ranta, I. Aho, U. Nieminen, H.-A. Unnérus, and E. Saksela, Cancer 38:2065 (1976).
38. M. Seppälä, K. D. Bagshawe, and E. Ruoslahti, Int. J. Cancer 10:478 (1972).
39. J. G. Grudzinskas, M. J, Kitau, and P. C. Clarke, Lancet 2:1088 (1977).
40. S. W. Rosen, B. D. Weintraub, J. L. Vaitukaitis, H. H. Sussman, J. M. Hershman, and F. M. Muggia, Ann. Intern. Med. 82:71 (1975).

41. P. Gold, Cancer 42 (Suppl. 3):1399 (1978).
42. S. K. Khoo and E. V. Mackay, Cancer 34:542 (1974).
43. V. Barrelet and J.-P. Mach, Am. J. Obstet. Gynecol. 121:164 (1975).
44. M. Seppälä, H. Pihko, and E. Ruoslahti, Cancer 35:1377 (1975).
45. J. R. van Nagell, Jr., W. R. Meeker, J. C. Parker, and J. D. Harralson, Cancer 35:1372 (1975).
46. P. J. DiSaia, C. P. Morrow, B. J. Haverback, and B. J. Dyce, Obstet. Gynecol. 47:95 (1976).
47. J. R. van Nagell, Jr., E. S. Donaldson, E. G. Wood, R. M. Sharkey, and D. M. Goldenberg, Am. J. Obstet. Gynecol. 128:308 (1977).
48. D. M. Goldenberg, R. M. Sharkey, and F. J. Primus, J. Natl. Cancer Inst. 57:11 (1976).
49. J. Lindgren, T. Wahlström, and M. Seppälä, Int. J. Cancer 23:448 (1979).
50. T. Wahlström, J. Lindgren, M. Korhonen, and M. Seppälä, Lancet 2:1159 (1979).
51. E.-M. Rutanen, J. Lindgren, P. Sipponen, U.-H. Stenman, E. Saksela, and M. Seppälä, Cancer 42:581 (1978).
52. G. D. Braunstein, J. L. Vaitukaitis, P. P. Carbone, and G. T. Ross, Ann. Intern. Med. 78:39 (1973).
53. E.-M. Rutanen and M. Seppälä, Cancer 41:692 (1978).
54. H.-C. Chen, G. D. Hodgen, S. Matsuura, L. J. Lin, E. Gross, L. E. Reichert, Jr., S. Birken, R. E. Canfield, and G. T. Ross, Proc. Natl. Acad. Sci. USA 73:2885 (1976).
55. J. M. Lieblich, B. D. Weintraub, S. W. Rosen, N. K. Ghosh, and R. P. Cox, Nature 265:746 (1977).
56. C. R. Kahn, S. W. Rosen, B. D. Weintraub, S. S. Fajans, and P. Gorden, N. Engl. J. Med. 297:565 (1977).
57. M. E. Crowther, J. G. Grudzinskas, T. A. Poulton, and Y. B. Gordon, Obstet. Gynecol. 53:59 (1959).
58. H. Würz, Arch. Gynaecol. 1:227 (1979).
59. J. G. Grudzinskas, R. C. Coombes, J. G. Ratcliffe, Y. B. Gordon, T. J. Powles, and A. M. Neville, Cancer 45:102 (1980).
60. S. A. N. Johnson, J. G. Grudzinskas, Y. B. Gordon, and A. T. M. Al-Ani, Br. Med. J. 1:951 (1977).
61. G. I. Abelev, S. Perova, N. I. Khramkova, Z. A. Postnikova, and I. Irlin, Transplantation 1:174 (1963).
62. D. Gitlin, A. Perricelli, and G. M. Gitlin, Cancer Res. 32:979 (1972).
63. G. I. Abelev, I. V. Assercritova, N. A. Kraevsky, S. D. Perova, and N. I. Perevodchikova, Int. J. Cancer 2:551 (1967).
64. J. Masopust, K. Kithier, J. Rádl, J. Koutecký, and L. Kotál, Int. J. Cancer 3:364 (1968).
65. J. Kohn, A. H. Orr, T. J. McElwain, M. Bentall, and M. J. Peckham, Lancet 2:433 (1976).

66. B. Nørgaard-Pedersen, Scand. J. Immunol. 4:7 (1976).
67. K. D. Bagshawe, in Alpha-Fetoprotein in Clinical Medicine (H. Weitzel and J. Schneider, eds.), Georg Thieme Publishers, Stuttgart, 1979, pp. 105-108.
68. A. Talerman, W. G. Haije, and L. Baggerman, Int. J. Cancer 19:741 (1977).
69. E. Ruoslahti, E. Engvall, A. Pekkala, and M. Seppälä, Int. J. Cancer 22:515 (1978).
70. E. Ruoslahti and M. Seppälä, Adv. Cancer Res. 29:275 (1979).

14

PROTEIN MARKERS FOR ANDROGENIC
ACTIVITY IN RAT VENTRAL PROSTATE

RUEY MING LOOR,* CHUNSHING CHEN, and SHUTSUNG LIAO / The Ben May Laboratory for Cancer Research and The Department of Biochemistry, University of Chicago, Chicago, Illinois

I. Introduction   351
II. Androgen-Receptor Protein   351
III. Nuclear-Acceptor Protein   357
IV. Spermine-Binding Protein   360
V. Prostate α-Protein   363
VI. Concluding Remarks   365
    References   366

I. INTRODUCTION

The four rat prostatic proteins which we will describe here are (1) a high-affinity, low-capacity, androgen-binding protein which is believed to be a functional receptor; (2) a nuclear (acceptor) protein, which may promote binding of the androgen-receptor complex to the nuclear chromatin· (3) a specific spermine-binding protein which appears to be one of the earliest proteins induced by androgen; and (4) a major secretory protein which is an end product of androgen stimulation but can inhibit the chromatin retention of the androgen-receptor complex.

Although their precise roles are not clear, these proteins, which are more abundant in the ventral prostate than in any other tissue of rats, can be useful in biochemical evaluation of the androgen responsiveness and viability of the prostate cells.

II. ANDROGEN-RECEPTOR PROTEIN

One of the most significant developments in the study of steroid hormone action in the last decade is the discovery of proteins that can selectively bind the active steroids in the target cells and can exhibit various properties

---

*Present affiliation: Department of Diagnostic Immunology and Biochemistry, Roswell Park Memorial Institute, Buffalo, New York.

which are likely to be characteristic of the functional cellular receptors. The search for the receptor proteins for steroid hormones follows a pattern that includes study of the uptake and retention of a radioactive hormone, identification of the hormone presumed to be the active form in the cell, and detection and isolation of a specific protein that binds the active steroid firmly.

Since the molecular function of the receptor proteins is not clear, investigators have used various criteria to determine whether the steroid-binding proteins are receptors. In general, the target cells are expected to have a high affinity for active forms of the steroid hormones; inactive steroids or their metabolites are not expected to bind to receptors to any significant extent at physiological steroid concentrations; target cells may contain more of the receptors than do the insensitive cells; and hormone antagonists can interfere with receptor binding or with some process that involves the receptor molecules. The steroid receptors thus characterized, both qualitatively and quantitatively, should be good markers for steroid sensitivity of the cells or tissues, although insensitivity may be caused by other factors.

One of the key events in the search for the androgen receptor was the observation that testosterone, the major testicular androgen circulating in the blood, is converted in the rat ventral prostate to 5α-dihydrotestosterone (DHT) and is retained selectively by the cell nuclei in vivo [1, 2] or in vitro when the prostate gland is incubated with radioactive testosterone [2]. These studies, carried out by biochemical techniques, were subsequently confirmed by autoradiography [3, 4].

Since DHT is a more potent androgen than testosterone in several androgen bioassay systems including prostate growth [5], and since androgens can rapidly affect nuclear RNA synthesis [6, 7], these findings strongly suggest that DHT is an active form of androgen, and that the prostate has a specific mechanism for concentrating DHT in cell nuclei. The biological importance of DHT retention was also suggested by the demonstration that cyproterone [8] and other antiandrogens can inhibit the nuclear concentration of androgen in vivo or in vitro [9-11].

Meanwhile, DHT in nuclei were found to associate with a nonhistone nuclear protein [12-14]. Since such an androgen-protein complex was not found in nuclei when isolated prostate cell nuclei were incubated directly with radioactive DHT, except when a cytosol fraction was also supplemented [13, 15], it was suggested that fromation of a complex between DHT and a cytosol receptor precedes retention of the androgen-receptor complex by the nuclear chromatin. The presence of a cytosol receptor protein for DHT was reported simultaneously in several laboratories [13, 16-18].

At least two prostate proteins (Table 1) can bind DHT [15, 19]. At low concentrations, DHT binds predominantly to a high-affinity ($K_a \sim 10^{11} \text{ M}^{-1}$) and low-capacity protein (β protein). The complex (complex II) formed can eventually be retained by nuclear chromatin. DHT can also form a complex

(complex I) with another low-affinity (Ka $\sim 10^7$ M^{-1}) and high-capacity protein (α protein), but chromatin does not retain complex I [15]. Whereas the high-affinity protein binds only natural and synthetic androgens, the low-affinity protein also binds estradiol, but not glucocorticosteroids.

There is general agreement that complex II represents a form of the cellular androgen-receptor complex. The interaction of DHT with the receptor is effectively antagonized by cyproterone acetate [8], flutamide [9, 10, 20], and other antiandrogens [11, 21-23]. In the fully grown prostate, as much as 30% of the cellular receptor protein is found in the cytoplasmic particulate (microsomal) fraction [24, 25]. Only a small quantity ($\sim$200 DHT-binding sites per nucleus) of the receptor is present in the cell nuclei of castrated rats. After injection of testosterone or DHT, the nuclear DHT-receptor content per nucleus is increased to 2000-6000 DHT molecules [15, 26]. The adult normal prostate has about 10,000-15,000 DHT-binding sites per cell (about 100 ± 50 fmol/mg cytosol protein), although higher numbers (54,000-65,000 per cell) have also been reported [33]. The receptor content may be considerably less ($\sim$1000 sites per cell) in castrated [27] or aged rats [28].

The cytosol DHT-receptor complexes have often been found to sediment as 7-12S and 3-5S units (see reviews in Refs. 29-33). Since the function of the steroid-receptor complex is not known, it is not possible to determine whether any of these are native forms. This difficulty is further complicated by the fact that almost all studies of sedimentation properties are carried out with receptor preparations which are much less than 0.1% pure, and without knowledge of the extent of the action of proteolytic enzymes on the receptor proteins and receptor-associated cellular components.

The 7-12S form of the DHT-receptor complex can be transformed to the 3-5S form by an incubation at 30°C, or by an increase in the salt concentration to 0.4 M KCl. Under some conditions, the small form can also be converted to the large form. In 0.4 M KCl or in 2 M urea, the cytosol DHT-receptor complex (3.8 ± 0.3S) sediments somewhat more quickly than the nuclear complex (2.9 ± 0.3S). At 0.1 M KCl and at a pH below 7.5, the 4S complex gradually aggregates to the 8S and larger forms. When the pH is raised to 9, the extent of aggregation is reduced, and the complex sediments in 8S and 3-4S forms. At pH 9.5, only a 7S form is clearly observed. The 8S and the aggregated form were not observed with hydroxyapatite-purified complex, suggesting that other cellular materials are involved in the formation of these large complexes [29, 34]. With the nuclear complex, extensive aggregation is observed at pH 6.4, but at a pH above 8.3 most of the complex sediments as a 3S form in 0.4 M KCl, or even in the absence of KCl.

The interaction of the androgen-receptor complex with divalent metal ions and nucleotides has been studied by gradient centrifugation [29]. At 1 to 5 mM, $MnCl_2$, $MgCl_2$, or $CaCl_2$ produced no significant effect in a medium containing 0.4 M KCl, but $CoCl_2$ facilitated aggregation of the 3.8S

TABLE 1 Comparison of Two Androgen-Binding Proteins and a Spermine-Binding Protein in the Rat Ventral Prostate

Protein	α Protein	β Protein	Spermine-binding protein
Representative ligands	DHT (Ka: 10^7 M^{-1}) testosterone, progesterone, estradiol	DHT (Ka: 10^{11} M^{-1}) and other androgens	Spermine (Ka: 10^7 M^{-1}) thermine
Ammonium sulfate precipitation (% saturation)	55–70	0–40	55–70
Approximate percentage of cytosol protein	20–30	0.002	0.5–1
DEAE-cellulose column chromatography	Eluted by 0.2 M KCl	Eluted by 0.05 M KCl	Eluted by 0.3 M KCl

Sedimentation property			
Low KCl	3–4S	3–4S, 7–8S, and aggregates	3S
High KCl	3–4S	3–4S	3S
Heat stability of the ligand-binding activity at 40°C	Stable	Unstable	Unstable
Subunit structure	Three or more components	Not known	Single polypeptide
Tight binding to chromatin	No	Yes	No
Inhibits complex II binding to chromatin	Yes	—	No
Spermine-binding activity			
At pH 8.5	Yes	No	Yes
At pH 7.5	No	No	Yes

form. With $ZnCl_2$, a shift in the sedimentation coefficient from 3.8S to 4.5S was observed after incubation at 0°C for 20 min, but the total amount of [^{3}H]DHT bound to the receptor did not change. At 20°C, the radioactive peak broadened (5 ± 2S), and eventually a considerable amount of the radioactive androgen dissociated from the receptor.

ATP and GTP can interact with and stabilize the DHT-receptor complex. Such an effect is observed most clearly when a freshly prepared receptor protein is incubated with 1-5 mM of these nucleotides at 20°C for 20 min. Under these conditions, the nucleotides appear to retard the release of DHT from the receptor as well as the transformation to the nuclear 3S form [29].

It is not clear whether the interaction of the receptor with divalent ions and nucleotides represents involvement of the receptor in a biologically important reaction, such as RNA synthesis. As reviewed elsewhere [29, 30], metal ions affect the properties of receptors for estrogens, mineral corticosteroids, and progesterone [35-38]. In some cases, this could be due to an indirect effect, such as Ca^{2+} activation of a protease that may transform the estradiol-receptor complex (8S and 5S) into a form (4S) that does not aggregate [38, 39].

The correlation between the androgenicity and the receptor-binding affinity of many steroids that have been tested is excellent [40]. With regard to receptor binding, the bulkiness and flatness of the steroid molecule, especially at the ring A area, appear to play a more important role than the detailed electronic structure of the steroid nucleus. For example, steroids with an A/B cis structure, such as the 5β-isomer of DHT, which is inactive in the rat prostate, are not bound by the prostate receptor. Other relatively flat steroids with rings A/B in the trans form also differ in their receptor-binding affinity, according to their bulkiness in the ring A/B area.

Testosterone binds to the receptor less firmly than DHT [40-44], apparently because testosterone dissociates more rapidly from the receptor. Since other Δ^4-3-ketosteroids can bind tightly to the receptor, the difference in dissociation rates may be due mainly to the steric property at the ring A/B area, rather than to the presence of the ring A double bond [40]. Potent androgens like 7α-17α-dimethyl-19-nortestosterone (DMNT), 2-oxo-17α-methyl-17β-hydroxyestra-4,9,11-triene-3-one, and 17α-methyl-17β-hydroxyestra-4,9,11-triene-3-one (R1881) are capable of binding to the androgen receptor more tightly than DHT. These androgens have conjugated double bonds which extend from rings A and B to C, and are indeed very flat molecules [40]. The removal of the angular methyl group between ring A and ring B also makes the area less bulky and possibly facilitates tight binding of these androgens to the receptor. This view is also supported by the fact that A-nor-17β-acetoxyestra-4,9,11-trien-3-one, with five carbons in ring A, is a very potent androgen, whereas A-homotestosterone, as well as A-homo-DHT, with seven carbons in ring A, are virtually inactive.

A unique structural feature of natural steroid hormones is that they contain an oxo group at the C-3 position. Whether this oxo group is absolutely required for androgen action is not clear, for many synthetic steroids without such a group have been shown to be androgenically active. These deoxy steroids generally show very low binding affinity for the prostate androgen receptor and have lower androgenicity than natural androgens. Some of the 3-deoxy androgens may be oxygenated and become androgenic, or various 3-oxygenated androgens may have different affinities toward different receptors and exhibit diverse biologic responses. The 17β-hydroxy group appears to be needed for high-affinity binding of androgens to the receptor, and for androgen action in the rat prostate. It is not clear, however, whether the 17β-hydroxy group is needed only for the formation of a tight androgen-receptor complex, or for the triggering process itself.

Antibodies against DHT and testosterone were found to be effective in removing steroids bound to nonreceptor proteins of the blood and prostate (steroid-metabolizing enzymes or blood steroid-binding globulin), since the rate of dissociation of the steroids from these nonreceptor proteins is much higher than the rate of dissociation from the receptor protein. The steroids bound far inside the receptor protein have very low rates of dissociation and are not readily removed by the antibody. The antisteroid antibodies can therefore be used for assays of the androgen receptor in the crude extracts containing nonreceptor steroid-binding proteins [45].

Purification of an androgen receptor has not been very successful [41-44, 46, 47]. Among the difficulties encountered are instability of the receptor, loss of the steroid ligand from the receptor protein being purified, and the limited quantities of tissues available for purification. Although the crude cytosol androgen receptors isolated from the seminal vesicles, testis, epididymis, kidney, uterus, ovaries, brain, and other organs appear to have similar physicochemical properties, such as sedimentation patterns, thermolability, and electrophoretic mobility, it would be premature to conclude that all of the receptors are identical or have the same genetic origin [48, 49].

III. NUCLEAR-ACCEPTOR PROTEIN

Steroid hormones appear to be taken up easily by various cells. The nuclear retention of steroid hormones, however, can be seen clearly only in tissues highly sensitive to these hormones. Since most steroids are present in cell nuclei in the protein-bound form(s), and since the free steroids incubated with cell nuclei do not form steroid-protein complexes, it is generally agreed that the nuclear retention of steroid hormones occurs in two steps: (1) an initial interaction of the hormone with the receptor protein, and (2) retention of the steroid-receptor complex by the nuclei. The precise intranuclear site where the steroid-receptor complex binds and functions is not clear; it is generally believed, however, that the complex is associated with the chromatin proper, and that it somehow modulates RNA synthesis. In sites such as the prostate, the cytosol receptor cannot be retained tightly by the prostate nuclei unless it first interacts with the hormone (DHT); therefore the first recognizable cellular action of the hormone is the alteration of the cytoplasmic receptor protein to a form that the nuclei can retain.

Autoradiographic studies [4] and experiments in cell-free media [15, 34, 46] showed that receptor retention by the nuclei occurred more readily at 20°C than at 0°C, apparently because the cytosol DHT-receptor complex must be transformed into a nuclear form by a temperature-dependent process before chromatin binding of the hormone-receptor complex. The receptor "transformation" seen in vitro appears to change the isoelectric point from 5.8 to 6.5 [46] and the sedimentation coefficient from 3.8S to 3.0S [29]; but whether these changes are artifacts is not clear. It is also not known whether the "transformation" process involves a conformational change of the receptor, a proteolytic action, or a bimolecular interaction.

The nuclear retention of the DHT-receptor complex appears to be tissue-specific. In cell-free systems, cell nuclei or nuclear chromatin isolated from tissues that are less sensitive to androgens than the prostate have low capacities for retaining the prostate DHT-receptor complex [15, 50, 51]. Since the nuclear receptor-binding activity is reduced if the nuclear preparation is heated to temperatures above 50°C, the receptor-binding activity may require a heat-labile factor [15]. In most prostate nuclear preparations that we have studied, each cell nucleus has 2000 to 6000 acceptor sites; this number may be higher ($\sim$10,000) in hyperplastic tissues and lower (200) in rats castrated many weeks previously.

Tymoczko and Liao [52] attributed the acceptor activity to a nonhistone protein in the prostate nuclear chromatin. In the presence of purified DNA, nonhistone proteins isolated from prostate nuclei can form a nucleoprotein complex that can bind the prostate DHT-receptor complex. The acceptor activity of the nuclear proteins (but not of histones) is similar to that of isolated prostate nuclei. The heat denaturation curves are identical and the bound DHT-receptor complex can be released in 0.4 M KCl medium. The heat-labile factors are not dialyzable; they can be fractionated by ammonium sulfate and ethanol or precipitated from the solution by acid (pH 4.5). Equivalent liver nuclear-protein preparations appear to contain a much smaller amount of acceptor-protein-like material, providing a reason for the low receptor-binding capacity of liver nuclei.

Klyzsejko-Stefanowicz and co-workers [53] studied chromatin acceptor molecules by sequentially removing urea-soluble chromosomal nonhistone proteins, histones, and DNA-associated nonhistone proteins from the chromatin of the rat ventral prostate and testis. The prostate [^{3}H]DHT-receptor complex was found to bind much more readily to the partially deproteinized chromatin, which still contains the DNA-binding nonhistone proteins, than to purified DNA. The receptor-binding activity of rat DNA was enhanced significantly by addition of the DNA-associated nonhistone protein of the prostate or testis, but not that of the liver.

Puca et al. [54] used nuclear proteins linked covalently to Sepharose to study the interaction of nuclear protein with the cytosol estrogen-receptor complex of the uterus, and they suggested that the nuclear acceptor molecule may be a basic protein. By the same technique, Mainwaring et al. [55] found that certain nonhistone basic proteins of the prostate nuclei, when bound to Sepharose, can retain the prostate cytosol [^{3}H]DHT-receptor complex. Androgens present in free form or bound to the sex steroid-binding globulin were not retained. The acceptor activity was higher in the nuclear preparations from the prostate than in those from the liver, spleen, or other tissues which are relatively insensitive to androgens. When various radioactive steroids were incubated with prostate cytosol as the source of steroid-receptor complexes, [^{3}H]DHT was about five times more active than labeled testosterone. Dexamethasone, progesterone, and androsterone were essentially inactive.

In cell nuclei, the function of the steroid-receptor complex may depend on its interaction with different forms of acceptor molecules. To explore such a possibility, we have studied the interaction of the DHT-receptor with ribonucleoprotein (RNP) particles extracted from prostate cell nuclei [56]. The ternary complex sedimented as heterogeneous components (60-80S) and could be analyzed by gradient centrifugation. The receptor-binding sites on the RNP particles could be saturated with excess DHT receptor, indicating that a limited number of binding sites exist. The apparent association constant calculated from such experiments is of the order of 10^{10} M^{-1}. Under saturating conditions with respect to the DHT-receptor complex, less than 5% of the isolated RNP particles can bind to the steroid-receptor complex. It is possible that only those RNP with heat-labile acceptor factors can associate with the DHT-receptor complex. Experiments with excess RNP demonstrated that only about 30-50% of the total DHT-receptor complex was capable of binding to the RNP particles, indicating that only certain forms of the steroid-receptor complex can interact with RNP.

The cytoplasmic polysome or 80S monosome forms of ribosomes do not bind the DHT-receptor complex, although the 40S and 60S subunit particles prepared from them can bind this complex readily [29, 57]. Radioactive steroids alone do not associate with any of the nuclear RNP or ribosomal particles. RNP or nuclear binding of the receptor complex is inhibited by aurin tricarboxylic acid (10^{-5} M), which dissociates the nucleic acid-protein complex and inhibits functions such as the initiation of protein synthesis.

Wang [58] reported that, when sequential extraction with 0.35 M NaCl and 2 M NaCl is used, the selective acceptor activity is also associated with the 2 M NaCl-insoluble residual fraction of the prostate chromatin, which contains DNA bound to nonhistone proteins. Nyberg and Wang [59] fractionated the androgen-labeled prostate chromatin after a single injection of radioactive testosterone into castrated rats. The quantity of radioactive androgen associated with a salt-soluble nonhistone protein fraction was high during the first hour, but declined rapidly during the second hour. The changes in the radioactivity of the salt-insoluble nonhistone protein and the DNA-histone complex fractions, however, exhibited the opposite pattern.

These studies and the others described above suggest that the androgen-receptor complex interacts with various nuclear components, possibly at different stages. Our suggestion [56, 57, 60] that the steroid-receptor complexes provide structural requirements not only for the formation of certain RNAs in the target cells, but also for their processing or functioning, or both, may be worth pursuing.

The control mechanism involved in the binding and release of the steroid-receptor complex from chromatin has not been studied in detail. The nuclear receptor-bound DHT may be metabolized [61] and released from the receptor, causing dissociation of the receptor protein from the nuclear acceptor site. As described above, we have also considered the possibility that some of the complex in the nuclei is associated with nuclear RNA or RNP particles and is recycled back to the cytoplasm in the RNA-bound form. Klyzsejko-Stefanowicz et al. [53] also reported that the capacity of the prostate chromatin to bind the DHT-receptor complex can be enhanced by phosphorylation of the chromatin in vitro. They suggested that phosphorylation may allow the acceptor protein to become more rapidly accessible to the receptor complex.

As will be discussed below, the target cells of steroid hormones may contain certain protein factors that can regulate the interaction of the hormone-receptor complex with the nuclear chromatin.

IV. SPERMINE-BINDING PROTEIN

The cytosol fraction of the rat ventral prostate contains an acidic protein which binds spermine rather selectively. The biological function of this protein is not known; however, its spermine binding activity is very sensitive to the androgenic manipulation of the animals. This protein may therefore be a good marker for evaluation of the androgen sensitivity of prostate cells.

We discovered the protein [62] during our study of the androgen enhancement of the initiation factor activity (Met-tRNA binding to the initiation factor eIF-2 which is needed in the protein synthesis) [63, 64], which is highly sensitive to certain polyamines, especially so spermine [65].

When the whole prostate cytosol preparation was incubated with [^{3}H]-spermine and the mixture was analyzed by gradient centrifugation, a distinct radioactive peak was found in the vicinity of 3S [62]. The macromolecular component of the complex was apparently a protein, since trypsin or chymotrypsin, but not pancreatic RNAse or DNAase, could reduce the formation of the 3S radioactive complex. The 3S protein is not present in blood. Although it has been observed in the cytosol fractions of various rat organs, this polyamine-binding component appears to be particularly abundant in the ventral prostate.

The spermine-binding protein can be precipitated quantitatively from the cytosol by the addition of ammonium sulfate to 50-80% saturation. When the precipitated protein was purified further by DEAE-cellulose column chromatography, the major spermine-binding protein was eluted from the column by a medium containing 0.3 M KCl. Further purification of the protein was achieved by Sephadex G-200 column chromatography, spermine-Sepharose affinity column filtration, chromatography on phosphocellulose and concanavalin A-sepharose columns, and gradient centrifugation [66]. The purified protein has a molecular weight of about 30,000, and it migrates, during electrophoresis, as a single band on polyacrylamide gels containing urea and sodium dodecylsulfate.

The radioactive ligand could be dissociated from the prostate protein when the preparation was acidified to pH 5 or lower or exposed in a 0.2 M or higher concentration of KCl. The recovered polyamine was identified as spermine by paper electrophoresis and thin-layer chromatography. Thus [^{3}H]spermine appears to bind to the protein noncovalently and without prior metabolism.

When we analyzed the ligand specificity by comparing the capacity of various nonradioactive amines to compete with [^{14}C]spermine for binding to the protein, we found that only spermine and thermine, a tetramine with one less carbon than spermine, were highly competitive. Considerably higher concentrations of spermidine were required before a similar competition was observed. Diamines of various sizes (putrecine, cadaverine, 1,10-diaminodecane, 1,12-diaminododecane) were generally not very effective. The precise measurement of the affinity constant of the ligands is complicated by polyamine adsorption by the tube wall; we have estimated the dissociation constant for spermine, however, to be approximately 0.1 μM, which is much lower than the spermine content in many animal tissues (0.5-5 mM) [67]. Divalent cations also interfered with spermine binding, but only at 100- to 1000-fold concentrations [62].

The spermine-binding activity of the purified protein was stable at temperatures below 30°C. At 50°C or higher temperatures, there was significant loss in binding activity. About 50% of the activity was destroyed within 1 hr at 37°C in 0.25 N NaOH. Treatment of the binding protein with alkaline phosphatase from calf intestine resulted in significant loss of spermine-binding activity. This loss was not observed when the phosphatase preparations were heated at 90°C for 10 min to destroy the enzyme. Beef heart protein kinase enhanced the spermine-binding activity of the phosphatase-treated binding protein. This effect was maximal in the presence of both cyclic AMP and ATP. These findings indicated that the spermine binding protein is a phosphoprotein, and that the ligand binding activity is dependent on the presence of the phosphate group.

The spermine-binding activity of the prostate cytosol or protein fractions can be analyzed by gradient centrifugation, or by polyacrylamide gel electrophoresis in which free spermine is separated from the glutaraldehyde-fixed spermine-protein complex. For gradient centrifugation in our laboratory, the protein sample was mixed with [^{14}C]spermine in 40 mM glycine, pH 8.7, layered on a linear sucrose gradient (5-20%), and centrifuged with an SW-56 rotor in a Spinco ultracentrifuge. For gel electrophoresis, the protein sample was incubated with the radioactive spermine in the glycine buffer and then fixed by glutaraldehyde. The fixed mixture was subjected to electrophoresis on polyacrylamide gel [68].

The spermine-binding activity of rat prostate protein fractions at different purifications was reduced significantly when the rats were castrated. Such a reduction was also observed when normal rats were injected with cycloheximide, a protein-synthesis inhibitor. The half-life of the spermine-binding activity estimated from such an experiment was about 3.5 hr.

Injection of 5α-dihydrotestosterone into rats castrated 24 to 48 hr earlier significantly elevated the spermine-binding activity of the prostate protein fractions. With both of the assay techniques described above, the effect of the hormone could be seen within 30 min after its administration and lasted for at least 2 hr [68].

The spermine-binding activity can be reduced by treatment of normal rats with antiandrogens, such as cyproterone acetate or flutamide. The androgen effect was considerably less pronounced when castrated rats were injected with a large dose of cycloheximide or actinomycin D. The androgen effect may therefore be dependent on new RNA or protein synthesis. Whether this represents induction of the protein or of other factors (such as those involved in protein phosphorylation) needed for the binding activity is being examined.

The biological role of the spermine-binding activity is not clear. It is conceivable that the secretion as well as the intracellular translocation or compartmentalization of polyamines depend on their binding to certain cellular proteins. On the other hand, spermine may assist the intracellular relocalization of certain acidic or phosphorylated proteins. Thus the rapid androgen action may mediate a number of cellular processes which are affected by spermine or by the binding protein. Whether such a mechanism is involved in androgen-induced stimulation of protein phosphorylation, RNA synthesis, protein synthesis, and chromatin alteration is worthy of further study.

V. PROSTATE α-PROTEIN

In 1971 we found that the high-capacity, low-affinity steroid-binding protein (α-protein) can inhibit the DHT-receptor complex from binding to the nuclear chromatin [15]. Other investigators have since reported the presence of similar inhibitors in the cytosol of rat uterus [69], chick oviduct [70], rat liver [71], and rat hepatoma cells [72]. The inhibition has not been regarded as a highly specific phenomenon, since bovine serum albumin was also inhibitory in some of these systems. With the hepatoma and prostate systems, however, bovine serum albumin is totally inactive.

Separation of α protein from the receptor protein (β protein) is not difficult [73, 74]. By ammonium-sulfate fractionation, β protein (or complex II) precipitates out at 0-40% saturation of the salt, whereas α protein (or complex I) precipitates out at 55-75% saturation of the salt. Complex II is not tightly retained on DEAE-cellulose, but complex I is retained and can be eluted at 0.2 M KCl. α Protein is a major cytosol protein (20-30%); β protein is a minor component (0.002%) of the cytosol.

Inhibition of the nuclear retention of the [^{3}H]DHT-receptor complex can be demonstrated by mixing the prostate α-protein fraction with the androgen receptor complex before, or simultaneously with, the addition of prostate cell nuclei. The extent of inhibition was not increased by prior incubation of the radioactive complex with the inhibitor, suggesting that the inhibitor did not cause irreversible destruction of the receptor complex. Apparently the inhibitor did not damage the nuclear binding site; nuclei incubated with the inhibitor and then washed (to remove the inhibitor) had the same receptor-binding activity and the same capacity to respond to additional inhibitor as did nuclei incubated in the absence of the inhibitor. That the inhibitor causes dilution of the isotope is unlikely, since addition of large quantities of nonradioactive DHT or of highly radioactive DHT did not affect the results, and since the inhibitory activity could not be removed from the protein fraction by extraction with organic solvents such as ethyl ether, carbon tetrachloride, chloroform, n-butanol, or ethylacetate.

The extent of inhibition cannot be reduced by an increase in the amount of radioactive androgen-receptor complex, indicating that there is no direct competition between the inhibitor and the receptor complex for the nuclear site(s). Interaction between the inhibitor and the receptor is possible, but the inhibitor may also act at a chromatin or DNA and indirectly affect the acceptor site to which the receptor complex binds.

α Protein appears to be identical to the "prostatein" of Lea et al. [75] and to the "prostate-binding protein" of Heyns and De Moor [76]. The latter investigators [77] as well as Parker et al. [78, 79] have also isolated poly A-containing prostate mRNA for an in vitro study of the synthesis of this protein, which like that of other major prostate proteins, is under the stimulatory influence of androgens. They found that the quantity of this protein in the prostate is reduced within a few days after castration and increases within a day after castrated rats are injected with androgens.

Polyacrylamide gel (7%) electrophoresis of the DEAE-cellulose-chromatographed α protein fraction gives, besides the α-protein band, two Coomassie-Blue-stainable bands which we designated as A and B bands. When the α protein and the two protein components were extracted from the gel and again subjected to electrophoresis on polyacrylamide gel, the A and B components were not contaminated with each other, but α protein could again generate A and B bands. This result suggested that α protein has two major subunits. The A subunit also appears to be present in free form in the crude cytosol and in the prostate fluid. Both subunits, as well as α protein, can be retained by a Con A-Sepharose column and eluted with α-methyl-D-mannoside, indicating that they are glycopolypeptides [74].

The A or B subunits individually can yield two bands after treatment with mercaptoethanol or dithiothreitol and analysis by polyacrylamide gel electrophoresis in the presence of sodium dodecyl sulfate. One of the two bands from the individual A and B subunits migrates to the same position in both cases, indicating that A and B subunits may have a common component. This reasoning is supported by an immunological analysis which we performed: Antibodies raised in the rabbits were found to cross-react with α protein and with both A and B subunits [74]. Our A and B subunits appear to correspond to the F and S subunits described by Heyns et al. [80]. Based on the protein bands obtained after polyacrylamide gel electrophoresis in the presence of sodium dodecyl sulfate, these investigators also suggested that there is a common component in the two subunits of the prostate-binding protein.

Besides its steroid-binding activity, α protein is also capable of binding radioactive spermine, and of sedimenting as a 3S complex during centrifugation in a 5-20% sucrose gradient medium [74]. This 3S spermine-protein complex can form only when the A and B subunits are mixed together. Clearly, α protein is different from the highly androgen-sensitive spermine-binding protein described above. Whereas α protein binds spermine well in a cell-free system at pH 8.5, but not at pH 7.5, the highly androgen-sensitive protein binds spermine at both pHs. Since the [^{3}H]spermine-binding activity can be detected with 10-30 μg of α protein, whereas the steroid-binding activity requires more than 500 μg of protein for its detection, polyamine binding may be used as a sensitive assay for α protein.

Of the two subunits, only the A subunit was inhibitory; the B subunit was totally inactive. This may suggest that the protein component (molecular weight about 9000) which is unique to the A subunit may be responsible for the inhibition. Preliminary experiments suggest that the inhibitory activity may be related to the high content of aspartic acid and glutamic acid in the protein.

Since α protein is a secretory glycoprotein, it is probably made by the endoplasmic-reticulum-bond ribosomes and stored in the cisternal space of this cellular structure. If α protein plays a regulatory role in androgen action, one may visualize the following sequence of events in the prostate: Castration causes disappearance of the androgen receptor from the cell nuclei, reduction in the cellular synthetic activity, and atrophy of cytoplasmic organelles. Androgen injection reverses this castration effect by promoting binding of the androgen-receptor complex to the nuclear chromatin with the help of an acceptor protein. This results in increased RNA and protein synthesis and facilitates reconstruction of the endoplasmic reticulum structure where the secretory proteins, including α protein, are synthesized. As α protein is accumulated, a component of the A subunit may enter the nuclei or may be released into the cytoplasmic compartment to limit nuclear chromatin binding of the androgen-receptor complex, thus allowing the cell to maintain the synthetic activity of the prostate cell in a balanced state. Loss of such a cellular organization-linked control of nuclear RNA synthesis may cause prostate cells to grow abnormally.

VI. CONCLUDING REMARKS

Androgenic steroids exhibit diverse effects on many vertebrate glands or tissues. There is no compelling reason why all the effects should be carried out by the same cellular process common to all of them. Generalization of the findings in the rat ventral prostate to other tissues must therefore be undertaken with caution.

Receptor-like proteins which bind DHT or testosterone proteins have been found in many androgen-sensitive rodent organs such as seminal vesicles, kidney, uterus, ovary, levator ani muscle, thigh muscle, sebaceous and preputial glands, coagulating glands, spermatozoa, hair follicles, skin, bone marrow, heart, specific areas of the brain (hypothalamus, pituitary, preoptic area, brain cortex, pineal gland, and certain androgen-sensitive tumors (see reviews in Refs. 30, 81-83). Yet it has been suggested that the androgenic action in the dog prostate depends on a receptor for 3α-17α-dihydroxy-5α-androstane rather than DHT [84], and that the induction of benign prostatic hyperplasia in dogs involves the action of certain 5α-androstanediols [85, 86]. In the vagina, 3α-hydroxy-androstanes and 3-keto-androstanes can stimulate the production of mucus by the superficial cells, whereas 3β-hydroxysteroids can affect deeper layers and 3β, 17β-dihydroxy-androst-5-ene and estrogens can cause keratinization of the epithelium [87]. Some of these actions may rely on receptors for hydroxylated androsta(e)nes in the cell nuclei of the vagina [88].

Quantitation and characterization of the androgen receptor in the human prostate have been very difficult, mainly due to the presence of large quantities of nonspecific steroid-binding proteins, and due to the complexity of the prostate itself. Although considerable progress has been made [89-101], further studies are necessary for the development of suitable receptor assay methods which can be used in clinical evaluation of the steroid-hormone responsiveness of normal and diseased prostate.

Three other proteins described above have not been studied in the human prostate. Our preliminary study showed that the human prostate contains no protein that can cross-react with the antibody against rat prostate α protein. A species-specific protein that resembles α protein, however, may exist in the human prostate.

ACKNOWLEDGMENT

This work was supported by Grants AM-09461 and CA-09183 from the U.S. National Institutes of Health, and by Grant BC-151 from the American Cancer Society, Inc.

REFERENCES

1. N. Bruchovsky and J. D. Wilson, J. Biol. Chem. 243:2012-2021 (1968).
2. K. M. Anderson and S. Liao, Nature (Lond.) 219:277-279 (1968).
3. K. J. Tveter and A. Attramadal, Endocrinology 85:350-354 (1969).
4. M. Sar, S. Liao, and W. E. Stumpf, Endocrinology 86:1008-1011 (1970).
5. S. Liao and S. Fang, Vitamins Hormones 27:17-90 (1969).
6. S. Liao, K. R. Leininger, D. Sagher, and R. W. Barton, Endocrinology 77:763-765 (1965).
7. W. I. P. Mainwaring, F. R. Mangen, and B. M. Peterken, Biochem. J. 123:619-628 (1971).
8. S. Fang and S. Liao, Mol. Pharmacol. 5:428-431 (1969).
9. E. A. Peets, M. F. Henson, and R. Neri, Endocrinology 94:532-540 (1974).
10. S. Liao, D. K. Howell, and T. M. Chang, Endocrinology 94:1205-1209 (1974).
11. J. L. Tymoczko and S. Lio, J. Reprod. Fertil. Suppl. 24:147-162 (1976).
12. N. Bruchovsky and J. D. Wilson, J. Biol. Chem. 243:5953-5960 (1968).
13. S. Fang, K. M. Anderson, and S. Liao, J. Biol. Chem. 244:6584-6595 (1969).
14. W. I. P. Mainwaring, J. Endocrinol. 44:323-333 (1969).
15. S. Fang and S. Liao, J. Biol. Chem. 246:16-24 (1971).
16. W. I. P. Mainwaring, J. Endocrinol. 45:531-541 (1969).

17. E. E. Baulieu and I. Jung, Biochem. Biophys. Res. Commun. 38:599-606 (1970).
18. K. J. Tveter, O. Unhjem, A. Attramadal, A. Aakvaag, and V. Hansson, Adv. Biosci. 7:193-207 (1971).
19. S. Liao and S. Fang, in Some Aspects of Aetiology and Biochemistry of Prostate Cancer (K. Griffiths and C. G. Pierrepoint, eds.), Alpha Omega Alpha Publishing, Cardiff, Wales, 1970, pp. 105-108.
20. W. I. P. Mainwaring, F. R. Mangen, P. A. Feherty, and M. Freifeld, Mol. Cell Endocrinol. 1:113-128 (1974).
21. F. R. Mangan and W. I. P. Mainwaring, Steroids 20:331-343 (1972).
22. W. I. P. Mainwaring, in Androgens and Antiandrogens (L. Martini and M. Motta, eds.), Raven Press, New York, 1977, pp. 151-161.
23. K. J. Tveter and A. Aakvaag, Endocrinology 85:683-689 (1969).
24. S. Liao, J. L. Tymoczko, T. Liang, K. M. Anderson, and S. Fang, Adv. Biosci. 7:155-163 (1971).
25. P. Robel, J. P. Blondeau, and E. E. Baulieu, Biochim. Biophys. Acta 373:1-14 (1974).
26. E. Van Doorn, S. Craven, and N. Bruchovsky, Biochem. J. 160:11-12 (1976).
27. J. N. Sullivan and C. A. Strott, J. Biol. Chem. 218:3202-3208 (1973).
28. S. A. Shain, K. W. Boesel, and L. R. Axelrod, Arch. Biochem. Biophys. 167:247-263 (1975).
29. S. Liao, J. L. Tymoczko, E. Castaneda, and T. Liang, Vitamins Hormones 33:297-317 (1975).
30. S. Liao, in Biochemical Actions of Hormones (G. Litwack, ed.), Academic Press, New York, 1977, pp. 351-406.
31. W. I. P. Mainwaring, Vitamins Hormones 33:223-245 (1975).
32. G. Verhoeven, W. Heyns, and P. DeMoor, Vitamins Hormones 33:265-281 (1975).
33. J. P. Blondeau, C. Corpechot, C. LeGoascogne, E. E. Baulieu, and P. Robel, Vitamins Hormones 33:319-345 (1975).
34. S. Liao and T. Liang, in Hormones and Cancer (K. W. McKerns, ed.), Academic Press, New York, 1974, pp. 229-260.
35. M. R. Sherman, S. B. Atienza, J. R. Shansky, and L. M. Hoffman, J. Biol. Chem. 249:5351-5363 (1974).
36. I. S. Edelman, J. Steroid Biochem. 3:167-172 (1972).
37. P. I. Brecher, A. Pasquina, and H. H. Wotiz, Endocrinology 85:612-614 (1969).
38. E. R. DeSombre, G. A. Puca, and E. V. Jensen, Proc. Natl. Acad. Sci. USA 64:148-154 (1969).
39. G. A. Puca, E. Nola, V. Sica, and F. Bresciani, Biochemistry 10:3769-3779 (1971).
40. S. Liao, T. Liang, S. Fang, E. Castaneda, and T.-C. Shao, J. Biol. Chem. 248:6154-6162 (1973).

41. M. Krieg and K. D. Voigt, J. Steroid Biochem. 7:1005-1012 (1976).
42. M. Krieg and K. D. Voigt, Acta Endocrinol. Suppl. 214, 85:43-89 (1977).
43. E. E. Baulieu, I. Jung, J. P. Blondeau, and P. Robel, Advan. Biosci. 7:179-191 (1971).
44. P. DeMoor, G. Verhoeven, and W. Heyns, J. Steroid Biochem. 6:437-442 (1975).
45. E. Castaneda and S. Liao, J. Biol. Chem. 250:883-888 (1975).
46. W. I. P. Mainwaring and R. Irving, Biochem. J. 134:113-127 (1973).
47. A. L. Hu, R. M. Loor, L. Chamberlin, and T. Y. Wang, Arch. Biochem. Biophys. 185:134-141 (1978).
48. A. Attramadal, S. C. Weddington, O. Nass, O. Djoseland, and V. Hansson, in Prostate Diseases (H. Marberger, ed.), Alan Liss, New York, 1976, pp. 189-203.
49. E. M. Ritzen, L. Hagenas, V. Hansson, S. C. Weddington, F. S. French, and S. N. Nayfeh, Vitamins Hormones 33:283-295 (1975).
50. W. I. P. Mainwaring and B. M. Peterken, Biochem. J. 125:285 (1971).
51. A. W. Steggles, T. C. Spelsberg, S. R. Glasser, and B. W. O'Malley, Proc. Natl. Acad. Sci. USA 68:1479-1482 (1971).
52. J. L. Tymoczko and S. Liao, Biochim. Biophys. Acta 252:607-611 (1971).
53. L. Klyzsejko-Stefanowica, J. F. Chiu, Y. H. Tsai, and L. S. Hnilica, Proc. Natl. Acad. Sci. USA 73:1954-1958 (1976).
54. G. A. Puca, V. Sica, and E. Nola, Proc. Natl. Acad. Sci. USA 71:979-983 (1974).
55. W. I. P. Mainwaring, E. K. Symes, and S. J. Higgins, Biochem. J. 156:129-141 (1976).
56. S. Liao, T. Liang, and J. L. Tymoczko, Nature (New Biol.) 241:211-213 (1973).
57. T. Liang and S. Liao, J. Biol. Chem. 249:4671-4678 (1974).
58. T. Y. Wang, Biochim. Biophys. Acta 518:81-88 (1978).
59. L. M. Nyberg and T. Y. Wang, J. Steroid Biochem. 7:263-273 (1976).
60. S. Liao, T. Liang, T.-C. Shao, and J. L. Tymoczko, Adv. Exp. Med. Biol. 36:232-240 (1973).
61. K. Nozu and B. Tamaoki, J. Steroid Biochem. 6:1319-1323 (1975).
62. T. Liang, G. Mezzetti, C. Chen, and S. Liao, Biochim. Biophys. Acta 542:430-441 (1978).
63. T. Liang and S. Liao, Proc. Natl. Acad. Sci. USA 72:706-709 (1975).
64. T. Liang, E. Castaneda, and S. Liao, J. Biol. Chem. 252:5692-5700 (1977).
65. S. C. Hung, T. Liang, L. M. Gluesing, and S. Liao, J. Steroid Biochem. 7:1001-1004 (1976).
66. R. M. Loor, G. Mezzetti, C. Chen, and S. Liao, Fed. Proc. 39:959 (1979).

67. A. E. Pegg, D. H. Lockwood, and H. G. Williams-Ashman, Biochem. J. 117:17-31 (1970).
68. G. Mezzetti, R. Loor, and S. Liao, Biochem. J. 184:431-440 (1979).
69. G. C. Chamness, A. W. Jennings, and W. L. McGuire, Biochemistry 13:327-331 (1974).
70. R. E. Buller, W. T. Schrader, and B. W. O'Malley, J. Biol. Chem. 250:809-818 (1975).
71. E. Milgrom and M. Atger, J. Steroid Biochem. 6:487-492 (1975).
72. S. S. Simons, H. M. Martinez, R. L. Garcea, J. D. Baxter, and G. M. Tomkins, J. Biol. Chem. 251:334-343 (1976).
73. C.-I. Shyr and S. Liao, Proc. Natl. Acad. Sci. USA 75:5969-5973 (1978).
74. C. Chen, R. A. Hiipakka, and S. Liao, J. Steroid Biochem. 11:401-405 (1979).
75. O. A. Lea, P. Petrusz, and F. S. French, Endocrinology Suppl. 100:217 (abstract) (1977).
76. W. Heyns and P. DeMoor, Eur. J. Biochem. 78:221-230 (1977).
77. W. Heyns, B. Peeters, and J. Mous, Biochem. Biophys. Res. Commun. 77:1492-1499 (1977).
78. M. G. Parker and W. I. P. Mainwaring, Cell 12:401-407 (1977).
79. M. G. Parker and G. T. Scrace, Eur. J. Biochem. 58:399-406 (1978).
80. W. Heyns, B. Peeters, J. Mous, W. Rombauts, and P. DeMoor, Eur. J. Biochem. 89:181-186 (1978).
81. J. L. Tymoczko, T. Liang, and S. Liao, in Receptors and Hormone Action (B. W. O'Malley and L. Birnbaumer, eds.), Vol. II, Academic Press, New York, 1978, pp. 121-156.
82. C. W. Bardin, L. P. Bullock, N. C. Mills, Y. C. Lin, and S. T. Jacob, in Receptors and Hormone Action (B. W. O'Malley and L. Birnbaumer, eds.), Vol. II, Academic Press, New York, 1978, pp. 83-103.
83. W. I. P. Mainwaring, in Receptors and Hormone Action (B. W. O'Malley and L. Birnbaumer, eds.), Vol. II, Academic Press, New York, 1978, pp. 105-120.
84. C. R. Evans and C. G. Pierrepoint, J. Endocrinol. 64:539-548 (1975).
85. P. C. Walsh and J. D. Wilson, J. Clin. Invest. 57:1093-1097 (1976).
86. G. H. Jacobi, R. J. Moore, and J. D. Wilson, Endocrinology 102:1748-1755 (1978).
87. C. Huggins, E. V. Jensen, and A. S. Cleveland, J. Exp. Med. 100:225-240 (1954).
88. T. C. Shao, E. Castaneda, R. L. Rosenfield, and S. Liao, J. Biol. Chem. 250:3095-3100 (1975).
89. M. Krieg, W. Bartsch, S. Herzer, H. Becker, and K. D. Voigt, Acta Endocrinol. 86:200-215 (1977).

90. M. Menon, C. E. Tananis, M. G. McLoughlin, and P. C. Walsh, Cancer Treatment Rep. 61:265-271 (1977).
91. M. Menon, C. E. Tananis, M. G. McLoughlin, M. E. Lippman, and P. C. Walsh, J. Urol. 117:309-312 (1977).
92. M. Snochowski, A. Pousette, P. Ekman, D. Bression, L. Andersson, B. Hogberg, and J. A. Gustafsson, J. Clin. Endocrinol. 45:920-930 (1977).
93. R. F. Morfin, I. Leav, J. F. Charles, L. F. Cavazos, P. Ofner, and H. H. Floch, Cancer 39:1517-1534 (1977).
94. E. F. Hawkins, M. Nijs, and C. Brassinne, Clin. Chim. Acta 75:303-312 (1977).
95. D. A. N. Sirett and J. K. Grant, J. Endocrinol 77:101-110 (1978).
96. C. Bonne and J. P. Raynaud, Steroids 26:227-232 (1975).
97. P. Davies and K. Griffiths, Mol. Cell. Endocrinol. 3:143-164 (1975).
98. V. Rosen, I. Jung, E. E. Baulieu, and P. Robel, J. Clin. Endocrinol. 41:761-990 (1975).
99. B. G. Mobbs, I. E. Johnson, and J. G. Connolly, J. Steroid Biochem. 6:453-458 (1975).
100. R. A. Cowan, S. K. Cowan, and J. K. Grant, J. Endocrinol. 74:281-289 (1977).
101. R. S. Hsu, R. C. Middleton, and S. Fang, in <u>Normal and Abnormal Growth of the Prostate</u> (M. Goland, ed.), Charles C Thomas, Springfield, Illinois, 1975, pp. 663-675.

INDEX

Acid hydrolases, 143-144
Acute lymphoblastic leukemia, 82, 90, 139
Acute myelomonocytic leukemia, 82
Adenosine deaminase, 90, 139-140
Adenosine kinase, 140
Adrenocorticotropin hormone (ACTH), 15, 276
Albumin, 28
Alkaline phosphatase, 93-109
  clinical application, 106-108
  developmental, 95, 98-100
  isoenzyme analysis, 94-96
  isoenzymes
    Kosahara, 104-105
    Nagao, 103-104
    non-Regan, 105-106
    of human cancer, 100-106
    of normal tissue, 96-98
    Regan, 104-105
Allotype, 313
Alpha fetoprotein (AFP), 12
  amino acid composition, 28-29
  biosynthesis, 33
  carbohydrate composition, 29-30
  DAB hepatoma carcinogenesis, 46
  effect on
    AFP-producing cells, 45-46
    fetus in pregnant animals, 47
    hepatoma growth, 46

[Alpha fetoprotein (AFP)]
  electrophoretic microheterogeneity, 29-30
  embryonal cells, 13
  endodermal sinus tumor cells, 13
  estrogen binding, 36-39
  F(ab'), 47
  hepatitis and liver cirrhosis, 39-40
  in early diagnosis of hepatoma, 40-41
  in gastrointestinal tumors, 43
  in germ cell tumors, 42
  in gynecological cancer, 345
  in hepatoma, 36-39
  in non-malignant disease, 44
  in pregnancy and fetal disease, 44-46
  methods of detection, 32-33
  mice AFP, 27-28
  Rαf gene, 27, 34
  rat AFP, 28
  seminomas, 13
Androgen-receptor protein, 351-357
Antibody-enzyme complexes, 125
Antigen-antibody complexes, 5, 19-20
Antisera, 12
APUD cell system, 89, 273, 288
Arginine, 276
Aryl sulfatase, 89
Asparaginase, 45

Barium enema, 67
Benign prostatic hypertrophy, 122
Bence Jones protein, 302-308
Biclonal gammopathies, 312-314
Bladder transitional cell sarcoma, 87
Bone marrow, 17, 121, 129-131
Breast cancer, 86-88

Calcitonin, 20, 271, 276, 280
  carcinoma of the thyroid, 15-16
  radioimmunoassay, 16
Carcinoembryonic antigen (CEA)
  bladder tumor, 76
  breast carcinoma, 14, 73-74
  CEA-M, 18
  CEA-S, 17
  chemotherapy, 72
  colorectal carcinoma, 14-15
  diagnostic adjuvant, 66-67
  differential diagnosis of cancer, 14
  gastrointestinal cancer, 14, 64-73
  gynecological malignancies, 74-75, 317-331
  lung carcinoma, 14-15, 74
  monitoring, 73
  radioimmunoassay, 14, 63, 65-73
  screening, 64-66
  second look surgery, 70-73
  staging, 67-70
Cervical carcinoma, 88
Choriocarcinoma, 15-16
Colorectal carcinoma, 89
Corticotropin-releasing activity, 284
CSAp, 18
Cyproterone acetate, 353
Cytidine deaminase, 140
Cytochemical staining, 138

Developmental alkaline phosphatase, 100-106 (see also Alkaline phosphatase)
Differential antigens, 17
5α-Dihydrotestosterone, 356
  receptor complexes, 353
p-Dimethylaminoazobenzene, 35
DNA polymerase, 140-141
DNase, 162
Dukes classification, 66-67

Ectopic hormone, 15-17
  biological significance, 287-290
  clinical syndromes, 276
  clinicopathological application, 284-287
  definition, 268
  evidence, 268-280
  multiple production, 283
  structure, 277-280
  tumors, 271-277
Enteroglucagon, 276
Enzymes, 81
Erythropoietin, 276
Esterases, 144-145
Estrogen binding protein, 200-204
Estrogen binding sites, 188-190
  inactivation, 197
  ligand specificity, 192-195
Estrogen receptors, 191-192
  activation, 204-211
  human breast carcinoma, 217-223
  stability, 197-200

F(ab'), 7, 305
Fetal sulfoglycoprotein, 18
Fetus, 12
Fucosyltransferase, 88

Galactosyltransferase, 86-88
  isoenzyme, 87

Index

Gaucher's disease, 123
Germ cell tumors, 13, 42
Glycoprotein, 146, 279
Glycosyltransferase, 85-88
β-Glucuronidase, 143
Glutamyl transpeptidase, 83-84, 142
Gonadotropins, 276 (see also Human chorionic gonadotropins)
Growth hormone, 184, 271

Heavy chain, 302
  disease, 302
  proteins, 314-315
Herpes simplex antigen, 75
Hepatoma, 13-14, 83
  AFP, 36-37
  AFP-producing and non-producing, 37
Hepatocellular carcinoma (see Hepatoma)
Hexosaminidase, 90, 144
Histaminase, 89
Hormone receptors, 147
Human chorionic gonadotropin (hCG), 13, 15-16
  AFP, 42-43
  α-subunit, 322-323, 337-340
Human placental lactogen, 276
Human seminal plasma, 122
Hypothalamic-type releasing factors, 283

Idiotypic determinants, 313-314
Isoenzymes
  galactosyltransferase, 87
  hexosaminidase, 90
  lactic dehydrogenase, 88
  Nagao, 103-104
  prostatic acid phosphatase, 118
  Regan, 94, 104-105
Immune complexes, 283

Immunocytochemical method, 272, 286
Immunoelectrophoresis, 18, 304, 310
Immunoglobulins
  atypical, 315
  classification, 303
  monoclonal, 302
Immunohistochemical technique, 323
Immunoperoxidase staining technique, 8, 13, 327-329, 335
Isoferritins
  amino acid composition, 230
  serum, 233-234
  structures, 229-230
  tumors, 231-233

Lactic dehydrogenase, 88-89, 146
Leukemia, 17, 20
Light chain, 302
  disease, 307
  fragment, 308
Lung tumor associated antigen, 19
Lysozyme, 82, 147

Mammary gland
  development, 184-185
  morphology, 185-187
  steroid receptors, 187-188
Medullary carcinoma of the thyroid, 15-16, 20
  histaminase, 89
Monoclonal gammopathies, 310-312
mRNA, 270, 279
Multiple markers, 321
Myeloma, 12, 308

N-alkaline phosphatase, 147
"Naked" DNA, 205
Nuclear-acceptor protein, 357-360

Nuclear chromatin, 352
5'-Nucleotidase, 83, 141-142

Oncofetal antigen, 2, 12
Oncodevelopmental gene expression, 26
Oncodevelopmental protein, 26
Ornithine decarboxylase, 243
Ovarian cancer, 86
Ovarian cancer associated antigen, 19

Pancreatic carcinoma, 84
Pancreatic oncofetal antigen, 18
Parathyroid hormone, 271, 276
Placental lactogen, 15
Placental protein five, 326-328
Plasma cell dyscrasia, 312
Pregnancy specific α-1-glycoprotein, 323-326, 340-345
Prolactin, 84, 276
Prostate α-protein, 353, 363-365
Prostate cancer, 121
Prostate proteins, 352, 354-355
Prostatic acid phosphatase, 18-19
  biochemical nature, 118-120
  bone marrow, 129-131
  clinical evaluation, 127-129
  counterimmunoelectrophoresis, 122-126
  fluorescent immunoassay, 126
  immunoadsorbent assay, 126-127
  isoenzymes, 118
  prostate cancer, 118-131
  radioimmunoassays, 120-122
Prostate-binding protein, 363
Prostatein, 363
Protease, 162
Protease inactivator, 328
Polyamines, 12, 241
  cancer patients, 248-259
  mammary turmor growth and regression, 245-246

[Polyamines]
  rat hepatoma, 246-247
  tumor cells, 243-244
  tumor regression, 245
  urinary, 259-260
Putrescine, 241
  growth process, 242
Purine nucleoside monophosphate kinase, 140
Purine nucleoside phosphorylase, 140

Radioimmunoassay, 324
  AFP, 32-33
  CEA, 14, 63-64
  immunoglobulins, 311
  prostatic acid phosphatase, 120-122
Rαf gene, 27
Ribonuclease, 84-85, 146
Receptors
  activated, 205
  charged, 205
Releasing factor-type activity, 283-284

S-Adenosyl-L-methionine decarboxylase, 242-243
Screening
  diagnostic test, 6-7
Second look surgery, 70-73
Sensitivity
  diagnostic test, 4
Sialyltransferase, 85-86
Somatomedin, 276
Specificity
  diagnostic test, 4
Spermidine, 241
  cell division, 242
  DNA synthesis, 242
  spermidine/spermine ratio, 242, 244
Spermine, 241

Index

Spermine-binding protein, 360-362
Steroid receptors
   characteristics, 188-211
   C3H mouse mammary adenocarcinoma, 213
   DMBA-induced mammary tumors, 214-217
   mammary gland, 187-188
   MCCLY mammary adenocarcinoma, 213
   NMU-induced mammary tumors, 211-213
   R3230AC mammary adenocarcinoma, 214

Testes
   germ cell tumors, 13
Terminal deoxynucleotidyl transferase, 90, 148-162
Testosterone, 352
Thyrocalcitonin (see calcitonin)
Trophoblastic tumors, 322-330

Tumor antigens, 2, 61
Tumor associated antigen, 1-2
   cervix cancer, 19
   detection of metastasis, 10-11
   diagnosis, 7
   histopathological evaluation, 7-8
   immunodiagnosis, 9
   localization of tumor, 10-11
   monitoring of therapy, 11-12
   organ or tissue, 17-19
   ovarian cancer, 19
   staging of cancer, 8-9

Vasopressin, 271, 276

Waldenstrom's macroglobulinemia, 12

Zinc glycinate marker, 18